VOLUMETRIC THREE-DIMENSIONAL DISPLAY SYSTEMS

VOLUMETRIC THREE-DIMENSIONAL DISPLAY SYSTEMS

BARRY G. BLUNDELL
ADAM J. SCHWARZ

A Wiley-Interscience Publication

JOHN WILEY & SONS, INC.

New York • Chichester • Weinheim • Brisbane • Singapore • Toronto

Library of Congress Cataloging-in-Publication Data:

Blundell, Barry, 1956-
  Volumetric three-dimensional display systems / Barry Blundell, Adam Schwarz.
      p.   cm.
   "A Wiley-Interscience publication."
   Includes bibliographical references and index.
   ISBN 0-471-23928-3 (cloth : alk. paper)
   1. Three-dimensional display systems.   2. Imaging systems.
I. Schwarz, Adam, 1969- . II. Title.   III. Title: Volumetric 3-dimensional display systems.
   TK8315.B58 2000
   621.36′7′02856693 — dc21   99-30124
Printed in the United States of America.

10 9 8 7 6 5 4 3 2 1

To
A.R.
sunlight that casts no shadow
and Mandy Waterman
for her courage

The University is a Paradise, Rivers of Knowledge are there, Arts and Sciences flow from thence. Counsell tables are *Horti conclusi* (as it is said in the Canticles) *Gardens that are walled in*, and they are *Fontes signati, Wells that are sealed up*; bottomless depths of unsearchable Counsels there.

— John Donne

CONTENTS

FOREWORD

There is a growing need to present increasingly complex and voluminous three-dimensional information in such a way that it may be interpreted rapidly, naturally, and with great accuracy. Consequently, it is appropriate to investigate and develop novel display techniques that are able to depict the three-dimensional form, spatial distribution, and dynamic properties of three-dimensional image data in a natural manner. One approach involves the use of volumetric displays (also called direct volume display devices) which permit three-dimensional image data to be presented within a transparent volume. Innate depth cues inherent in the three-dimensional objects are then automatically present, and in principle, the three-dimensional images may be viewed from any arbitrary direction. The proper development of displays of this type is particularly challenging from both the hardware and software perspectives. Furthermore, in some cases, concomitant advances in the physical mechanisms underlying the three-dimensional image creation process are also required. However, the optimum manner in which these displays may be utilized in applications such as medicine and design must still be established.

This is the first book to be devoted exclusively to this type of display methodology. The authors have provided an analytical insight into the capabilities of these systems and developed essential terminology, mathematical formalisms, and a systems-level approach that will be of particular assistance to those involved in or joining this area of research and development. They outline the operation of a number of volumetric display architectures that have been the subject of past work and describe areas of current activity. Interesting proposals are made for the next generation of volumetric displays, which they refer to as high-definition systems. These illustrate the challenging and diverse nature of this work. Furthermore, the development of new autostereoscopic display techniques is an exciting area of activity.

I trust that this book will be of interest not only within the computer science and engineering communities, but also to those who are involved in any way with the interpretation of volume data, in disciplines such as medicine, science, and engineering.

ARIE E. KAUFMAN

Leading Professor of Computer Science
State University of New York
Stony Brook, New York
April 1999

PREFACE

Volumetric display systems provide a transparent volume in space within which animated images may be depicted. Since these images occupy three physical dimensions, a number of depth cues are automatically satisfied without the need to perform extensive computation. Furthermore, images may be observed naturally without the need to wear special glasses. A variety of volumetric images are illustrated in the color plates included in this book.

The impetus for the research that has been undertaken in this area during the twentieth century has often been provided by small numbers of creative people who have perceived the benefits that could ultimately be gained from a novel and intuitive method of visualization in applications such as air traffic control and medicine. The full extent of the research that has been carried out may be gauged by reviewing the relevant scientific literature, patent applications, and other published works. This material also provides an insight into the considerable ingenuity directed to the design and construction of prototype systems. A French patent filed in 1912 by Luzy and Dupuis provides what we believe to be the first description of a three-dimensional display architecture that today would be considered as volumetric.

Volumetric display systems developed to date are generally best suited to the depiction of relatively sparse data sets and permit the three-dimensional form, spatial separation, and dynamic properties to be discerned with great clarity. In principle, many of these systems impose very little restriction on viewing position — an observer is able to move around the physical volume within which the images are cast and examine features of interest from any chosen orientation. However, imperfections in display techniques frequently cause image clarity to vary considerably with viewing position, and this naturally restricts viewing freedom. Most volumetric architectures give rise to translucent images and this reduces the effectiveness of the depth cue of occlusion (although the natural way in which the accommodation depth cue is satisfied ameliorates this to some extent). However, perhaps as a consequence of their ghostly appearance, the images are particularly dramatic and provide an insight into the benefits that may ultimately be derived from volumetric visualization techniques.

We believe this to be a fascinating area of research that involves the close integration of a diverse range of technologies. During the preparation of this book, we have corresponded with a number of people who have previously worked upon volumetric or other three-dimensional display systems. Without exception they have expressed continued enthusiasm and confidence for this approach to visualization. Volumetric display systems present a considerable intellectual challenge. Furthermore, as a consequence of the diversity of technologies employed in their implementation, research in this area provides an ideal vehicle for the training of both undergraduate and postgraduate students.

As will become apparent during the course of this book, volumetric systems are likely to complement the already wide range of two-and three-dimensional visualization techniques, being particularly suited to enabling both the overall three-dimensional structure and smaller details of a data set to be appreciated. Each visualization technique is best suited to a range of applications. This book represents a first attempt to formalize and thereby structure the subject of volumetric display system engineering. We have attempted to provide a critical and unbiased discussion on many of the facets to this approach to visualization. In preparing the book we have set ourselves three broad objectives. First, we wish to establish a technical and mathematical formalization in relation to this area of research. Our second intention is to record some of the work that has been carried out to date in devising volumetric architectures. In this way we hope to provide a sound basis for our discussions and ensure that researchers will benefit from the effort that has already been expended on this subject. As a consequence of the considerable time during which volumetric research has been carried out and the diversity of approaches adopted in the implementation of these systems, these objectives have represented a major undertaking. Our third aim is to propose various areas of investigation that we believe may lead to the advancement of this type of display. In particular, we emphasize á systems-level approach to the design of volumetric display systems, including both display hardware and control software components. We also propose techniques that may lead to the next generation of high-definition volumetric display systems.

Unfortunately, there is often insufficient time to read a textbook in its entirety. We have therefore cross-referenced the sections of the book and included some limited repetition in the hope that the reader may comfortably peruse particular areas of interest without necessarily following the order in which we have chosen to present the subject matter. In formalizing aspects of this work, it has been necessary to introduce various terminologies. To assist the reader with this, a glossary is included toward the end of the book. We have also included a list of symbols that will be encountered in the mathematical expressions that we present. This distinguishes between symbols that we consider to be global and which are encountered throughout the work, and symbols that are specific to certain chapters.

We trust that this book will be of interest not only to physicists, computer scientists, and engineers, but also to those who are users of display technologies and creative media. We anticipate that the book will also be relevant to people working in the areas of medical and scientific visualization, computer-aided design, air-traffic control, and entertainment.

Despite our best endeavors during the three years in which we have been occupied with the preparation of this book, it will certainly contain mistakes and omissions. We should be extremely grateful to hear from readers who may be able to assist us in piecing together the history of this area of research or who may find errors and inconsistencies. Above all, we sincerely hope that this book will be of interest to those who read it — we should indeed be pleased to hear from you.

BARRY G. BLUNDELL
barry.blundell@physics.org

ADAM J. SCHWARZ
adam@icr.ac.uk

April 1999

ACKNOWLEDGMENTS

And some were kind, and some were crusty.
And some would give and some would not.
It is rather difficult work asking for things,
even for other people.... .*

This book begins in a vibrant and highly chaotic laboratory in New Zealand where the first version of a volumetric display system that became known as the Cathode Ray Sphere (CRS) began to take shape, passes through a haze of congenial conversation in front of bright log fires with good friends and fine wine there in the Staff Club, and for a short time, settled in the sands of the United Arab Emirates. In between is a blur of faces, lands, oceans, and deserts but always, shimmering like a mirage within the mind's eye, are those volumetric images that we last saw in New Zealand before embarking on these travels.

In June 1996 the authors and a small research team traveled from New Zealand to work for one sabbatical year on volumetric systems and to establish a visualization laboratory (the AVL) in southwest France. Its primary function was to be the development and application of all types of three-dimensional display technology. At the eleventh hour, funding for the venture fell through and sadly the development team dispersed. We are grateful to the many people in France who assisted with the planning of the AVL and who provided a means for work to be undertaken on this book. We should particularly like to thank Frederique Henri, Marie-Anne Fontenier, Paul Finzi, and Sharon Cacheux in Valenciennes, and Professor Patrick Meyreuis of the Université Louis-Pasteur in Strasbourg. At that time we moved to a château set amid the tranquil rural beauty of France and,

---

* E. Nesbit, *The Railway Children*, Puffin Books, 1994.

to restore our spirits, started work on this book. Before you is the culmination of ten years of intensive work that has, to say the least, affected many lives.

Of those connected with the CRS in New Zealand, and consequently this book, we should like to begin by expressing our particular thanks to all those in the Department of Electrical and Electronic Engineering at the University of Canterbury who provided both facilities and technical support and the graduate and postgraduate students who worked on the development of the technology and on the related establishment of the AVL in France. Particular mention should go to Warren King, who had the dubious honor of being there at the beginning of the project and also to Art Vernon. The late Professor Richard Bates was a source of great encouragement to a newcomer working late into the night. We acknowledge the hard work undertaken in subsequent years by Damon Horrell, Richard Newstead and Kenneth Teo, which helped to advance succeeding prototypes of the CRS system. Later, Iwan Pekerti kept B.G.B. sane, while Rosemary Russo and Paul King provided hours of friendship and light relief as well as invaluable administrative support at a crucial time. Keith Lewis provided sound advice and circuit diagrams.

As we started work upon this book in 1996 in the solitude of the French countryside, we had not imagined the scale of the task before us. Despite our best efforts, B.G.B.'s allotted time in Europe expired with only a skeletal framework of the book in place. The return to New Zealand denoted a period in which the chaos of our arrival in France appeared in hindsight to be a time of relative peace and tranquility.

Milo and Jasper were the focus of the problem. All airlines firmly refused to transport our two cocker spaniels from France to New Zealand, so B.G.B. and his family embarked on a cargo ship voyage intended to last twenty-eight days, but which in fact took far more than eleven weeks. There can be few books written in so many diverse locations. The sea voyage could occupy a book in itself: a computer on the deck of a ship next to a wooden kennel where Milo and Jasper lived; weeks without seeing land; the cold Indian Ocean and the relative calm of the Australian Bight. It was an ideal environment to write a book (provided that the necessary books, papers, patents, etc., are at hand). In the event, it regrettably proved necessary to return to Europe. We therefore remained on the ship and completed a large portion of Chapter 5 while crossing the South Pacific and rounding Cape Horn. During that time, many of the crew — from the bo'sun down — assisted in various ways with the production of this book. Their generosity of spirit (and with spirits) is well remembered.

So B.G.B. (and family) traveled around the world, arrived back in Europe twenty miles from the original point of departure, and set off for a new American university in the United Arab Emirates (by air this time). Regular e-mails between B.G.B. and A.J.S. (who had by then moved to England) supplemented by occasional meetings in France, assisted the book in its rather shaky progress. Since that time, Professor Rüdiger Hartwig has given invaluable assistance and we are deeply indebted to him for the preparation of some of the diagrams and for his written contributions used in Sections 5.5.2, 5.6.6 and 11.3 His constant support and sound advice have been a source of great encouragement.

In the United Arab Emirates, B.G.B. should like to thank Venkat Reddy Surabhi, who provided unstinting support and assistance. I acknowledge also the many Indian workers who have constructed the university under such terrible conditions. The quiet dignity and courage of these men has been a source of great inspiration, although their situation troubles me profoundly.

The patronage of Richard Wood facilitated the development of the numerous CRS prototypes. Without his involvement and visionary purpose many of the events of the last few years could not have happened. He should be very proud of all his achievements.

Although work on the book has been carried out in England, France, Germany, the larger oceans, and amid the slowly shifting sands of the United Arab Emirates, there is a special place in these acknowledgments for the people in our tiny corner of France who continue to welcome us and our halting French with interest and friendship. They are too many to mention, but they have our grateful thanks for their unfailing warmth, interest and humor. They will be amused to hear that the first volumetric system was most probably proposed by a Frenchman. Our thanks go in particular to Madame de V, who opened her home to us all at a time when we needed solitude to begin our work; Jean-Pierre, who walks the dogs in the grounds; and to Jeannine, who provides croissants (not forgetting Lionel).

A book is the sum total of its parts, and there are many people who have contributed to this work directly and indirectly. Along with those we have already mentioned, we should like to thank the following, who have contributed by their discussions and friendly support over the years: Professor John Willmott CBE, Dr. John Baker, and Dr. Scott Hamilton (all of the University of Manchester, UK) and Dr. V. Kerdemelidis, Professor A. Williamson (University of Canterbury). For their interest in the work, and in many cases for contributing or checking material, we should like to thank Mr. Robert Batchko (Stanford University), Dr. Richard Ketchpel, Mr. Doug Ketchum (Thomas Electronics, New York), Dr. Duncan MacFarlane (University of Texas at Dallas), Dr. William Simon (University of Rochester, New York), Mr. Dennis Solomon (Holoverse), Mr. Parviz Soltan (NRaD), Dr. John Szilard (CSIRO, Australia), Dr. Adrian Travis (University of Cambridge, UK), and Mr. Peter Wallis FIEE. Thanks also to Andrew Smith, Lina, George, and Rosalyn at Wiley Interscience for producing the book.

B.G.B. would like to extend sincere thanks to Adam Schwarz for the years during which he has suffered for his science. Thanks also to Alistair Scott, for his long-standing support and patience.

A.J.S. would like to thank Barry Blundell for his vision and determination to finish the book, and in particular for his patience as other work commitments arose. The friendship and hospitality of him and his family, especially during the last two years on the odd occasions when we managed to find ourselves in the same country (France), is deeply appreciated. I am particularly grateful to my friends and family for their understanding as I have receded from normal life in recent months. In particular, I thank my fiancée, Emmanuelle, for her love, extreme patience and continual support, and her family for their kindness and hospitality.

Our involvement with volumetric systems began in New Zealand. The pioneering spirit, ingenuity and individuality of many New Zealand people made it possible to embark upon research which is sometimes viewed as not falling within the framework of currently accepted scientific endeavour. The University of Canterbury provided the environment for this work and facilitated its progress. We are indebted to all the people in New Zealand who eased the way by giving their time, knowledge, friendship, and financial assistance to the CRS research project.

Finally I (B.G.B.) would like to thank my family particularly Jandy, my wife. Without her help it is unlikely that this book would have been completed. She has not only been a source of inspiration but has also devoted a great deal of time to the production of the book. Her work has included the correction of text and the tracking down of patents and publications. She has also located and corresponded with many of those who have worked previously in this field and liaised with the publishers. I never understood her allergy to ships — perhaps she told me but at the time I was probably busy writing this book.

B.G.B.'s special thanks go to Milo and Jasper, who are currently in France, for their courage and unquestioning faith. Last but not least, how do I begin to thank Fluffy's Gang and its Director, Adriana?

BARRY G. BLUNDELL
ADAM J. SCHWARZ

The authors wish to acknowledge permissions received to reprint figures from the following papers and some text from the Parker and Wallis paper:

E. Parker and P.R. Wallis, "Three-dimensional display systems," *Journal of the IEE*, **95**, 371–390 (1948); © 1948 IEE.

R.D. Ketchpel, "Direct-view three-dimensional display tube," *IEEE Transactions on Electron Devices*, September, 324–328 (1963); © 1963 IEEE.

H. Mark and F. Hull, "3-D viewing of tomographic data: the Tomax system," *SPIE*, **120**, *Three-dimensional imaging*, 192–194 (1977); © 1977 SPIE.

R. Yamanaka, K. Yamamoto, N. Handa and H. Yoshikura, "A 3D display with a linearly moving mirror to reflect a series of 2D cross-sections and its application to noninvasive angiography," *IEEE Transactions on Medical Imaging*, **7**, 193–197 (1988); © 1988 IEEE.

W. Simon and T. Walters, "A spinning mirror autostereoscopic display," *SPIE Proceedings*, **120**, 180–183 (1977); © 1977 SPIE.

K. Kameyama, K. Ohtomi and Y. Fukui, "Interactive volume scanning 3D display with an optical relay system and multidimensional input devices," *Stereoscopic displays and applications IV, SPIE Proceedings*, **1915**, 12–20 (1993); © 1993 SPIE.

D.L. MacFarlane, *Applied Optics*, **33** (31), November, 7453–7457 (1994); © 1994 OSA.

R. Zito and A.E. Schraeder, *Applied Optics*, **2** (12), December, 1323–1329 (1963); © 1963 OSA.

LIST OF SYMBOLS

The following list summarizes the symbols used in the book. Those that are used in many chapters are grouped under "Global Symbols." Those whose use is more restricted are listed under the chapter in which they appear.

GLOBAL SYMBOLS

$\mathbf{x} = (x, y, z)^{\mathrm{T}}$	Position of a point in a Cartesian coordinate frame defined within the image space (the *image space frame*)
T	Voxel time, the total time required to create single voxel: $T = T_S + T_V$
T_S	System adjustment time
T_V	Voxel activation time
D	Distance between origins of beam source and image space frames (applicable for beam addressed display units)
t	Continuous time variable
P	Parallelism of voxel activation, the number of voxels that can be activated simultaneously
f_r	Image space refresh frequency [Hz]
τ_r	Image space refresh period, the time available for each refresh of the image space: $\tau_r = 1/f_r$
N_a	Voxel activation capacity, the maximum number of voxels that can be activated within the image space each refresh

N_l	Voxel location capacity, the number of possible voxel locations within the image space each refresh
$\hbar$	Planck's constant (h) divided by 2π : $h = 6.626176 \times 10^{-34}$ J · s, $\hbar = 1.05458866 \times 10^{34}$ J · s · rad^{-1}
c	Speed of light in vacuum: $c = 299,792,458$ m · s^{-1}

LOCAL SYMBOLS

Chapter 1

L	Side length of the elementary system example
N	Total number of voxel elements within the elementary example image space
ϕ	Visual angle subtended by an object of size x at distance s
s	Distance of an object from the eye
x	Size of an object in view
ϕ_c	Approximate maximum resolving power of the human eye, equaling approximately 1 minute of arc
f_c	Critical flicker fusion frequency [Hz]
L_{av}	Average luminance of an image

Chapter 2

l	Length of each side of a square two-dimensional display screen, or of each edge of a cubic three-dimensional image space
n	Number of pixels along any row or column of a two-dimensional screen or a three-dimensional image space
ρ	Voxel density
ψ	Fill factor; the ratio of the voxel activation capacity to the voxel location capacity: $\psi(\%) = N_a/N_l \times 100$

Chapter 3

f_s	Cyclic motion frequency of screen [Hz]: also, angular frequency is denoted $\omega_s = 2\pi f_s$, and the motion period by τ_s
τ_s	Screen motion period [s]: $\tau_s = 1/f_s$
ω_s	Angular frequency of screen motion [rad · s^{-1}] : $\omega_s = 2\pi f_s$
N_a^{max}	Maximum voxel activation capacity in a swept-volume system
$\Delta\theta$	Angular range of screen positions

$\delta\theta$	Incremental screen angle moved in a single voxel time T
n	A given number of voxels
r	Radial distance from the rotation axis
Δd	Distance in direction of screen motion
k	Thickness of screen supporting structure (assumed constant)
$\delta\tau$	Minimum time between consecutive activations of a voxel in a translational motion display unit
A	Peak-to-peak amplitude of a translational screen motion
ξ	Number of slices into which the volumetric data are divided, for output to a translational motion display unit
$z(t)$	Position of a translating surface in direction of motion
$z^{\text{approx}}(t)$	Linear approximation to a sinusoidal screen motion
Δz	Slice thickness in translational motion system. $\Delta z = A/\xi$
z_j	Position of jth image slice in translational motion example
Δt	Time between start of output of consecutive slices
t_j	Time at which output of jth slice begins
j	Counter through the ξ slices
e_i	Positioning error of ith voxel having the same z coordinate in the example system
i	Counter through voxels having ideal locations at the same instantaneous surface position
m	Mass of the screen in a swept-volume display unit
α	Slope parameter for the linear approximation to the sinusoidal position profile of a reciprocating screen
q	Fraction of reciprocating screen amplitude occupied by the image space in a linear approximation ($0 < q \leq 1$)
V_d	Voxel diameter
V_s	Separation between centers of adjacent voxels
E_k	Kinetic energy acquired by a screen moving in sinusoidal motion
P_s	Power required to accelerate/decelerate screen moving with sinusoidal translational motion
(L_x, L_y, L_z)	Dimensions of translational motion image space, where z is the direction of screen motion

Chapter 4

ϕ	Incidence angle of a beam onto a screen
b	Unit vector in direction of a beam
s	Unit inward normal vector to the local tangent to the screen

W	Diameter of the incident beam
V	Voxel length (greatest extent) in the plane of the screen
e	Eccentricity of an elliptical voxel in the plane of the screen
$(x_s, y_s, z_s)^T$	Coordinates of the beam source in the image space frame
B	Distance between the beam source and a voxel location within the image space
H	Half-height of the cylindrical image space
R	Radius of the cylindrical image space
η	Angle between the rotation axis of the screen (z axis of the image space) and a line connecting the beam source and the origin of the image space frame
β	Instantaneous angle of the screen about rotation axis (z-axis of image space frame)
r	Radius of a point (x, y, z) from the rotation (z) axis of the image space [$r = (x^2 + y^2)^{1/2}$]
ψ	Angle of a radial line segment of a helical surface at height z, relative to that at $z = 0$
f_s	Frequency of screen motion in swept-volume display units [Hz]
S	Number of sectors into which each revolution of the screen is divided for voxel-ordering purposes in beam-addressed swept-volume units

Chapter 5

r	Radius of an Archimedes spiral surface (also: r_{max}, r_{min})
a	Coefficient of angular dependence of an Archimedes spiral
b	Constant term in equation for the surface of an Archimedes spiral
θ	Angular coordinate of an Archimedes spiral

Chapter 6

ΔE	Energy difference between two electronic states of a quantum system: specifically, of a fluorescent center employed the stepwise excitation of fluorescence [J]; for two energy states, $|i\rangle$ and $|i + 1\rangle$, $\Delta E_i = E_{i+1} - E_i$; also, energy of each photon in the activation beam resonant with this transition
ω	Angular frequency of radiation [rad $\cdot$ s^{-1}]
b_i	Coefficient of absorption rate for a transition between two energy states, $|i\rangle \rightarrow |i + 1\rangle$, of a fluorescent center; defined as $b_i = \sigma_i I_i / \Delta E_i$ [s^{-1}]

σ_i	Resonance absorption coefficient for transition $\|i\rangle \to \|i+1\rangle$ [m^2]
I_i	Peak intensity of activation beam resonant with transition $\|i\rangle \to \|i+1\rangle$ [W $\cdot$ m^{-2}]
τ_i	Average decay time constant for all spontaneous transitions from state $\|i\rangle$ to lower-energy states [s]
η	Fluorescence efficiency; this is the fraction of decays from state $\|3\rangle$ in a three-state stepwise excitation model giving rise to visible radiation and thus contributing to the voxel intensity ($0 \le \eta \le 1$)
N_i	Population of state $\|i\rangle$; most commonly, this is an electronic energy state
N_{tot}	Total population of fluorescent centers involved in stepwise upconversion process
$\mathbf{N}(t)$	Vector representation of energy state populations in a three-state system: $(N_1(t), N_2(t), N_3(t))^T$
N	Total number of voxel elements in a static, directly addressed, active matrix display unit (hence it equals the voxel location capacity N_l of such a system)
$p_j(t)$	Temporal profile of activation pulse j, normalized to unity (for a two-step excitation process, $j = 1, 2$)
a	Cross-sectional area of each activation beam at the voxel location [m^2]
W	Diameter of each activation beam
P_j	Power of activation beam j[W] : $P_j = aI^j$
$\mathbf{A}(t)$	Coefficient matrix for the vector representation of the rate equations, as defined in Eq. (6.1)
t_j	Time at which the pulse from activation beam j arrives at the fluorescent center of interest
T_p	Characteristic duration of the radiation intensity pulse from an activation beam
Δ	Normalized relative timing between the two activation pulses in a two-step excitation process, $\Delta = (t_2 - t_1)/T_p$
Δ_{opt}	Value of Δ that optimizes the voxel brightness
S_3	Total number of fluorescent centers excited to the fluorescent state $\|3\rangle$ by the two-pulse excitation combination
E_V	Total energy radiated from each voxel during each activation sequence [J]
E_{fl}	Energy of each fluorescent (visible) photon [J]
D_{fc}	Density of fluorescent centers in the image space [m^{-3}]
V	Voxel volume [m^3]
P_V	Power radiated from each voxel due to a single activation pulse sequence [W]

(r, β, ϕ)	Spherical polar coordinates of the image space, as defined in Figure 6.11
β_{12}	Angle between beam sources 1 and 2 in the $\phi = 0$ plane
R	Radius of spherical image space for dead zone analysis in Section 6.6
D	Distance of beam source from the center of the image space
ℓ	Linear distance between the two beam sources in Section 6.6
d_j	Distance between beam source j ($j = 1, 2$) and a given point (voxel location) in the image space; when $d_1 = d_2$, we define $d \equiv d_1 = d_2$
ψ	Angle of intersection between the beams intersecting at a point within the image space
s	Voxel elongation factor, being the ratio between the longest and shortest voxel dimensions; hence $s_{\max}$ is the maximum permissible elongation
α	Ratio of beam source distance to image space radius, $\alpha = D/R$
α_{180}	Value of α at which the beam intersection angle at the near face of the image space is $180°$
$\alpha_{\text{near}}, \alpha_{\text{far}}$	Value of α corresponding to near and far solutions
c	Defined as shorthand for $\cos(\frac{1}{2}\beta_{12})$
A	Defined as shorthand for $\cos(\psi_{\min})$
$^{*}c_{\text{far}}$	Value of c for which second derivative of Eq. (6.27) is zero
$^{*}c_{\text{near}}$	Analogous to $^{*}c_{\text{far}}$ for near condition

Chapter 7

ΔE	Energy gap between ground and fluorescent electronic states in a material exhibiting two-photon absorption (Section 7.5.4)
I_1, I_2	Intensities of two activation beams pumping a two-photon absorption in a volumetric display application
ω_1, ω_2	Angular frequencies of the radiation in the two activation beams [$\text{rad} \cdot \text{s}^{-1}$]
Δt	Duration of overlap between pulses from the two activation beams in a two-photon absorption in a volumetric display application
n	Number of voxels to be activated along the activation beam paths through the image space
d	Distance between adjacent voxels along the beam paths within the image space

λ_1, λ_2	Wavelength of activation radiation beams for use with volumetric display units employing photochromic material (Section 7.5.5)
N^2	Number of pixels upon a two-dimensional panel employed to address an active matrix display unit (Section 7.7.1)

Chapter 8

(u, v, w, i)	Three-dimensional spatial coordinates and intensity of voxels in the original data (application) coordinate system
(x, y, z, i)	Three-dimensional spatial coordinates and intensity of voxels in image space frame
t_{op}	Time coordinate governing when a voxel descriptor is to be output to the image space
(x', y', i)	Generalized coordinates and intensity value for voxel descriptors arranged in memory so that the remaining degree of freedom (the third spatial position) is governed by the voxel descriptor output time
(X, Y, Z)	Coordinates of a right-handed Cartesian coordinate system fixed to each beam source (beam source frame); the origin is defined as being coincident with the deflection point of the beam,* the Z axis along the optical axis of the beam source (path of undeflected beam), and the X and Y axes parallel to pure horizontal and pure vertical deflections, respectively $(\theta_X, \theta_Y, t_{op}, I)$ voxel descriptor expressed as horizontal and vertical deflection angles, output time, and voxel intensity
f_s	Cyclic motion frequency of screen [Hz]; also, angular frequency is denoted $\omega_s = 2\pi f_s$, and the motion period by τ_s
τ_s	Screen motion period [s]: $\tau_s = 1/f_s$
ω_s	Angular frequency of screen motion [rad $\cdot$ s^{-1}] : $\omega_s = 2\pi/\tau_s$
L_X, L_Y, L_Z	Image space dimensions of the example system
D	Distance between electron gun and center of image space for example system
n	Number of slices into which the example image space is divided
δZ	Incremental distance moved by the surface of emission in the TM (translational motion) example system in a single voxel time T

---

* Horizontal and vertical deflections are assumed to occur at the same point in space.

T_G	Gating time of the electron beam employed in the TM example system
T_D	Dwell time of the beam employed in the TM example system, this being the time it is required to impinge on the surface of emission in order to achieve the required voxel brightness

Chapter 9

N_a	Voxel activation capacity
η	Number of independent voxel activation mechanisms; hence the number of subspaces
N_j	Number of voxels that may be activated within a single subspace
N_{sub}	Number of voxels that may be activated within each subspace when they all may contain an equal maximum number of voxels
n_i	Voxel activation capacity of the ith subspace
L	Linear dimension of an image space in the direction of screen motion
ξ	Separation of adjacent voxels on the surface of emission of an active swept-volume display unit
χ	Separation of adjacent voxels along the one-dimensional subspace created by each voxel generation center in an active swept volume display unit as the screen sweeps through the image space
r	Distance of voxel from rotation axis in a rotating screen display
i	Integer counter through subspaces ($1 < i < \eta$)
j	Integer counter through groups of subspaces lying at the same distance r from the rotation axis in a rotating screen display
V_i	Volume of ith subspace
V_e	Effective image space volume
V_d	Volume of dead image space
σ_i	Temporal fraction of the image update period τ_u for which the ith subspace may be addressed
σ_{sub}	Value of σ_i when all subspaces occupy the same temporal fraction
P	Parallelism of the display unit
T_i	Time duration corresponding to σ_i: namely, $T_i = \sigma_i \tau_u$
T_{sub}	Value of T_i when all subspaces occupy the same temporal fraction
R_{sub}	Voxel throughput rate (voxels/second) into each subspace
B_D	Digital bandwidth (bytes/second)

n_p	Number of replicated voxel processors, each responsible for addressing one or more subspaces
φ	Angle between horizontal (xy) plane of image space frame and line connecting the origin of the beam source and image space frames [equal to $(\pi/2 - \eta)$, where η is used in Chapter 4]
(r, z, β)	Cylindrical polar coordinates within the image space frame, with coincident origin and z axis to frame $\mathbf{x}'$ above; in rotating screen displays, β thus parameterizes the screen position
β_0	Angular position of beam source frame about rotation (z) axis of image space frame
$\mathbf{X} = (X, Y, Z)$	Coordinates of a right-handed Cartesian coordinate system fixed to each beam source (beam source frame); the origin is defined as being coincident with the deflection point of the beam,* the Z axis along the optical axis of the beam source (path of undeflected beam), and the X and Y axes parallel to pure horizontal and pure vertical deflections, respectively
$\mathbf{T}$	Homogeneous (4×4) transformation matrix describing the rotations and translations required to map $\mathbf{X}$ into the image space frame $\mathbf{x}$
$\mathbf{t} = (t_x, t_y, t_z)^{\mathrm{T}}$	Translation vector linking origins of the image space and beam source frames
$\mathbf{n} = (n_x, n_y, n_z)^{\mathrm{T}}$	Normal unit vector, comprising part of rotation submatrix linking $\mathbf{X}$ and $\mathbf{x}$ coincident with X axis
$\mathbf{o} = (o_x, o_y, o_z)^{\mathrm{T}}$	Orientation unit vector, comprising part of rotation submatrix linking $\mathbf{X}$ and $\mathbf{x}$ coincident with Y axis
$\mathbf{a} = (a_x, a_y, a_z)^{\mathrm{T}}$	Approach unit vector, comprising part of rotation submatrix linking $\mathbf{X}$ and $\mathbf{x}$ coincident with Z axis
$\mathbf{R}$	3×3 rotation submatrix describing relative orientations of $\mathbf{X}$ and $\mathbf{x}$

Chapter 10

t_{op}	Time during an image refresh frame at which a voxel is to be activated; for swept-volume systems, this must correspond to the screen passing through the voxel location within the image space

Chapter 12

m	Number of beam sources supported by each scanning mechanism

---

* Horizontal and vertical deflections are assumed to occur at the same point in space.

n	Number of voxel generation centers in a column at a distance r from the axis of rotation
r	Radial distance of the selected column of voxel generation centers from the axis of rotation
L	Length of column of voxel generation centers
ω	Angular frequency of motion of image medium: $\omega = 2\pi f$
Δt	Time between activation of adjacent voxels within a single subspace

Appendix

$\mathbf{x} = (x_1, x_2, x_3)^{\mathrm{T}}$	3-vector describing the position of a point in a Cartesian coordinate frame
$\mathbf{x}'$	Transformed 3-vector
$\mathbf{A}$	3×3 matrix encapsulating shear, rotation, and scaling operations
$\mathbf{t} = (t_{x1}, t_{x2}, t_{x3})^{\mathrm{T}}$	Vector describing a translation in three-dimensional space
θ_{x1}	Angle of rotation about the x_1 axis
θ_{x2}	Angle of rotation about the x_2 axis
θ_{x3}	Angle of rotation about the x_3 axis
$\mathbf{A}_{\mathrm{rotn},x1}(\theta_{x1})$	Matrix describing a rotation of θ_{x1} about the x_1 axis
$\mathbf{A}_{\mathrm{rotn},x2}(\theta_{x2})$	Matrix describing a rotation of θ_{x2} about the x_2 axis
$\mathbf{A}_{\mathrm{rotn},x3}(\theta_{x3})$	Matrix describing a rotation of θ_{x3} about the x_3 axis
s_{x1}	Scaling factor parallel to the x_1 axis
s_{x2}	Scaling factor parallel to the x_2 axis
s_{x3}	Scaling factor parallel to the x_3 axis
$\mathbf{A}_{\mathrm{scale}}(s_{x1}, s_{x2}, s_{x3})$	Matrix describing scaling parallel to the three coordinate axes
$\mathbf{X} = (X_1, X_2, X_3, X_4)^{\mathrm{T}}$	4-vector invoked in homogeneous transformation formalism
$\mathbf{X}'$	Transformed 4-vector
$\mathbf{M}$	4×4 homogeneous transformation matrix
s	Overall scaling factor

VOLUMETRIC
THREE-DIMENSIONAL
DISPLAY SYSTEMS

CHAPTER 1

VOLUMETRIC SYSTEMS AND THE PROCESS OF VISUALIZATION

We shall not cease from exploration
And the end of all our exploring
Will be to arrive where we started
And know the place for the first time[*].

1.1 INTRODUCTION

Volumetric displays have been the subject of almost continual research activity for more than 50 years; during this time numerous systems have been proposed and some have been implemented in prototype form (Blundell and Schwarz 1995; Blundell et al 1993a).[†] Although simple in concept, volumetric displays have in general turned out to be complex when constructed, often taking the form of large systems of experimental apparatus surrounded by bundles of cables, optical components, and electromechanical devices operating in semidarkness. The construction of each prototype has generally represented a major undertaking in terms of the time devoted to the production and integration of mechanical, electronic and software subsystems.

The fascinating history of the research undertaken on volumetric display systems is outlined, perhaps for the first time, in this book. While reviewing historical aspects of volumetric display development we learn from the technical progress (and mistakes) of others and may also begin to understand why volumetric systems have failed to progress from the experimental prototype phase

---

[*] T.S. Eliot, "Little Gidding," in *The Four Quartets*, Harcourt Brace, NY, 1974.
[†] A very small number of systems have been developed as commercial products mainly for evaluation purposes.

to that of significant commercial realization. It is, after all, rather surprising that despite such a protracted period of research, the immense effort directed toward their development and the relative simplicity and availability of the technologies involved in their implementation, there has not yet been any major breakthrough in their commercialization.

Unfortunately, scientists and engineers working on the development of volumetric systems have tended to work in isolation. This has led to a major duplication of effort and a failure to adopt standardized terminology. This situation is reflected in scientific publications and patents where even the most basic components in display systems are often referred to by different names. Despite the extensive period during which work on this area has been carried out, we are forced to conclude that this area of research has not yet been adequately formalized and that past research has largely been conducted on an ad hoc basis. This situation may have occurred as a consequence of the perceived commercial potential of the technologies under study and a natural wish on the part of researchers to protect their commercial interests.

In this work we intend to review historical and current aspects of volumetric display system research, and perhaps more importantly, to establish standardized terminology and formalism which we hope will form a foundation for those studying this subject in the future. By means of these foundations, it is possible, in later chapters, to perceive with greater clarity the potential of volumetric displays as an aid to visualization and anticipate the next generation of high-definition volumetric systems.

In this chapter we introduce the most elementary principles of volumetric display system techniques and present a simple display model that will be used during the course of the book for the purpose of reducing some of the more abstract concepts to a simple form. Section 1.4 provides a brief description of the processes through which we perceive the three-dimensional (3-D) nature of our surroundings. Since many excellent texts exist concerning this fascinating subject, our discussions will, in general, be limited to matters that are especially relevant to volumetric visualization.

Each type of 3-D display technique is able to depict image scenes in such a way as to satisfy a number of the physiological and psychological depth cues. The inability of display techniques to satisfy all the various depth cues in a natural manner coupled with the varying emphasis that we place on these cues and the variety of situations in which displays may be operated precludes the possibility of developing a single universal 3-D display system. Volumetric displays will therefore complement a range of visualization systems. Although some display techniques have gained more widespread popularity and acceptance, there can be no question of direct competition between them, as each has a unique role to play within the spectrum of our visualization needs. In Section 1.5 we briefly outline some other types of 3-D display technologies. Rather than attempting a comprehensive discussion on each of these types of system, we provide references to some of the many books and journal articles that describe them in detail.

The use of specialized terminology in this chapter has been minimized so as to allow those visiting this area of research for the first time to concentrate on general concepts. Expressions have therefore been chosen for their descriptive merit and are not necessarily the same as those introduced in subsequent chapters which are more precisely defined.

1.2 VOLUMETRIC DISPLAY

Common to all volumetric display systems is a volume or region occupying three physical dimensions within which image components may be positioned, depicted, and perhaps manipulated. Some examples of simple volumetric images are illustrated in the set of color plates included in the book. Since these images are generated within a region of 3-D space, and may be observed directly therein, a number of depth cues (see Section 1.4) are automatically satisfied and their three-dimensionality is naturally perceived. The volume may therefore be considered to provide a 3-D tableau and in contrast to many other 3-D display techniques, the sensation of depth occurs automatically and is an intrinsic property of volumetric image depiction. Depth cue conflict may therefore be avoided and this denotes a considerable benefit of this type of display methodology. In principle, the computational overheads are small in comparison to other approaches, furthermore, images displayed within the volume may be observed naturally without the need for an operator to wear special glasses.

A number of the various volumetric display technologies that have been the subject of past research or which are currently under investigation impose very little restriction in the viewing angle. An operator may therefore move around the volume and, in the case of some volumetric architectures, view an image scene from practically any orientation. For those visiting this area of research for the first time, the concept and characteristics of the display volume may be appreciated by watching fish moving in an aquarium. The fish (which we may think of as image objects) can position themselves at any location within a clearly defined volume. As a consequence, the image scene is automatically 3-D and the spatial separation of the fish is readily apparent.

It is interesting to note (and will be a matter for further discussion later in this work) that the physical dimensions of the volume in which the fish move (as measured) differ from the apparent volume (as observed) due to the refraction of the light as it passes from the water and containing glass vessel to the surroundings. So although we consider that a volumetric display must, by definition, employ a physical volume within which images may be constructed, the apparent dimensions of this volume may not correspond to its physical measurements. Should we generate images within a volume whose refractive index differs from that of the surroundings, we may anticipate a bending of the light as it crosses the boundary of the volume. The shape (and curvature) of the volume will then become a matter for important consideration. In the case of a spherical volume with a high refractive index, the lensing effect may cause considerable image distortion, and it is suggested that the reader examine the consequences for volumetric imaging by observing fish swimming within a spherical bowl. As we will see, optical effects at the volume boundary may, for certain applications, be advantageous and for others quite the opposite.

In the case of all volumetric display systems known to the authors, the generation of images occurs within a containing vessel from which the observer(s) is excluded. Volumetric systems therefore provide a "God's-eye" view of any image scene. However, it is possible to envisage the potential of a display device able to project volumetric images into a region of space through which one or more observers are free to pass. The development of such an immersive environment may one day completely change the nature of volumetric display systems.

Finally we believe it to be appropriate at this point to provide a definition that characterizes the essential nature of any volumetric display system and which may therefore be used to differentiate between this type of display and other 3-D visualization techniques:

> *A volumetric display device permits the generation, absorption, or scattering of visible radiation from a set of localized and specified regions within a physical volume.*

1.2.1 Display System Classification

For many years, systems able to provide a 3-D tableau within which images may be constructed have been referred to as *volumetric displays*. The origins of this term are unclear, and it is perhaps unfortunate that the word *volumetric* was adopted in the first instance for the classification of these display devices. In the *Oxford English Dictionary** we find *volumetric* defined as follows:

volumetric: adj. Of, pertaining to, or noting measurement by volume.

Volumetric therefore clearly implies measurement by means of spatial information. Since all types of physical measurement lead to some variety of quantitative information, the term *volumetric display* may be considered to refer to display apparatus that permits quantitative measurement of visual information. This is misleading, as volumetric display systems are best suited to the qualitative visualization of spatial information and, as we will see in Section 8.2, are least suited to the depiction or provision of quantitative information.

Volumetric display systems have also been referred to as *multiplanar display systems* (Williams and Garcia 1988, Bains 1993) and *direct volume display devices* (DVDDs) (Clifton and Wefer 1993a, 1993b). The latter is perhaps a more appropriate title. However, it has not been adopted for use in this book, for two reasons: First, the use of this expression is not yet widespread, and second (and perhaps more importantly), the word *volumetric* is now firmly associated

---

*The *Oxford English Dictionary*, 2nd edition, Volume XIX, prepared by J.A. Simpson and E.S.C. Weiner (Oxford, Clarendon Press, 1989).

with certain types of image data sets. Although such data may be depicted on all types of display systems, in the case of volumetric displays they form an essential source of image information. Consequently, the common use of the word *volumetric* in the description of image data has led to our continued adoption of this word as a means of classifying the display systems that are the primary subject of this book.

1.2.2 Volumetric Display System Family

When we examine various display system implementations, it is generally very easy to distinguish between those that are volumetric and those that fall outside this classification. However, for a small number of display system implementations, classification is more difficult. For example, varifocal mirror technologies employ a reflective surface, the curvature of which is rapidly varied. The surface therefore acts as a mirror of varying focal length. By means of an appropriate control system, portions of a 3-D image projected thereon appear to originate from different depths. The image thus occupies a virtual volume whose dimensions exceed the amplitude of the surface's motion. By the definition provided earlier in this section, such displays fall outside the scope of volumetric displays as considered in this book (since they employ a virtual rather than a physical display volume). Varifocal mirror techniques are discussed in a number of excellent publications [e.g., Traub 1967; Harris et al. (1986); U.S. patent 4,130,832; Sher (1993) and references therein]. The matter of their proper classification is left to future debate.

1.3 AN ELEMENTARY SYSTEM MODEL

Many issues relating to the implementation of a volumetric display may best be illustrated by reference to a simple example. In this section we therefore introduce an elementary model of a display device within which volumetric images may be generated. This model will be referred to during the course of the book and has been selected to reduce the underlying concepts and properties of an image depiction volume to their most basic and therefore the most comprehensible level. In the form described here, the model has a number of both obvious and subtle deficiencies and is therefore not intended to represent a feasible volumetric implementation. In subsequent sections we highlight some of the deficiencies associated with this model.

Graphical images are typically constructed from a set of image elements. In the case of conventional cathode ray tube (CRT)–based displays or those employing a liquid crystal display (LCD) panel, images are formed by the activation of picture elements (pixels) at appropriate locations within 2-D space. Each of these elements has a unique location (x, y) within a two-dimensional plane and a color and gray-scale value may be assigned to each. Volumetric display systems employ a similar method for image construction; elements [referred to as voxels (see Section 2.3)] are activated at the appropriate locations within the 3-D space contained within the volume. As is the case for conventional displays, each voxel has a unique position. However, unlike conventional display devices, this is defined within a 3-D volume (x, y, z). In common with conventional display systems, each voxel may be assigned a color and gray-scale value and perhaps for future display systems it will also be possible to assign to each voxel an opacity descriptor (see Sections 7.5.5 and 10.2.3).

Any volumetric display system must provide a mechanism for the generation of voxels within a volume, and in the case of our elementary model we employ for this purpose a simple 3-D array of light bulbs (filaments). The volume will be assumed to take the form of a cube with sides of length L containing N bulbs. Since each filament will be responsible for the generation of a single voxel, the display will be capable of generating up to a total of N spatially separated voxels. This example display model is illustrated in Figure 1.1. In view of the simple technique employed for the generation of each voxel, our elementary display will clearly support the depiction of only single color images. Controlling the current passing through each of the bulbs may accommodate gray scale.

Clearly, should we wish to generate small voxels, the display device will need to employ a large number of bulbs (and electrical connections) to achieve a useful display volume and so be able to depict images of an appropriate size. Furthermore, we can see that voxels may only be generated at unique locations within the volume, and as a consequence aliasing (Foley et al. 1990, Hill 1990) will occur as we attempt to generate lines that are neither horizontal nor vertical, nor which form a diagonal across the cubic image space.

Also worth consideration is the fraction of the total number of available bulbs that we may wish to activate simultaneously. Clearly, as more bulbs (and hence

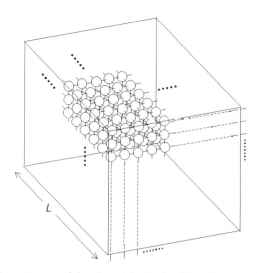

FIGURE 1.1   Elementary model volumetric display.The display volume has sides of length L and contains N regularly spaced lamps (filaments). Since this type of display does not incorporate any mechanical motion of components within the volume, it is classified as a static-volume display device (see Sections 3.1 and 6.2).

voxels) are illuminated, the display volume will become increasingly cluttered and it will become increasingly difficult to discern the spatial separation and form of the translucent image components contained therein. As will become clear in subsequent chapters, the number of bulbs illuminated at any one time should represent only a small fraction of the total number of available bulbs. This may be demonstrated by reference to our previous comparison between a display volume and tank containing fish. As we add increasing numbers of fish to the tank, their spatial separation becomes less apparent and some fish will become obscured by others that are closer to the observer and cross the observer's line of sight. If we continue adding fish, those closest to the observer will eventually obscure completely the motion of those located farther within the tank. In a similar way, if we increase the fraction of activated voxels in a volumetric display device, we will eventually reduce the benefits associated with the display volume's third dimension.*

1.4 PERCEPTION AND THREE-DIMENSIONAL SPACE

Human perception of the 3-D world is an extremely complex process that continues to be the subject of research. To depict images that may be viewed in a natural manner, electronic display devices take advantage of, and are perhaps limited by, the way in which our visual system perceives its surroundings. For example, the required brightness and speed at which images need to be updated for the realistic illusion of motion are both determined by parameters associated with the visual system. A study of 3-D display techniques must clearly encompass a knowledge of *depth cues*, the visual-based mechanisms through which we obtain depth information about the surrounding environment. An understanding of depth cues and the human visual system should form an essential starting point for all working on and developing display system technologies. Many excellent texts have been written on this fascinating subject. In this section we discuss some of the visual characteristics and depth cues that will be of importance in our subsequent discussions. For further information we refer the reader to some of the comprehensive works published in this area (Kaufman 1974, Dember and Warm 1979, Schiffman 1990, Wade and Swanston 1991, Yeh 1993).

1.4.1 Visual Acuity

Visual acuity is a measure of the level of detail the eye can distinguish. Most electronic displays present their images as a 2-D array of pixels, and volumetric systems do so as a 3-D distribution of voxels. In order that images appear sufficiently smooth, the required size and separation of these image elements, is dictated by the visual acuity of the eye in combination with the probable viewing

---

* This analogy is not quite accurate since the majority of volumetric display technologies give rise to translucent rather than opaque images. In this case considerable reliance is placed on the accommodation depth cue.

distance.* Moreover, the smallest size of image detail that can reliably be detected by an observer is also important when considering the likely or intended application(s) of a display system. This will govern the required spatial displacement of an image component relative to others in order that it may be observed reliably. It will also indicate the level of image detail that may be discerned.

The acuity, or maximum resolving power, of the eye's foveal region,[†] is generally considered to be a visual angle ϕ_c approximately equal to 1 minute of arc. However, this acuity is broad and very approximate, encompassing a number of acuities involving different visual tasks (Poole 1966, Schiffman 1990). The resolving power of the eye is generally different for each acuity subtype and varies widely from person to person. Five types of visual acuity are relevant to the study of display systems: minimum visible, minimum perceptible, minimum separable, vernier, and stereoscopic acuity. These are discussed by Poole (1966) and are outlined briefly below in the context of volumetric displays:

- *Stereoscopic acuity.* This describes the minimum difference in distance (from the observer) of two objects that may be discerned. It depends not only on the absolute depth of the objects from the observer, but also on their spatial separation perpendicular to the line of sight. Consider two objects located at slightly different distances and offset from each other perpendicular to the line of sight, as illustrated in Figure 1.2. Each eye alone will perceive a slightly different angular separation of the two objects, denoted β_1 and β_2 in this diagram. The minimum difference $\eta = |\beta_1 - \beta_2|$ between the angular separation seen by each eye between two such objects provides a measure of the stereoscopic acuity. This is, in effect, a quantification of the binocular parallax depth cue (discussed in Section 1.4.4.). The stereoscopic acuity is typically about 10 arc seconds but may vary between observers from 2 to 100 arc seconds. This measure is of considerable importance to 3-D display systems, as it indicates the magnitude of relative depth displacement that may be distinguished reliably. In the case of volumetric (and autostereoscopic) systems, the viewing distance may be variable, but the availability of other depth cues will aid in appreciation of the 3-D structure of the displayed data. However, applications requiring a rapid response from the observer may rely more critically on stereoscopic acuity.
- *Minimum visible acuity.* This refers to the minimum size of an object that can be seen when it is brighter than its background[‡] and is a function of intensity rather than size; it is detected by the triggering of one (or more) cones in the retina. Interestingly, this means that a bright object against a dark background may be visible even though it subtends an angle smaller

---

* This is because the acuity depends on the feature size as registered on the retina—the acuity is expressed in terms of a visual angle.

[†] The region containing a high concentration of cones in the center of the retina.

[‡] In the context of volumetric displays, this could correspond to a luminous voxel representing, for example, aircraft position in air traffic monitoring applications.

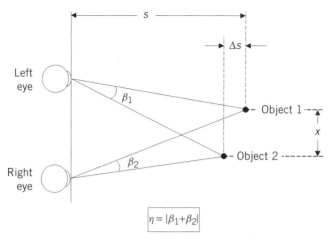

FIGURE 1.2   Stereoscopic acuity. Two objects located at distances s and $s - \Delta s$ from the observer are discerned as being at different depths if the difference between angles β_1 and β_2 subtended at each eye by the two objects is greater than the stereoscopic acuity. These angles depend on the absolute distance s, the relative distance Δs, and their separation x perpendicular to the line of sight. [After Poole (1966, p. 277) with permission.]

than the eye's minimum resolvable spot size (1 arc minute, determined by the size of the cones in the fovea) but it will *appear* to subtend 1 arc minute. This is because once a cone is triggered, the object appears as large as the cone.

- *Minimum perceptible acuity.* This is analogous to minimum visible acuity,[*] but for the case of a dark object against a bright background. This generally has a larger value than the minimum visible acuity, as a dark object will not be noticed if the surrounding bright region triggers all the cones in the corresponding area of the retina. The eye is thus better able to discern isolated bright points against dark backgrounds than vice versa.

- *Minimum separable acuity.* This corresponds to the ability of the eye to distinguish alphanumeric characters and is thus of considerable importance in the context of 2-D display systems. However, in the case of volumetric displays, the depiction of textual information may be considered at the present time to be of less importance, as discussed in Section 8.2. We refer the reader to Poole (1966, Chapter 15) for further discussion.

- *Vernier acuity.* This is the ability of the eye to distinguish an offset between two nearly colinear lines. The eye may distinguish such offsets at scales finer than 1 minute of arc; this is due to the random placement of the cones in the retina and the fact that a line will overlap to varying degrees a large number of cones.

---

[*] The minimum visible and minimum perceptible acuities may be considered together as minimum detectable acuity (Poole 1966).

An important consideration for volumetric displays is the role played by the background as well as the foreground in most acuity tasks. Recognition and interpretation of displayed information may thus be affected, in display units offering an all-round view, by objects and other people situated on the opposite side of the image space from the observer as well as by the ambient lighting conditions.*

Our ability to distinguish the physical dimensions of an object or detail depend on the distance from which it is viewed — the more distant an object, the smaller the angle it subtends at the eye. Consider an object of size x a distance s from the eye; as illustrated in Figure 1.3, the visual angle ϕ subtended by the object is given by 2 arctan $(x/2s)$. For small angles, standard approximations yield the dependence of the visual acuity on object size and distance as

$$\phi = \frac{x}{s} \qquad (1.1)$$

Therefore, any detail of size x perpendicular to the line of sight can be resolved only while the viewing distance is less than $\approx x/\phi_c$.

Such considerations affect the design of electronic display systems, where the intended and possibly worst-case viewing positions may be considered. In the specific case of volumetric displays, the size, shape, and separation of individual voxels drawn within image space will determine to what extent, and for which viewing positions, the visual acuity limit is reached for a particular technique. This in turn may influence likely applications for the system. Although the discussion above has highlighted some of the issues governing the size and relative placement of image features, a sounder understanding of the precise implications of the various acuities as they apply to volumetric displays awaits empirical data from well-defined experimental trials.

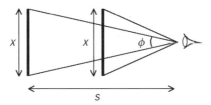

FIGURE 1.3 An illustration of Euclid's law (Burton 1945) in 2-D. An object of given size x subtends a smaller solid angle ϕ as its distance s from the observer increases.

---

*If the envelope surrounding the volume within which images are depicted is chosen so as to be semitransparent (rather than fully transparent), the image contrast relative to visual interference from background objects within the room may be improved. This is because external light crossing the volume must pass through this envelope twice and thus is attenuated by twice the amount compared to light originating from within the volume. However, the absolute image brightness is, of course, diminished.

1.4.2 Temporal Perception: Flicker

Most electronic display systems utilize transient luminescent phenomena for the creation of visible images. Images must therefore be refreshed at a frequency sufficient to ensure that the observer(s) perceives the visual information as continually present. For example, in the case of cathodoluminescence (von Ardenne 1939, Ozawa 1990), the light intensity emitted from an excited phosphorescent layer may decay substantially in a time period of perhaps 100 μs after the removal of the source of excitation. Any display employing this technique must therefore be refreshed at a rate sufficiently frequent that the continual variations in intensity are not apparent to the visual system. The minimum refresh frequency at which such an image appears to be continually present is known as the *critical flicker fusion frequency* (f_c). However, a precise value for this quantity is not easily determined, as it is dependent on the angular motion response of the eye (and hence the direction of gaze relative to the stimulus) and on the spatial and temporal nature of the flickering pattern. It also varies with the logarithm of the average luminance (L_{av}) of the stimulus according to the Ferry–Porter law (Sherr 1992):

$$f_c \sim a \log L_{av} + b \tag{1.2}$$

where a and b are constants. However, due to the other factors mentioned above, this equation does not in general provide an accurate value for f_c. The effects of flicker are more visible when the brightness levels of the stimulus are greater — this may be advantageous for displays operating in subdued lighting conditions. Since many volumetric display system technologies produce images of low intensity that are viewed in semidarkness and for only short periods, it has generally been possible to reduce the refresh frequency to levels that would otherwise be unacceptable.

An image that is in direct gaze begins to appear free of flicker at refresh frequencies in excess of 25 Hz.* The higher the refresh frequency, the smaller is the perceived variation in brightness. The human eye is more sensitive to motion at the periphery of its vision (i.e., outside the fovea centralis); flicker unobservable when looking directly at an image is often noticed when the gaze is elsewhere. It is interesting to note that although an image may appear comfortably flicker-free to both the direct gaze and peripheral vision, it may still exhibit subliminal flicker that may lead to discomfort after prolonged viewing. To minimize the effects of flicker, computer monitors now commonly employ effective update frequencies of 70 Hz or greater, and time-sequential stereoscopic displays may have refresh frequencies as high as 120 Hz (MacDonald and Lowe 1997).

1.4.3 Illusion of Motion

The minimum refresh frequency required to ensure freedom from flicker is somewhat greater than that required to provide the illusion of motion. For a sequence

---

* This is an approximate value. Under favorable conditions it be may reduced.

of images to appear to change continually and smoothly, animation frames need only be changed at a frequency of approximately 10 Hz. Thus, while a higher image refresh rate is required to minimize flicker, it is not necessary to alter the displayed image frame at each refresh. For example, in the case of a display system in which the output device is updated at a frequency of 30 Hz, the image content need (in principle) change only after each third refresh cycle. As a consequence, during a continual animation sequence, the buffer that contains the image data for this display need only be updated each one-tenth of a second.*

1.4.4   Summary of Some Depth Cues

The 3-D space around us is, visually speaking, mostly void. To perceive and comprehend the spatial distribution of objects within these surroundings, we employ (at a subconscious level) various features of the 2-D projection of the objects in the visual field on our retinas. Furthermore, we also derive information from the physiological behavior of the eyes during the process of vision. The projection of a 3-D scene onto the retinas (or any 2-D surface) will at every point on the surface include information about the nearest opaque object in the line of the projection.

Elements of the visual perception of depth are known as *depth cues*. It is important to note that these cues do not act in isolation — visual input is augmented continually by tactile and auditory sensations. In addition to these external stimuli (which depend on the nature of the scene being observed), "internal" input from the brain, such as prior knowledge (experience), associations, or expectations, may also affect the process of visual perception. Past experience provides a calibration of the visual input and is an important sense of absolute depth information; most depth cues by themselves indicate only relative distance (Ittelson 1960, Kaufman 1974).

The depth cues outlined briefly below may be grouped according to several methodologies (Dember and Warm 1979). One grouping scheme divides them into two categories (Figure 1.4): the *physiological cues* comprise those that rely on feedback from muscle groups, and *retinal image cues* comprise the remainder, in which features of an image scene observed by the visual system provide the source of input. Retinal image cues may be further subdivided into static monocular (or "pictorial") cues, which function with a single retinal projection, and parallax cues, which make use of information from more than one projection.

- *Binocular parallax (or stereopsis).* Each human eye occupies a slightly different position in space (the interocular distance is, on average, about 6.5 cm), and hence each receives a slightly different perspective of any scene. The brain fuses the images from each eye and interprets the differences in the apparent relative positions of objects in each view as relative

---

* This is a generalization that makes assumptions regarding the dimensions of the display and the speed of the motion of any image components depicted therein.

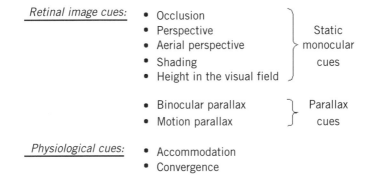

FIGURE 1.4   One possible classification of depth cues, in terms of physiological cues (those based on feedback from muscle groups in the eye) and retinal image cues (those based on the visual projections received on the retinas).

depth (Wheatstone 1838; Julesz 1964, 1971.) In practice, for objects located at distances greater than approximately 10 m, the differences between the two perspectives are negligible, so beyond this distance the cue of binocular parallax provides no depth information. However, it is of interest to note that approximately 5% of the population may not possess this depth cue (Richards 1970, Yeh 1993).* Bardsley and Sexton briefly discuss benefits that may be derived from the use of stereoscopic video display techniques (MacDonald and Lowe 1997) and comment:

An element of visual noise may be present in the video image due to environmental factors or poor quality image reception. The human visual system is particularly adept at filtering out uncorrelated visual noise from binocular scenes to give greater picture quality. This ability is invaluable in determining not only *where* objects are in unfamiliar or complex scenes, but frequently *what* they are. (Merritt 1983).

- *Motion parallax (or temporal parallax).* Just as each eye receives a different perspective of the 3-D environment, relative motion between the head and the 3-D scene provides a continually changing perspective. This provides depth information in a manner that is analogous to stereopsis. Objects at different distances will alter their position in the 2-D retinal projection(s) at different rates.
- *Linear perspective.* Objects at greater distances from the observer subtend a smaller angle in the visual field than do nearby objects, and hence appear smaller (Edgerton 1975) (cf. Section 1.4.1); this is known as *Euclid's law* (Figure 1.3). Thus the relative sizes of similar objects gives an indication of their spatial arrangement, as evidenced by the convergence to an

---

* Of the 150 subjects comprising the study sample of Richards (1970), 4% were unable to use disparity, and 10% did so only with great difficulty.

FIGURE 1.5   The linear perspective depth cue. The parallel straight railway lines appear to converge with distance from the camera, as they are not perpendicular to the line of sight. Also, similar-sized objects appear smaller as they become more distant.

apparent vanishing point of a railway track, fence, or road extending into the distance (Figure 1.5). When the objects in question are texture elements on the surfaces of other objects, the effect is sometimes referred to as *texture perspective* (Gibson 1950).

- *Occlusion (or interposition).* An opaque object interposed between the observer and a second object will render at least part of the second object invisible. By means of this cue, the brain interprets partially occluded objects as lying farther away than interposing ones (Figure 1.6).

- *Shading.* The shading and shadowing patterns of an object provide an indication of its form, surface texture, and position relative to its surroundings. Since light propagates in straight lines, the shading gradients on the surface of an object enable the topography of its surface to be predicted by the brain; a gradual change in the darkness of a surface indicates a curved surface, whereas an abrupt change signals the presence of a sharp edge. In the absence of information to the contrary, the brain generally assumes a light source incident from above (Figure 1.7). In conventional computer graphics, this cue is heavily exploited in rendering techniques (Foley et al. 1994).

- *Aerial perspective.* Due to light scattering by the interceding atmosphere (including water and dust particles), distant objects tend to appear hazy.

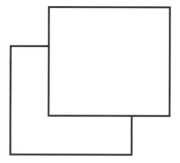

FIGURE 1.6 Occlusion depth cue. The square at the top right is perceived as being in front of the other because it partially occludes the latter.

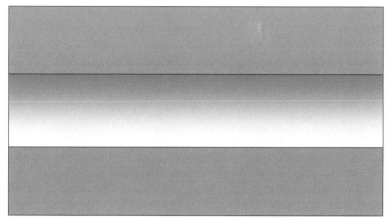

FIGURE 1.7 The central bar is generally seen as a concave trough, due to the brain assuming a light source from above. If the figure is viewed upside down, the same (subconscious) assumption leads to the pattern appearing convex.

As the distance of an object increases, it is seen less clearly, its coloration dulls, and it takes on a bluish tinge. Thus, assigning these characteristics to features of a rendered scene provides a sensation of distance. The aerial perspective depth cue may cause distant geographical features to appear to be in closer proximity on days when the air is clearer than usual. Under such circumstances, the effects of aerial perspective are minimized.

- *Height in the visual field.* Objects appearing higher in the retinal image are generally interpreted as being farther away, as the vertical positioning of objects in the visual field generally increases toward a horizon line (at eye level) as their distances approach that of the horizon. Similarly, objects above the eye level (e.g., clouds) appear to be farther away as they descend toward the horizon.
- *Accommodation.* This represents a physiological depth cue. To focus on the plane of attention, exertions of the ciliary muscle alter the shape of the lens

of the eye and so modify its focal length. The magnitude of the tension applied to these muscles is used by the brain to provide information relating to the depth in the field of view of the object under scrutiny. Beyond about 10 m, the lens shape is constant, so this cue is applicable only for objects that are closer than this distance. This cue is of particular importance in the case of volumetric displays which depict translucent images.

- *Convergence.* As with accommodation, this is also classified as a physio-logical cue in which the brain makes use of the muscular forces exerted on the two eyes so that they swivel inward by an amount necessary in order that they both center their gaze on an object of interest. As with binocular parallax and accommodation, this cue is effective only for distances up to approximately 10 m from the observer; as beyond this point the eyes are essentially parallel.

- *Familiarity, expectation, and context.* In the interpretation of a 3-D scene, the role of prior knowledge, experience, and the expectation of the attributes of an object in a particular context should not be underestimated. It has been demonstrated, for example, that assumptions about the size of a familiar object, based on the viewer's experience, can dictate the perceived distance of the object (Schiffman 1990).

- *Color.* The color of an object can also affect its perceived distance — colors toward the red end of the spectrum are perceived as being closer than those toward the green/blue. Although this effect has little consequence when regarding a real-world scene where a range of colors and shadings are present and other depth cues dominate, it is easily noticed when only a few discrete colors are present. The effect may be noticed on electronic text displays at some airports and other public buildings, where alternating lines of, say, red and green text are presented. From a distance, a clear depth offset between the text of the two colors may be noted. A similar situation applies to the current generation of volumetric displays, where only a few discrete colors of voxel may be generated. Volumetric displays are designed to be viewed sufficiently closely that cues such as motion and binocular parallax may be employed to their best advantage, but conflict with the color cue may be a potential source of eye fatigue. The color effect is, however, most noticeable in 2-D photographs or slides of volumetric images, where the additional depth cues do not apply.

1.4.5   Considerations Relating to Volumetric Displays

Many 3-D display techniques seek to add a sensation of depth by simulating either one or both of the parallax depth cues (see Section 1.5). In the case of volumetric display systems the situation is somewhat different. Rather than computing and presenting (via a 2-D medium) different perspectives of a 3-D data set, the entire 3-D data set is depicted simultaneously within a volume. Thus, an observer's eyes will each automatically see the image from a slightly different perspec-tive. Furthermore, by moving relative to the volume, the observer changes the viewing position and hence the perspective. Both parallax cues are thus intrinsically satisfied in a natural manner. This is advantageous for viewers who do not possess the cue of stereopsis (see Section 1.4.4). The physiological cues are also satisfied, alleviating a possible source of eye fatigue. Furthermore, the depth cue information is presented without the need to undertake complex and often time-consuming computation.

Conventional high-resolution display devices are more able to satisfy some of the pictorial depth cues (such as shading) and better represent the surface detail of opaque objects than are current volumetric systems. Furthermore, the image space of a volumetric display represents directly the data displayed — apparent depth in a particular direction cannot be amplified by transforming the data using perspective and relative size, as is the case on a 2-D screen. The data displayed should be scaled uniformly in all spatial dimensions to ensure that it appears accurate from all viewing directions (see Section 10.7). The inability of most volumetric architectures to depict opaque objects results in a failure to properly support the cue of occlusion (see Section 10.2.3).

1.5   METHODS OF DISPLAYING 3-D INFORMATION

Any display system able to provide an observer with the sensation of depth can greatly enhance the impact and qualitative understanding of 3-D information. Rendering techniques used on conventional computer terminals utilize retinal image depth cues such as occlusion, perspective, shadowing, and possibly motion parallax (Fuchs et al. 1989, Coatrieux et al. 1990). The added realism that may be achieved by simulating the binocular parallax depth cue has spurred research into displays capable of replicating the slightly different views seen by each eye. [An excellent outline of some of the current research in this area may be acquired from McAllister (1993) and the *SPIE Proceedings*: for example, Fisher and Merritt (1990), Merritt and Fisher (1991, 1992, 1993), Fisher et al. (1994, 1995, 1996, 1997) and Bolas et al. (1998)]. Research is also being carried out (Okoshi 1976, 1980; Travis 1997) on display systems that permit a modified binocular view to be seen as the observer's head is moved. Examples of such systems include computer-generated holography (Tricoles 1987, Benton et al. 1993, Onural et al. 1994) and parallax barrier techniques (Akiyama et al. 1991). The majority of these autostereoscopic systems operate over a limited range of viewing angle.

1.5.1   Display Systems and Depth Cues

Any display system that attempts to depict spatial information creates images that satisfy a certain subset of the depth cues discussed in Section 1.4. Although cues such as binocular parallax and occlusion are often considered to be dominant in the perception of relief, realism is enhanced by satisfying as many cues as possible. In particular, it is important to avoid prolonged *conflict* between different depth cues. Situations in which two or more cues conflict and therefore provide the brain with differing depth estimates can lead (at the very least) to fatigue (Hart and Dalton 1990). For example, in the case of some stereoscopic techniques, the information provided by binocular disparity is inconsistent with that provided by accommodation. If the display system is not of sufficiently high resolution, the plane in which the image is generated may be perceived subliminally. The accommodation and convergence information may then be in conflict with that provided by the retinal image cues. Although volumetric displays automatically satisfy a number of depth cues in a natural manner and so avoid conflict, we will see in subsequent sections that they may exhibit other problematic characteristics.

Conventional 2-D display systems satisfy only the monocular (retinal image) depth cues. In the case of 3-D display techniques, some or all of the remaining depth cues (i.e., convergence, binocular parallax, accommodation, and motion parallax) are satisfied. Of these, binocular and motion parallax are most frequently employed to enhance the sensation of depth. Ideally, all the retinal image cues would also be satisfied; however, this is not always the case. If a display device cannot present certain depth cues to the viewer, its application tends to be restricted to areas in which the absent cues are not a major disadvantage. It is important to note that the relative emphasis that we assign to the various depth cues varies according to the nature of the image scene under observation. Current volumetric display systems satisfy most of the nonretinal cues but in general do not accommodate shading and full occlusion. However, for certain applications the absence of these cues is not a major disadvantage.

1.5.2 Monocular Displays

Following the development of mathematical prescriptions enabling objects and scenes to be accurately recreated according to the rules of linear perspective, it became possible to utilize 2-D renditions of 3-D objects within scientific works. The studies of Viator and Dürer in the fifteenth and sixteenth centuries were particularly influential in this regard (Ivins 1973). Coupled with the invention of printing (ca.1430), it became possible to include within scientific studies realistic sketches of objects ranging from abstract mathematical concepts and technical drawings to biological specimens. Even when a perspective reconstruction was not necessarily required, the philosophy of portraying objects "as seen by the eye" was extremely influential and marked a significant step forward for science in general (Edgerton 1991).

The invention of photography (ca. 1839) further facilitated the recording of items such as biological specimens. Perhaps the most significant development was the discovery of evidence for the existence of cathode rays (Plücker 1858, Hittorf 1869) using tubes in which an electron beam was generated by a high-voltage discharge in a gas (usually air) at a pressure of about 0.5 torr (Martin 1986). Crookes (1879) demonstrated that these "rays" comprised negatively charged particles that could be deflected by an electromagnetic field. Braun

(1897) invented the first CRT for use as an oscillograph (Blondel 1891). This also consisted of a gas discharge tube containing a cold cathode, requiring a high voltage (≥ 10 kV) applied to an accelerating anode so as to maintain the discharge. Having passed by the anode the beam's spread was limited by an aperture plate before it impinged on a fluorescent screen. External coils were used initially to provide magnetic deflection, but later models incorporated additional electrodes within the tube so as to permit electrostatic deflection (Martin 1986).

Wehnelt (1904) invented the triode gun, which utilizes a control grid, so establishing the foundation for the modern CRT. Between 1904 and 1940 the fundamentals of electron optics were formulated by Busch (1926, 1927), Davisson and Calbick (1932), Knoll and Ruska (1932), Brueche and Scherzer (1934), and von Ardenne (1935, 1939, 1940). The cathode ray oscillograph enabled electrical signals to be depicted in real time. The development of computer systems from the 1940s onward provided further impetus for display systems[*] and it was soon realized that interaction with computers would be more efficient if printed output was augmented with devices able to display both text and graphics.

Initially, CRT-based computer graphics were generated primarily through the use of vectors; that is, the beam was directed from one visible point on the screen to the next and so moved in an irregular path; the path depending on the geometry of the image being depicted. The subsequent refinement of raster-scanned CRTs, the provision of full color, and the development of the associated hardware (including dedicated processing chips) have further increased graphics visualization capabilities. Ray tracing provides a very useful technique for increasing realism (in the photographic sense) of monocular computer images. This is a computationally expensive process in which the position and characteristics of one or more light sources are assumed and the luminance and color properties reflected in the direction of a chosen viewpoint are calculated (Foley et al. 1994). In essence, the more realistic the image, the greater the computation overhead, and this is reflected by the frequent inability of even today's high-performance workstations to render high-quality images in real time. Monocular displays are particularly suited to the depiction of surface detail. Furthermore, the ability to view data in cutaway form facilitates an appreciation of internal structure.

1.5.3 Stereoscopic Displays

The depth cue of binocular parallax, whereby the slightly different views of a scene obtained by each eye are combined in the brain to give an impression of relief (see Section 1.4.4), provides a powerful depth effect. There have been many successful attempts to incorporate binocular information into visualization hardware and a considerable amount of research continues in this area. In the case of stereoscopic displays, each eye is artificially presented with a slightly different view of an object or scene, so as to provide a strong sensation of depth.

---

[*] The earliest electronic (digital) computer systems also employed special-purpose CRTs for information storage and switching.

Although this effect was not noted until the nineteenth century; it had long been realized that each eye sees a different view of the visual field. Previously, philosophers had been more concerned with explaining why we do not see double and had assumed that the only depth cue arising from two eyes was that of convergence (Wade 1987). Wheatstone (1838) appears to have been the first to describe the phenomenon of stereopsis. Furthermore, he was able to demonstrate the powerful nature of this cue by using a device he termed the *stereoscope* which artificially presents each eye with its correct perspective of a scene or object. The object is then seen in relief and no longer appears contained within a 2-D plane. It is interesting to note that there was a certain amount of controversy during the nineteenth century regarding the true inventor of the stereoscope (Wade 1983, 1987). Brewster (1856), in particular, maintained not only that Wheatstone did not originate the device, but that the effect of relief was due to convergence alone. Brewster attempted to attribute the invention of the stereoscope to Elliot in 1834, whereas Wheatstone drew attention to earlier work by Mayo (1833). The relevant excerpts are quoted in Wade (1983, pp. 149, 173).

The stereoscope rapidly became extremely popular, and several different implementations were produced (Brewster 1856). With these devices, physically separate right and left eye views are presented to the eyes via a device comprising mirrors or lenses. A barrier is used to separate the visual fields of the two eyes. The Wheatstone and Brewster stereoscopes are illustrated in Figures 1.8 and 1.9. From the outset the educational and scientific applications of stereoscopic visualization equipment were emphasized (Brewster 1856, Helmholtz 1909).

Other stereoscopic techniques do not require the stereopair to be physically separated, so the viewer need only wear special glasses. This engenders significant freedom of head motion and enables the image to be viewed in a more natural manner. One such method, the *analglyph*, originating with the work of Rollman in 1853 (Southall 1962, p. 356), requires glasses comprising eyepieces of a different

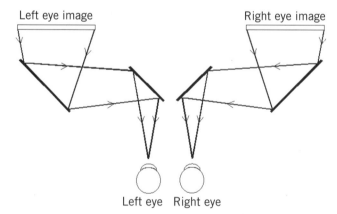

FIGURE 1.8   Arrangement of the Wheatstone (reflecting) stereoscope. Light from each view travels to the intended eye by the system of mirrors.

Left eye image   Right eye image

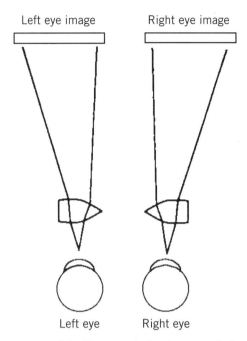

Left eye   Right eye

FIGURE 1.9   Arrangement of the Brewster (refracting or ocular) stereoscope. The two image projections, corresponding to the views seen by the left and right eyes, are directed to each eye by means of prisms.

color (e.g., red and blue). If a stereopair of images are overlaid in red and blue, each eye will then only see the image of the opposite color. This process has been applied to cinematography at various times since the beginning of the twentieth century (Dudley 1951) and is still used today for an economical 3-D effect in scientific visualization and entertainment. One problem with this technique is the provision of natural color to the stereoscopic images. Color may be retained to a certain degree by ensuring the colored filters on the eyepieces are wideband and admit a range of colors rather than a narrow range of wavelengths. Thus, one of the images would contain the low-frequency (yellow, orange, red) color components, and the other the higher-frequency (blue, green) colors. When viewed through corresponding filters, the colors are blended with the fusion of the two images, and a reasonable color reproduction may be obtained (Dudley 1951).

In an alternative embodiment the eyepieces of the viewer's glasses take the form of perpendicularly polarized glass or plastic. An image stereopair is displayed in which the light of one of the pair is perpendicularly polarized with respect to the other. According to Dudley (1951), Anderton first proposed this process in 1890, although difficulties in obtaining the necessary materials delayed practical implementation for nearly half a century. In the context of modern scientific visualization, consecutive frames on a computer monitor may form stereopairs and be polarized alternately. In this instance the members of the stereopair are temporally separated (frame sequential), but since visual information received within about 100 ms is processed simultaneously, the consecutive left/right eye images are, in effect, concurrent.

1.5.4 Autostereoscopic Displays

A limitation of stereoscopic displays is that motion parallax is not provided automatically; moving one's head from side to side does not necessarily produce a different viewpoint of the scene depicted. Display devices that are able to provide the correct stereoscopic perspective to each of the viewer's eyes over a range of viewing positions may be defined as *autostereoscopic*. Volumetric displays fall into this category. Below, two other broad classes of autostereoscopic display techniques, which do not require special glasses or headgear, are outlined.

The stereoscopic displays described above, in combination with powerful computer systems, can also provide an updated stereoscopic view in response to motion of the observer's head if an appropriate position or motion sensor is provided [see Travis (1997) and references therein]. However, the processing of the sensing device feedback and recomputation of the image so as to update the stereoscopic views of the virtual scene continuously and smoothly greatly increase the burden on the associated graphics engine. Nevertheless, increasing computational speed coupled with more efficient image data-processing algorithms and feedback from applications is steadily improving this technology. When the viewing apparatus takes the form of a head-mounted display, providing an immersive environment that the operator "enters", such systems are often known as *virtual reality* or *virtual environment* systems. Despite the computational overheads, the realism associated with this approach (which may be enhanced by the incorporation of tactile feedback in sensor equipment (such as gloves) ensures the continued development of this type of display methodology.

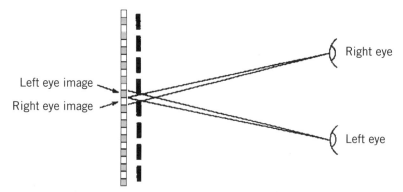

FIGURE 1.10 Parallax barrier technique for presenting binocular images. The viewer's head position needs to be quite constant relative to the display, or the left and right eye images swap and the relief of the figure is inverted.

Parallax Barrier and Lenticular Screen Techniques

The parallax stereogram, illustrated in Figure 1.10, was originally devised by F. E. Ives in 1903 (Okoshi 1976). A sheet containing fine vertical slits is placed in front of a specially prepared picture containing alternate left and right eye views of an image. The slits control the direction of the light emitted from the picture so that each eye sees a different perspective of the image. However, this original method allowed little freedom in viewing position. In 1918, Kanolt improved the system in this regard by decreasing the ratio of the image strip width s to the slit width p from about $\frac{1}{2}$ to $\frac{1}{10}$ (Figure 1.11).

A difficulty associated with this technique was the creation of the picture sheets with the alternate-view strips. Various ingenious schemes to achieve this, utilizing one or more cameras, were devised by H. E. Ives (the son of F. E. Ives) in the late 1920s (Ives 1928, 1929, 1930a, 1930b; Okoshi 1976). However, several major drawbacks to the parallax barrier method resulted in the technique losing favor by the end of the 1950s. The darkening of the images due to the presence of the barrier increases as the slit-to-strip ratio decreases, and hence as motion parallax range increases. Moreover, the slits must be sufficiently narrow that they are not observable. But this then can lead to unacceptable spreading of the image light due to diffraction.

The technique of integral photography, invented by Lippmann (1908), provides a method of recording a spatial image on a single photographic plate. The process

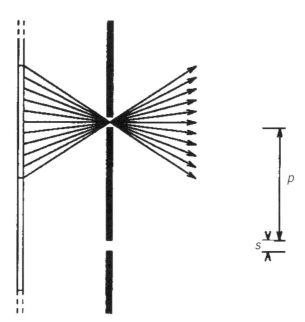

FIGURE 1.11 Parallax panoramagram, in which the slit/strip width ratio s/p is decreased from about $\frac{1}{2}$ to $\frac{1}{10}$. This gives more freedom of lateral movement to the viewer.

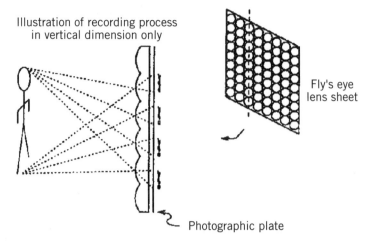

FIGURE 1.12 Technique of integral photography. The array of lenses provides an array of photographic replicas of the target object. When developed appropriately and viewed through an equivalent lens array, a 3-D image ensues as each lens projects a slightly different view of the object to the observer. [after Okoshi (1976, p. 22) with permission.]

utilizes a large array of tiny convex lenses, known as a *fly's-eye lens sheet.* An object in the focal plane of this sheet will be imaged by each lens from a slightly different viewpoint. A photographic plate behind the sheet records a large number of small images of the object, each viewed from a different direction. By developing and reilluminating the photograph, and viewing through the fly's-eye sheet, a 3-D image of the object, satisfying binocular and motion parallax (both horizontally and vertically) is observed (Figure 1.12). However, the resulting image will be *pseudoscopic*, that is, reversed in depth. This difficulty may be overcome by taking a second integral photograph of the first image (Ives 1931), although this may result in resolution being degraded (Okoshi 1976). The first experiments with this technique were performed by Sokolov in 1911, using an array of pinholes to approximate the small lenses (Valyus 1966, Okoshi 1976); lens sheets of sufficient quality were not available until after World War II.

Interest in a simplified version of the technique, dating back to the work of F. E. Ives and H. E. Ives, underwent a resurgence in the United Sates in the 1960s (Burckhardt 1968a). The 2-D array of lenslets was replaced by a linear array of cylindrical lenses, known as a *lenticular sheet* (Figure 1.13). In this case, the parallax is obtained in the horizontal direction only. Arrays of diffraction gratings may be employed in place of lenslet arrays (Nordin et al. 1994, Sakamoto et al. 1996, Toda et al. 1996). These provide a field of view of greater angular extent than the approximately 15° available with a multilens array. However, both of these methods require a pixel resolution corresponding to the product of the resolution of each view and the number of views (Travis 1997). This places strong demands on both the manufacture of the display screen and on the computation and data throughput capabilities required of the graphics

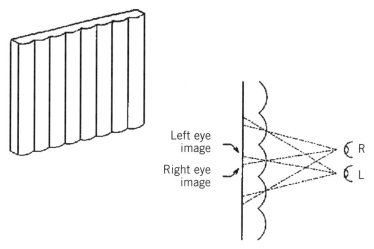

Left eye
image

Right eye
image

R

L

FIGURE 1.13 Principle of lenticular sheet imaging. The principle is the same as that of integral photography, except that parallax is retained in the horizontal direction only. The array of strip lenses is more easily fabricated than an array of fly's-eye lenslets (Figure 1.12).

engine. A further possibility for autostereoscopic display is that of spatially or temporally multiplexed video images with the aid of a lens array (Travis 1997).

Lenticular sheet–based electronic displays still form an active area of research interest. An LCD panel illuminated from the behind and equipped with a lenticular sheet permits the presentation of autostereoscopic images across a range of viewing positions (Isono et al. 1993, Sheat et al. 1993).

Computer-Generated Holography

The method of holography was originally proposed by Gabor (1948) to overcome resolution limitations in electron microscopy. The potential application of Gabor's holographic technique to 3-D imaging was heralded by the work of Leith et al. (Leith and Upatnieks 1964, 1965; Leith et al. 1965), which pioneered the two-beam method of hologram formation. This required a stable optical bench, on which the coherent light of a laser source was split into two beams, one of which was scanned across an object and subsequently interfered with the other beam to produce an interference pattern, characteristic of the object, and which could be captured on film. When reilluminated by laser light of the same wavelength, the interference pattern diffracts the light in an identical fashion to the object, so that a 3-D image of the object is made visible and appears suspended in the air between the observer and the holographic plate. Images generated in this manner satisfy both the binocular and motion parallax depth cues. Moreover, the cue of accommodation is present — the eye must change focus depending on the depth of the area of attention within the hologram. An important advance of this technique was the development of white-light holograms (Leith 1976, Caulfield 1979).

In employing holograms as an electronic display, it is necessary to compute the holographic interference pattern for each frame, which requires a large amount of information. Each point in the hologram may be calculated using the sum of the partial waves from all points in the object. For the computation of holographic images to be practical on a near real-time basis, the information requirements must be reduced substantially (Burckhardt 1968b). To this end, a number of simplifications and efficiencies have been adopted (Benton et al. 1993, Lucente 1993) and advanced prototypes are under development.

1.6 DISCUSSION

Volumetric displays provide 3-D images that are in many ways unique among visualization techniques. Being cast within a physical volume, the images naturally satisfy depth cues such as binocular and motion parallax without the need for additional viewing apparatus, and generally, there is very considerable freedom in viewing orientation. However, as will become apparent during the course of this book, a bewildering variety of technologies and physical mechanisms have been proposed and/or developed (in prototype form) with the aim of achieving these goals. In this book, we attempt to formulate concepts that are common across the wide range of the technologies and embodiments to be found in the volumetric field.

Although the ideal volumetric image space may be imagined as a uniform, isotropic, high-resolution tableau, this represents a formidable and as yet unrealized technical challenge. A number of aspects of the image space, including its dimensions, voxel size, and voxel spacing, affect the degree of detail and spatial structure that may be depicted therein. Conversely, at the design stage, the intended application will place requirements on these and other parameters of the image space and display system.* The manner in which the observer perceives the displayed data plays a crucial role in this interrelationship.

Almost all of the research that has been carried out to date regarding human depth perception has concerned our perception of the world around us (being the environment in which our visual system has evolved) and more recently the perception of images on 2-D or stereoscopic displays. Volumetric display units provide quite a different medium for the presentation of computer-generated data, and a number of perception issues relevant to volumetric display systems remain open. For example, the spatial acuity (and viewing distance) of the eye affect the required voxel size and separation. Furthermore, the relationship of the spatial acuity to the contrast between a feature of interest and its surroundings and to its degree of motion is complex. Sound knowledge of the spatial acuity in

---

* The application dependence of the image space characteristics is an important issue (see Chapter 10).

relation to volumetric images would therefore be invaluable in the development of volumetric systems, but to the authors' knowledge, has not yet been established.

Another important issue is that of conflict between depth cues. The physically 3-D nature of volumetric images provides consistency between the information obtained from a number of the depth cues (in particular, the physiological and parallax cues). However, the perception of volumetric image components comprising different discrete colors in systems providing a very limited color palette may provide some confusion. As the color of the component itself can provide a cue to depth, this may conflict with the information provided by other cues. This effect is stronger in 2-D representations of volumetric images (such as photographs), where the important parallax and physiological cues have been removed. The perception of images depicted on volumetric systems remains a largely unexplored area of research. It is to be hoped that some of the questions in this area will be addressed, and resolved, in the not too distant future.

CHAPTER 2

BASIC CONSIDERATIONS ON VOLUMETRIC DISPLAY UNITS

Take now this meatless penance,
It shall later be amended,
The Knight laughed and grew merry,
As wine to his head ascended.*

2.1 INTRODUCTION

In this chapter we introduce some fundamental concepts relating to volumetric displays and define essential terminology needed to describe their operation properly. Our purpose is therefore to lay essential foundations for the subsequent text. To describe the general principles of volumetric display systems research, it is often advantageous to consider display architectures in an abstract manner. Therefore, when possible, we emphasize concepts that are independent of the physical mechanisms employed within particular display system architectures. However, so as to provide particular instances of these principles, reference will be made to the elementary model introduced in Section 1.3 and to the various architectures described in Chapters 5 and 7. We recommend that those studying this topic for the first time not only peruse this chapter but refer regularly to Chapters 5 and 7. In this way, the reader may quickly gain insight not only into some of the abstract aspects of volumetric image depiction, but also into some of the more practical matters relating to their implementation.

The wide variety of techniques that may be used in the implementation of a volumetric display often make it impossible to state concepts and provide generalized technical descriptions that apply equally and without exception to

---

* *Sir Gawain and the Green Knight* (trans. Marie Borroff), WW Norton and Co., NY, 1967.

all. For example, it is assumed that all volumetric images are composed of a suitable arrangement of voxels (image "particles"), each of which acts as a source of visible light. Although this is generally the case, possibilities exist for the creation of opaque voxels that are visible as a consequence of the reflection of ambient light (see Section 7.5.5). For the sake of clarity, exceptional cases are dealt with elsewhere in the book. The reader may find the glossary toward the end of this book to be of assistance in providing brief descriptions of technical terms introduced in this chapter and elsewhere in this work.

2.2 TERMINOLOGY

The expression *volumetric display unit*, or simply *display unit*, is assigned to the apparatus responsible for the physical generation of the visible 3-D image. An analogous 2-D display unit would be, for example, a cathode ray tube (CRT). Just as an integral part of a CRT is the screen upon which images are cast, an *image space* forms the most visible component within a volumetric display, and in this volume 3-D images may be depicted. For example, in the case of the elementary display unit introduced in Section 1.3, the image space is defined by the extent of the filament matrix.*

We distinguish between the display unit and the *volumetric display system* (or simply *display system*), which will be assumed to encompass not only the display unit but also the hardware and software subsystems responsible for the generation, manipulation, and throughput of image data. By simple analogy, the display unit would in the case of a television receiver be equivalent to the CRT and the display system to the television set as a whole.

A volumetric display system comprises two essential subcomponents: the display unit and the *graphics engine*, as illustrated in Figure 2.1. It will be assumed that the graphics engine contains the image-processing portion of the display system, together with any associated electronic control and feedback mechanisms necessary to generate volumetric images within the display unit. These mechanisms are required by some technologies to achieve synchronization between the display unit and the image data stream and also to permit the computational hardware to dynamically calibrate components within the display unit.

During the operational life of a display system prototype, modification is likely to be necessary to obtain improvements in performance. For example, one may wish to permit the depiction of more detailed images or improve the overall visual quality. Unfortunately, the detailed architecture of a volumetric display unit tends to be complex, comprising a number of interdependent subsystems and incorporating a variety of technologies (see Section 2.4). The physical intricacy of these subsystems makes the construction of display unit prototypes very time consuming, and the resulting systems are often difficult to modify and upgrade.

---

* In principle, the ultimate goal of the volumetric display system engineer should be to develop architectures that permit an all-round, unobscured, view of the image space and hence of the image displayed. This is seldom possible, however, and may in fact be problematic in some applications.

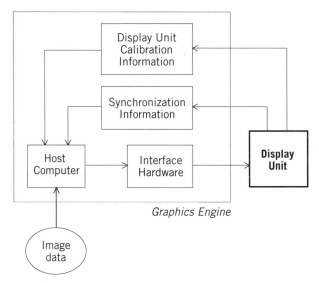

FIGURE 2.1   A volumetric display system consists of two major components, the display unit and the graphics engine. The display unit is responsible for the creation of the visible voxels within its transparent image space. The graphics engine is responsible for the processing of image data into an appropriate form and the passage of these data to the display unit. It encompasses any control hardware that may be needed to maintain the correctly synchronized flow of data to the display unit and feedback systems responsible for measuring and monitoring display unit characteristics.

Modifications made, for example, to one part of a display unit may affect many other aspects of it—sometimes in the most unexpected manner.

To minimize the need for, and impact of, subsequent modifications, it is important to have a clear understanding of the precise way in which a volumetric display system will operate and perform prior to its implementation. In short, it is highly desirable to adopt an engineering design approach in the development process. During the years of volumetric system research, approaches have often been limited to the initial experimental verification of techniques and ideas, as may be seen from some of the displays illustrated in Chapters 5 and 7. Unfortunately, it would appear that at times insufficient attention has been given to the detailed performance characteristics of proposed volumetric architectures prior to construction.

The construction of a volumetric system able to depict simple images that may be viewed from certain directions is a relatively straightforward undertaking. However, it is extremely difficult to produce a system that is able to display, in a reliable manner, complex high-quality images that may be viewed from a wide range of orientations without undue variation in image quality.

2.3   FUNDAMENTAL IMAGE PARTICLE: THE VOXEL

Images created by all but the most elementary volumetric display systems are composed of voxels. A *voxel* (volume element) represents the 3-D equivalent of a *pixel* (picture element) and forms the fundamental "particle" from which volumetric images are constructed. A voxel should take the form of a sharply defined source of visible radiation, and ideally is spherical in shape. Furthermore, the voxel attributes (which we describe later) should be invariant with respect to viewing direction.

It is unfortunate that most volumetric display unit implementations lack the ability to generate identical, regularly shaped voxels throughout the entirety of the image space. Where voxel attributes such as shape and intensity are affected by the position at which voxels are located within the image space, or by the direction from which they are viewed, a degree of image degradation will most certainly occur.* For this reason it is important to develop techniques to ensure that voxel attributes remain invariant in these respects.

The production of a voxel within an image space corresponds to effecting a visible change in the optical properties of some localized and defined location. For the purposes of this book we refer to this process as *voxel activation*. In the case of most volumetric display technologies, an activated voxel emits light, although alternatively, the change from the inactive to the active state may also correspond to the absorption or scattering of ambient light. The transition from an inactive to an active state is caused by the application of one or more activation stimuli, which are directed to the appropriate location within the image space.

The production of voxels within an image space may occur at a discrete set of locations defined by the system architecture. Alternatively, in some embodiments, voxel production may in principle take place at any position within the volume. For example, in the case of the elementary system model introduced in Section 1.3, voxel production can occur only at a certain set of locations within the image space, each corresponding to a site occupied by a lamp. Alternatively, should we imagine an image space comprised of a homogeneous material, which may be stimulated by means of external beam sources, voxel production may, in principle, take place at any position desired.† In the case of an image space comprising a set of discrete devices, each able to cause the production of one or more voxels, these will be referred to as *voxel generation centers*.

Following the removal of the stimulus responsible for causing the activation of a voxel, the optical change will often reverse automatically, returning the region of image space to its inactive (ideally invisible) state. For example, in the case of

---

* It is important to place equal emphasis not only on the generation of voxels having appropriate characteristics, but also the transmission of light from each voxel location across the image space and hence to the observer. This theme (which has often been ignored) is frequently referred to in this book.

† As we will see, in the case of such display unit architectures, aspects of both the graphics engine and display unit often ultimately determine the freedom in voxel positioning.

a system employing the two-step excitation of fluorescence (see Sections 6.5 and 7.5.1), light may be emitted (and therefore a visible voxel produced) within the region at which two nonvisible beams intersect. Following the removal of these beams, the light output diminishes rapidly and the material returns to its inactive state. Should such a process be used for the production of voxels, then in order that the perceived level of light emitted by each voxel remain constant, regular activation signals must be applied. This reactivation process is referred to as an *image refresh*. Refresh also permits image scenes to be changed and therefore enables the depiction of animated image sequences.

It is also important to consider the optical characteristics of each voxel generation center when in the inactive state. Transparency is of particular significance, as light emitted by activated voxel generation centers may pass through those that are inactive as it propagates through the image space and so to the observer. This marks a particular difference between pixel generation on a conventional display (e.g., a cathode ray tube) and voxel generation within a volumetric system. In the case of the CRT we are interested in the emissive characteristics of each pixel and need not concern ourselves with the transmissive nature of pixels that are in an inactive state. In the case of volumetric display systems, the active and inactive states of each voxel generation center are of equal importance.

During the course of this book we outline a variety of mechanisms by which visible voxels may be created. As a consequence of the great diversity of approaches that may be adopted, the foregoing discussion fails to encompass all techniques. It is, however, sufficient for our present purposes.

2.3.1   Voxel Attributes

In designing a volumetric display unit the characteristics that are to be associated with a voxel should be clearly identified. Although these characteristics are the subject of discussion in Section 10.2, it is instructive to summarize at this time those of particular importance:

- Geometrical shape of the voxel
- Size of the voxel
- Permissible variations in shape and size
- Color(s) that may be assigned to the voxel
- Maximum intensity
- Opacity of voxels while emitting (active), and their optical properties when inactive

Each of the above plays an important (and often complex) role in determining the quality of images that may be depicted by a particular display unit technology. However, it should be remembered that in the case of many volumetric systems, individual voxels are sufficiently small, or overlap to such an extent, that they cannot be resolved individually when combined to create image primitives (see

Section 1.4.1). The particular set of technologies employed within the display unit will define the possible range of voxel characteristics and therefore affect the types of applications for which the display unit is most suited. Conversely, the applications foreseen for the system will dictate the voxel characteristics required and so indicate the types of technologies that may best be employed within the display unit.

For example, let us consider a volumetric display system that is to be used to show the spatial separation of a number of small objects (e.g., aircraft or projectiles). An initial task for the design engineer is to consider the size and level of detail that is to be associated with each class of object and also to gauge the typical viewing distance. From this information an indication of the required voxel size may be obtained. This in turn has a direct impact on the required total number of voxels needed to portray a typical image scene and on the necessary intervoxel spacing. There would clearly be little merit in a display unit that permitted the creation of fine voxels of such a low density that when activated, adjacent nearest-neighbor voxels appeared as separate sources of light. Conversely, the production of large voxels may make it impossible to incorporate the required amount of detail within the image objects. In this particular application, the depiction of small objects containing fine detail would necessitate the use of small voxels able to be positioned at a high density within localized regions.

The use of color plays an important role in many aspects of scientific visualization. Artificial color mapping is often used to indicate variations of parameter values or to highlight different parts of an image. Most present volumetric systems can depict voxels with only a small number of discrete colors (e.g., red, green, and blue). In the case, for instance, of the visualization of a molecular structure, this would permit atoms to be represented in one or two colors and the bonds between them in another (see Plate iv). For such an application this might be sufficient. However, in general, the required range of the color palette must be determined by examining the types of image to be depicted by the display unit, and suitable techniques must then be identified for the generation of voxels of the appropriate color. As on conventional displays this may be achieved by the suitable mixing of component colors (Sproson 1983). Considerable research needs to be undertaken on the development of techniques that will increase the range of colors that can be depicted by the numerous volumetric display unit architectures.

The nature of the images to be portrayed by a display unit has an immediate impact not only on the required set of voxel attributes, but also on other issues relating to the display unit architecture, for example, the physical dimensions and shape of the image space. There is, therefore, a strong link between the architecture of a display unit and its suitability for various applications. This application dependence is introduced at the very outset of the design process when we define voxel attributes, image space dimensions, and so on. The prospects for the development of a general-purpose high-performance volumetric system are considered briefly in Section 10.6.

2.4   SUBSYSTEMS WITHIN A VOLUMETRIC DISPLAY UNIT

Any volumetric display unit may be considered to comprise three major interrelated subsystems, each of which can have a profound impact on the nature and quality of images depicted by the system. These three subsystems are described briefly below.

2.4.1   Image Space Creation Subsystem

Image space creation represents the physical process used to produce a transparent, addressable volume. Within this volume, voxels may be activated in a controlled manner and images can be formed. For example, in the case of the elementary model introduced in Section 1.3, the image space is created by the provision of a 3-D array of individually addressable voxel elements (voxel generation centers). Alternatively, an image space may be formed by the rapid motion of a surface. A variety of image space creation techniques are described in Chapters 5 and 7.

The technique employed for the creation of the image space has a strong impact on many issues, including the following:

- Voxel attributes, and therefore the image quality perceived
- Light propagation through the image space
- Image distortion occurring at the image space boundary
- Restrictions on the angle through which images may be viewed
- Maximum refresh frequency of the display unit
- Life expectancy of the display unit
- Possibilities of acoustic noise
- Suitability of the system for concurrent voxel activation

This last property is considered to be of major importance in the case of future high-definition volumetric systems, as it enables multiple voxels to be activated simultaneously and plays an essential role in determining the rate at which groups of voxels may be activated (see Section 2.6). The manner in which the image space is created will also determine the degree to which it provides a uniform (homogeneous and isotropic) 3-D tableau for the depiction of images. Should it fail to exhibit these properties to a satisfactory extent, it may be necessary to impose them artificially through manipulation of the image data within the graphics pipeline. In general, this is undesirable, as it results in an additional computational burden and so may cause a reduction in the overall performance of a display system (see Section 10.5).*

Finally, since it is generally desirable to maximize the volume of the image space, it is important to pay attention to its physical density, as this will have a

---

* Furthermore, this technique may enforce the minimum value of a performance characteristic throughout the image space.

major impact on the portability of the system. For example, a display unit whose image space is formed from a solid cube of glass with sides of length 30 cm would weigh approximately 70 kg. Consider the prospects for transporting an image space of this type with a volume of 1 m^3!

2.4.2   Voxel Generation Subsystem

Voxel generation represents the underlying physical process by which a visible change is effected within localized regions of an image space so as to give rise to the production of voxels. For example, in the case of the elementary display unit (see Section 1.3) the voxel generation technique corresponds to the emission of light from each hot filament. For each method of image space creation there are, in general, a number of possible voxel generation processes. For example, should the image space be created by the rotation of a surface, various voxel generation processes may be employed, including cathodoluminescence (see Section 5.4.1), light scattering (see Section 5.5.2), or electroluminescence (see Section 5.4.4). The voxel generation process employed within a display unit will greatly affect many of the voxel characteristics. For example, both the length of time for which the activation stimulus must be applied to each voxel location to achieve the required level of intensity and the requirements for image refresh will be strongly influenced by the voxel generation technique. The voxel generation process may also often affect the minimum achievable voxel size and together with the voxel activation technique will determine the maximum attainable voxel intensity.

For the purposes of this book we define the *voxel visibility lifetime* as the time duration for which an activated voxel continues to remain visible after removal of the activation stimulus.* The voxel generation technique is primarily responsible for determining the magnitude of this parameter. It is of interest to note that in the case of image space generation through the use of a rapidly rotating or reciprocating surface (Sections 3.4 and 3.5), the voxel visibility lifetime will directly affect the voxel asymmetry and may give rise to visible trails. This effect is illustrated in Plate vii.

When investigating a voxel generation technique, the active and inactive voxel states should both be considered. As mentioned previously, when in the inactive state voxel locations within the image space should not be visible (see Section 2.3). For example, in the case of the elementary display introduced in Section 1.3, the filaments are visible even when inactive, and this is clearly undesirable. Particular care must be taken in the choice of voxel generation technique when implementing display units that are required to depict multicolor or full-color images.

2.4.3   Voxel Activation Subsystem

Voxel activation represents the stimulus used to effect the transition of each voxel from its inactive to its active state. For example, voxel activation may be

---

* The perceived voxel visibility lifetime is subjective and will obviously be affected by the ambient lighting conditions.

achieved by means of electrical signals directed to individual voxel generation centers; through the use of nonvisible laser radiation; or by electron beams. A display unit's voxel activation subsystem represents the pathways through which information passes from the graphics engine into the image space. In the case of the elementary model introduced in Section 1.3, the electrical hardware responsible for the activation of the filaments could be constructed in such a way that a significant fraction of the filaments may be activated simultaneously. This would enable multiple voxels to be switched simultaneously from their inactive to their active state and would increase the total number of voxels that could be created during an image update period. In other display unit architectures (such as the system outlined in Section 5.3.1), voxels are activated sequentially. In this case, the total number of voxels that may pass from the inactive to the active state during a given image update period would be determined by the time required for the activation of each individual voxel.

2.4.4   Integration of the Subsystems

In the case of the elementary example introduced in Section 1.3, the image space is defined by the extent of the filament array. The voxel generation process equates to the emission of light from each heated filament, and the voxel activation technique is represented by the application of an electrical current to a combination of the filaments. It is important to note that individually and in combination, the three subcomponents of the display unit affect the various voxel attributes listed in Section 2.3.1. For example, the provision of voxels whose shape is symmetric and uniform throughout the image space and which appear identical from any direction of view (isotropic) is generally problematic. For instance, although the voxel generation process may result in isotropic radiation from each active voxel, the image space may not be equally transparent in all directions, and furthermore, the refractive index of the image space may be nonhomogeneous. In this case, voxels exhibiting isotropic characteristics could appear anisotropic to an observer.

In the case of the elementary display system model (Section 1.3), each individual filament is not transparent when inactive, and hence voxels viewed from, for example, the end of a "row" of filaments may appear dimmer due to this partial obstruction compared with their apparent brightness when viewed from other directions. Moreover, the wires traversing the image space, by which the voxels are activated, may also cause some visual obstruction. In such a situation, the reduction in intensity will also depend on the depth of an active voxel from the surface of the image space.

In summary, it is apparent that the attributes of each voxel, and hence the overall image quality, are affected not only by the voxel generation process itself but by the nature of the image space and the means by which individual voxels are activated. Furthermore, the method by which the image space is generated has a considerable impact on the physical portability of the display unit and its working life expectancy. Effects of this type arising in respect to various approaches to volumetric display unit implementations are discussed in greater depth in Chapter 10.

2.5 VOXEL TIME COMPONENTS

In the case of *all* volumetric systems, it takes a finite time to achieve the appropriate change in optical properties within a region of image space (i.e., to activate each voxel). This time duration will be called the *voxel time* and is denoted by T. It is convenient to subdivide this period into two components: the time required for the activation mechanisms to adjust before the creation of each voxel, and the time for which the voxel must be activated in order to achieve a satisfactory change in visibility. The former component will be termed the *system adjustment time* (T_S) and the latter the *voxel activation time* (T_V). Therefore,

$$T = T_S + T_V. \tag{2.1}$$

In the case of some volumetric technologies, the values of T_S and T_V may not be constant. For example, the required magnitude of T_S may depend on the position within the image space of preceding voxels.* By accommodating variations in T, it is possible, during the refresh period, to activate more voxels than would otherwise be the case. However, within our discussions, we assume (unless otherwise stated) that its value remains constant throughout any image refresh period and that it represents the sum of the worst-case system adjustment and voxel activation times.

For the elementary model described in Section 1.3, the system adjustment time would correspond to the time required for the control system to establish the appropriate electrical connectivity for each set of voxels that are to be activated simultaneously and initiate the flow of current. The voxel activation time would correspond to the time for which the current must be applied to each filament so as to produce a voxel of sufficient brightness.† In this example, a large number of the filaments may, in principle, be activated simultaneously. In the case of display units in which the activation of voxels is sequential, Eq. (2.1) will apply to each voxel in turn.

Example: A Display Unit Employing a Beam for Voxel Activation   Consider the display unit depicted in Figure 5.4, which employs a phosphor-coated surface

---

* This applies to display units employing deflectable beams for voxel activation.

† This provides an illustration of the variable voxel time referred to in the preceding paragraph. The magnitude of T_V could, in principle, be varied so as to provide voxel gray scale (fixed voltage, variable time), resulting in corresponding variations in T. Alternatively, the magnitude of the voltage applied to each element may be used to determine brightness, in which case a constant activation time is maintained (variable voltage, fixed time).

moving with a translational motion and addressed by an electron beam. The system adjustment time includes the time required for the beam to be deflected so that it will impinge on the surface at the appropriate location and so enable activation of a voxel at the required image space coordinates as well as the time required to turn the beam on and off. The voxel activation time may be defined as the duration for which the electron beam (operating at maximum current) must impinge on the surface so as to create a voxel of maximum brightness. In the case of this display unit technology, gray scale may be achieved by operating with a constant value of T_V and an adjustable beam current or, alternatively, by employing a constant beam current during the activation process and varying the voxel activation time.

In the remainder of this chapter we concern ourselves with issues relating to the number and spatial density of voxels that may be activated within an image space. When considered in association with voxel size, these attributes have a direct impact on the types of application for which a particular volumetric display unit architecture may be suitable.

2.6 VOXEL ACTIVATION CAPACITY

Consider a display unit in which voxels are activated sequentially. In this case the maximum number of voxels that can be activated within a single image update period is determined by the time required for the activation of each voxel and the duration of the image refresh period.* We denote this maximum number of voxels by N_a and define it to be the *voxel activation capacity* of the display system. For any volumetric display unit, the voxel activation capacity is limited fundamentally by the time (T) required to create each voxel (the voxel time), by the number of voxels (P) that can be generated concurrently[†] (i.e., the degree of parallelism), and by the time available for each image refresh (τ_r). Since τ_r is the reciprocal of the image refresh frequency (f_r), the voxel activation capacity may be expressed as:

$$N_a = \frac{P}{T f_r}. \tag{2.2}$$

The refresh frequency has, in practice, a minimum value due to flicker and/or animation considerations, so it is apparent from this expression that as the voxel activation capacity is increased, the time available to create each voxel must decrease (Figure 2.2). In turn, unless methods are found to decrease the system adjustment time, there will be less time available for voxel activation. This may result in a decrease in the image intensity and it is necessary to achieve a balance

---

* The maximum duration of this refresh period is determined either by the frame rate required to achieve smooth image animation or by the need to avoid image flicker.
[†] Should the voxels be activated sequentially, $P = 1$.

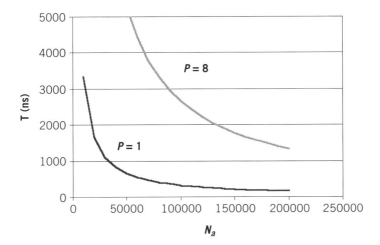

FIGURE 2.2 Variation of the time T (in nanoseconds) available to generate each voxel with the voxel capacity N_a, assuming an image space refresh frequency of $f_r = 30$ Hz, for two levels of parallelism, $P = 1$ and $P = 8$.

between the voxel activation capacity and overall image brightness.[‡] Further increases in N_a are *possible only by the inclusion of concurrency in the voxel activation process.* The degree to which display unit technologies support concurrency varies widely, and this matter is the subject of discussion later in the book.

2.7 THE INCLUSION OF THE THIRD DIMENSION

In the case of conventional display systems, such as those employing a CRT or LCD panel, the pixels are arranged in a closely spaced 2-D array. It is instructive to consider a volumetric image space as a 3-D generalization of this 2-D model. Let us therefore consider a simple 2-D display screen, such as a noninterlaced CRT monitor, whose resolution is indicated by its size and the number of pixels located horizontally and vertically. For simplicity we assume that the screen takes the form of a square with edges of length l comprising n by n pixels and permit any combination of these n^2 pixels to be activated during each refresh period.

Now let us suppose that a volumetric image space is to offer equivalent resolution, and consider the consequences of this assumption. The square array of pixels is therefore extended by an additional spatial dimension to form a 3-D cubic array of n^3 voxels. We will assume that this cube also has edges of length l. Flicker considerations give rise to a minimum image refresh frequency, and

---

[‡] In the opening paragraph of Chapter 1 we remarked on the frequent operation of volumetric display systems within semidarkened rooms. Should the voxel activation capacity be increased at the expense of the voxel time, image intensity will, in general, be compromised. Ambient lighting must then be lowered to maintain image visibility. It is interesting to note that a reduction in image intensity may permit a small reduction in f_r without any consequent increase in perceived flicker.

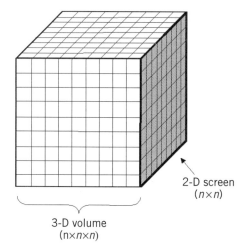

FIGURE 2.3   Simple extension of a 2-D screen consisting of n^2 pixels results in a 3-D image space of n^3 voxels. To be able to address the extra $(n-1)n^2$ points sequentially, each must be activated in a fraction $1/n$ of the time, an impossible task for values of n approaching those encountered in typical computer monitors.

therefore it is appropriate to assume that both the 2-D display and 3-D image space are refreshed at the same frequency.

The 3-D image space contains $(n-1)n^2 = O(n^3)$ *more* points than the 2-D model (Figure 2.3). If the pixels in the 2-D array are activated sequentially (e.g., in the case of a raster-scanned CRT monitor) the time available to activate each of the n^2 points is $T_{pixel} = 1/(f_r n^2)$, where f_r represents the refresh frequency of the display. By contrast, the time available to activate each of the n^3 voxels in the 3-D image space, if addressed sequentially, is $T_{voxel} = 1/f_r n^3 = T_{pixel}/n$. The time available to address each point in the 3-D array is therefore significantly less than in the 2-D case. Unfortunately, current technologies are not able to support the activation of voxels to an appropriate intensity in such a short time.

As a numerical example, consider the values $n = 10^3$ and $f_r = 60$ Hz. In this case, each voxel would have to be activated in a period that is 1000-fold shorter than that available for the activation of each pixel. In practice, most current volumetric display units employ a refresh frequency of less than 30 Hz. This affords over twice as much time per voxel activation. However, this benefit is of little consequence compared with the 1000-fold decrease in the available voxel time due to the inclusion of the extra dimension of the image space. We must therefore conclude that a 3-D image space, populated at the same density as a 2-D computer monitor, contains too many voxels to be addressed exhaustively during each image refresh *unless a very high degree of concurrency is included within the voxel activation process.*

Let us seek an alternative solution by taking the n^2 points depicted on the conventional monitor as indicative of the maximum number of points that can be activated within each image refresh period and redistribute these evenly throughout a cubic 3-D image space. That is, we demand no more time to create each voxel than it takes to create each pixel on a conventional 2-D screen. This redistribution results in either a reduction in the length of each side of the cube or in a lowering of the voxel density. Continuing with the numerical example employed above, let us assume that the 2-D screen has sides of length 300 mm. If the $n^2 = 10^6$ points are redistributed evenly throughout a cube, there will be only $(n^2)^{1/3} = 100$ voxels along each cube axis. Should the length of the edges of the cubic image space remain the same as those of the 2-D screen, the point density in each direction drops from 3.3 mm^{-1} in the 2-D case to 0.33 mm^{-1} (or one point every 3 mm) in the 3-D case. This is illustrated in Figure 2.4. Alternatively, to maintain the same linear point density, the size of the 3-D image space

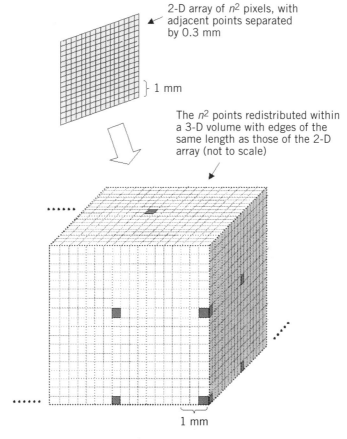

FIGURE 2.4 If a 2-D array of n^2 points is redistributed as a 3-D array in a cubic volume of the same linear dimension, the number of points per unit length along each axis is reduced by a factor of $n^{2/3}$. For example, if $n = 1000$ and the length of each side is 30 cm, the 3-D array contains 100 points along each axis, and the number of points per unit length is reduced from 3.3 points per millimeter to one point every 3 mm.

must be reduced to that of a cube measuring only 30 mm along each side. Once again, this simple calculation implicitly assumes that the refresh rate is the same for both 2-D screen and 3-D image space. However, as stated above, the effect of the extra time gained by lowering the refresh frequency is much smaller than the impact of the inclusion of the third dimension.

2.8 VOXEL DENSITY

In the preceding section we have seen that unless the voxel activation mechanism exhibits a high degree of concurrency, exhaustive addressing of all possible voxel locations within each image refresh period would result in one of the following:

- Insufficient time in which to achieve satisfactory voxel activation
- Inadequate image space volume
- Low voxel density

A reduction in the voxel time will generally cause a reduction of image intensity. Consequently, it may be necessary for the display unit to be operated in subdued ambient lighting.

Since volumetric systems are intended primarily for the visual presentation of spatial information, the image space volume should perhaps be as large as possible to maximize clarity. Should the voxel activation subsystem exhibit a high degree of parallelism, the voxel activation capacity may be large enough to support an image space with dimensions sufficient for most purposes. Conversely, if the concurrency in voxel activation is low, the image space may be too small to be of practical value. For example, if the voxel activation capacity is about 100,000 and voxels of diameter approximately 1 mm are packed at a density of 1 mm^{-3}, a cubic image space would measure only about 46 mm along each edge.

Increases in voxel separation will reduce the level of detail that can be incorporated within image components. For example, consider a volumetric system used to show the spatial separation of aircraft within 3-D space. Should a large intervoxel spacing be employed, it may only be possible to represent each aircraft icon as a cluster of activated voxels or by means of a vector (through the use of a directed line segment). However, by reducing the intervoxel separation, it becomes possible to depict each aircraft in considerable detail and perhaps associate textual information with each icon. Therefore, in the case of systems that do not employ parallelism in the voxel activation process, the consequences associated with the exhaustive addressing of an image space may be most undesirable. However, as we will see in subsequent chapters, any system that is properly able to support exhaustive scanning is likely to provide a more uniform and predictable image space.

Since voxels are positioned within 3-D space, it may be convenient to describe their spatial separation in terms of a voxel density (ρ). However, this term is not easy to define precisely when applied to systems in which the image space fails to exhibit homogeneity and isotropy with respect to voxel placement. Unfortunately, very few technologies offer such characteristics across the three dimensions of the volume, and as a consequence, use of the term *voxel density* should be avoided when possible. However, since density provides us with an intuitive and therefore convenient description of voxel spacing, we will on occasion make use of this expression and neglect anisotropic considerations. For the purposes of this book, the term *voxel density* will be loosely defined and assumed to relate to the maximum number of voxel locations, situated within a specified cubic region of image space, which may be activated during a single image refresh period.

In the case of the elementary display unit introduced in Section 1.3, configured to exhibit maximum concurrency and isotropy in voxel positioning, the voxel density is obtained simply by dividing the voxel activation capacity by the image space volume. Unfortunately, this represents an ideal case and has seldom been realized in the majority of volumetric technologies proposed to date.

2.9 VOXEL LOCATION CAPACITY

For systems exhibiting a low concurrency, one solution to the problems associated with the exhaustive addressing of an image space (which were highlighted in the preceding section) may be found by not addressing the entire image space during each image refresh period. A high density of potential voxel locations may be defined within the image space, but only a certain fraction of these may be activated during any one refresh period.

The number of possible voxel locations will be referred to as the *voxel location capacity* (N_l). For many technologies this is likely to be considerably greater in magnitude than the number of voxels that can be activated during a single refresh period (corresponding to the voxel activation capacity). Therefore, the voxel activation capacity is decoupled from, and so does not necessarily relate to, the dimensions of the display volume. For example, a display unit architecture may limit the voxel activation capacity to 100,000. However, these voxels may be distributed within an image space containing a far larger number of potential voxel locations. Ideally, this distribution would be free of constraints. Unfortunately, many volumetric architectures impose restrictions on the maximum voxel density, and therefore the distribution of voxels indicated by the voxel activation capacity cannot be undertaken with complete freedom.* As we will see, this affects the predictability of the image space.

It is appropriate to express the fraction of the total number of available voxel locations that may be activated during each image refresh as a percentage. In subsequent chapters this will be referred to as the *fill factor* (ψ) and defined as

$$\psi(\%) = \frac{N_a}{N_l} \times 100 \tag{2.3}$$

---

* The maximum density of voxels that may be achieved within an image space is determined by the complex interrelationship of the three subsystems that comprise the display unit.

An essential strength of volumetric display systems concerns their ability to depict spatial relationships in such a way that they may be perceived clearly. However, in order that both the form and spatial separation of objects may be observed with clarity, a large portion of any 3-D image space must be void. Should the image space be overpopulated by activated voxels, it will appear cluttered and the benefits that may be derived from this type of display technique will be negated. Consequently, there is no demand to be able to activate every voxel within a 3-D image space during each image refresh period. This denotes an important difference between a conventional 2-D display and a volumetric system.

2.10 DISCUSSION

In this chapter we have formulated some of the essential terminology that is necessary to describe properly the operation of volumetric display systems. The meaning of the various technical expressions that have been introduced are summarized in the Glossary. We have focused mainly on display units that exhibit a low concurrency in the voxel activation process and in doing so, have limited ourselves to the majority of display unit architectures proposed to date. Later in the book we discuss further the exhaustive addressing of an image space and examine advantages (associated with image space predictability) that may be derived by achieving a 100% fill factor. However, for the present time it is instructive to consider systems that exhibit low parallelism in voxel activation and so gain a proper appreciation of the basic techniques that may be applied to the development of a volumetric display system.

   In the next three chapters we concentrate on volumetric systems that employ periodic mechanical motion for the creation of an image space. Systems that place no reliance on mechanical motion are discussed in detail in Chapters 6 and 7.

CHAPTER 3

SWEPT-VOLUME DISPLAY UNITS

Even the most complex stereometric drawings, representing models of crystals, which are scarcely intelligible without a stereoscope, can be made perfectly clear and will look like figures in space.[*]

3.1 INTRODUCTION

An image space may be composed of a material or arrangement of materials or, alternatively, may be created by the rapid periodic motion of a surface. In the latter case, the image space comprises all or part of the volume swept out by the surface and a mechanism is provided for the production of voxels therein. Display units of this type that rely on the rapid mechanical motion of a surface[†] will be classified as *swept-volume systems*. The principles governing their operation are introduced in this chapter and discussed at length in subsequent chapters.

Volumetric systems that do not rely on mechanical motion for image space creation will be classified as *static-volume systems*. For example, the image space may in principle consist of a volume of transparent gas or may perhaps be formed from a fixed 3-D array of light-emitting elements. The elementary example system introduced in Section 1.3 clearly falls into this category. Static volume systems are discussed in some detail in Chapters 6 and 7.

The classification of volumetric architectures into swept- and static-volume systems provides a convenient and easily conceptualized scheme for grouping

---

[*] Hermann von Helmholtz, in *Helmholtz's Treatise on Physiological Optics* (trans. J. P. C Southall), Dover Publications, NY, 1962.

[†] The image space is swept out at frequencies equal to or greater than the critical flicker fusion frequency (see Section 1.4.2).

together many types of display units. It is, however, based only on the operation of a single display unit subsystem (pertaining to image space creation; see Section 2.4.1). Of equal importance are the techniques used for the processes of voxel generation and voxel activation. Within the static- and swept-volume classifications, we therefore create subclasses of display unit. This results in a hierarchical classification structure within which most volumetric display unit architectures may conveniently be accommodated. Such a classification scheme is developed for swept-volume systems in Section 3.3 and for static-volume systems in Section 6.2.

In Chapter 4 we continue to examine the nature of swept-volume systems and focus on display units employing one or more beams for voxel activation. Chapter 5 provides a discussion on a variety of display unit implementations and detailed references to published work.

3.2   CONCERNING THE DISPLAY UNIT

Within this section we introduce some essential ideas relating to the surface responsible for image space creation and consider a number of general issues concerning the characteristics of this type of display unit. Terminology needed for our subsequent discussions is also defined.

3.2.1   The Moving Surface

The structure within swept-volume systems that is responsible for sweeping out an image space and so defining its extent is referred to in the literature by a variety of names, including *screen, plate, display surface, panel*, and *matrix*. Within this work we assign the title *screen* to this structure. As we will see in the following pages, the precise nature and form of the screen greatly affect many image space characteristics and also the overall image quality. It is therefore most important to consider its design with the greatest of care.

We are able to identify two essential aspects of the screen. Firstly, there is the component responsible for the generation of voxels and secondly, the structural portion which is present in order to provide mechanical rigidity. The title *surface of emission* will be assumed to refer to the component of the screen responsible for the generation of voxels. Unfortunately, the need for mechanical rigidity does not permit this surface of emission to exist in isolation and so it is formed on, or attached to, a *supporting structure*. The screen therefore represents the combination of the surface of emission and its associated supporting structure.

In the case of display units that permit considerable freedom in viewing angle, the supporting structure should clearly be transparent, and therefore it is generally constructed from either glass or a transparent plastic. Both the physical nature and form of the supporting structure are extremely important, as its presence is likely to affect the propagation of light emitted in certain directions from the surface of emission. This can adversely affect image clarity and also cause images to appear distorted. These effects arise primarily as a consequence of the finite thickness of the supporting structure and the difference in refractive index between it and the surroundings. Light emitted by the surface of emission may undergo refraction as it enters or emerges from the supporting structure, so image distortion may occur. Furthermore, light emitted at certain angles from the surface of emission may experience total internal reflection within the supporting structure, and this can result in portions of an image appearing to be reduced in intensity.

Unless careful attention is given to the design of both the surface of emission and its supporting structure, different parts of displayed images are likely to have greater clarity and possibly appear to be more intense when viewed from certain positions. This will result in preferential viewing directions that vary according to the position at which an image is depicted within the image space. In this case, any advantages that may be derived by providing an unrestricted view of the image space are likely to be negated. Any detrimental impact that the supporting structure may have on the image quality must clearly be minimized. This matter is discussed further in Sections 3.4.3 and 4.4.2.

3.2.2   Swept-Volume Techniques

For the purposes of this book, display units in which the image space is defined by the region swept out by a screen moving with either rotational or translational motion, are classified as swept-volume systems.* Voxels may be activated on the screen (or more precisely on the surface of emission) in synchronization with its motion and may therefore be positioned in the 3-D image space through which the screen moves. A variety of swept-volume display unit architectures are described in Chapter 5.

Fundamental to the operation of a swept-volume display unit is the surface of emission which must sweep out the entire image space at rates in excess of about 25 Hz (1500 rpm).[†] The reliance that swept-volume systems place on mechanical motion gives rise to a number of potential problems, such as:

- The impact of the mechanical component(s) on the reliability and ultimate life expectancy of the display unit
- The possible production of vibration and noise by the mechanical components

---

* In Section 12.4 we propose a hybrid swept-volume display unit in which a *volume* of voxel generation centers sweep out an image space. The definition of a swept-volume display unit given in the text above is therefore not strictly correct. It should encompass any system in which a screen *or collection of voxel generation centers* move with a cyclic motion. However, for our present purposes, the definition of a swept-volume display unit presented above is sufficient.

[†] This frequency of motion arises from a need to refresh the image at a rate that is sufficient to eliminate the worst effects of flicker (see Section 1.4.2). However, in the case of some screen configurations, refresh may be achieved several times during each cycle of motion of the screen. When this is possible, the screen's frequency of motion may be reduced. For example, see Section 3.4.1.

- The effect of the inclusion of mechanical components on the portability of the display unit
- The ability of the display unit to withstand mechanical shock and rapid acceleration (this may affect its suitability for operation in certain environments)

Furthermore, to ensure that voxels are accurately positioned within an image space, it is essential that the surface of emission remains free of any distortion that may be caused by its motion.* These potentially problematic matters become more severe as the size of the mechanical components (particularly the dimensions of the surface of emission) are increased so as to provide image spaces of greater volume. As these difficulties are implicitly associated with any swept-volume system, it is natural to question the desirability of such an image space creation technique and in turn to examine any potential benefits that such systems may provide. Three of the most basic advantages of swept-volume display units over their static volume counterpart are:

- The image space posesses a low physical density (and therefore mass).
- Swept-volume architectures often facilitate parallel voxel activation. Unfortunately, numerous previously proposed and current implementations have failed to make use of this flexibility.†
- In the case of swept-volume systems employing a matrix of voxel generation centers, each element is responsible for the production of multiple voxels during the course of each cycle of motion. The total number of voxel generation centers positioned within the image space is therefore very much less than would be needed for an equivalent static-volume display unit.

During the course of this book we highlight other advantages and disadvantages associated with swept-volume display unit architectures.

3.3 CLASSIFICATION OF SWEPT-VOLUME DISPLAY UNITS

Within this section, swept-volume display units are classified according to the method by which each of their subsystems is implemented. Categories of image space creation, voxel generation, and voxel activation are therefore defined. This results in a simple hierarchical classification scheme that is able to accommodate all but a very few of the swept-volume display units that are either the subject of current activity or that have been proposed in the past.

This bottom-up classification scheme permits systems to be grouped according to the types of technologies employed in their implementation and is therefore

---

*Flexing of the screen will also cause optical distortion and so affect the general quality of the perceived image.
† This is an important consideration when maximizing the voxel activation capacity of the display unit.

particularly useful when referring to or comparing display unit architectures. We begin our classification hierarchy by grouping display units according to the method by which the image space is created. As we will see, the approach taken for implementation of this subsystem is likely to be influenced and ultimately determined by a number of factors, including the required freedom in the range of viewing direction and the physical characteristics of the image space (i.e, its shape and dimensions). In practice, therefore, the manner in which the image space is to be created often forms the first step in the design process. We defer the classification of static volume systems until Section 6.2 and consider only those display units that employ mechanical motion in the image space creation subsystem.

In relation to the image space creation subsystem, there are two important matters that must be considered:

- The *geometry* of the surface of emission. This is often planar or helical. Alternative shapes are possible: for example, see Section 5.6.1.
- The *type of motion* described by the surface of emission. The surface may move through the image space with a *rotational* or *translational* motion.

The next stage of the classification process allows us to distinguish between display units according to the type of voxel generation technique employed in their implementation. Therefore, we must consider the nature of the surface of emission — this may be either *passive or active*:

- A *passive surface of emission* may be addressed by one or more directed beams. In this case, visible light is emitted (and therefore voxels produced) as a consequence of the interaction between the beam(s) and the physical surface. This interaction process may, for example, involve the scattering of laser light by a semiopaque surface coating (see, e.g., Section 5.5.2). Alternatively, cross-sectional image sequences may be cast on a reflective surface (see e.g., Sections 4.7 and 5.6.4). In the latter case the source of such 2-D cross sections may be, for example, a conventional CRT employing a high scan speed and short persistence phosphor.
- An *active surface of emission* is equipped with a 2-D array of elements that can be individually addressed (activated) by, for example, an electrical or optical stimulus. When instructed to do so, each element is able to emit light. An example of this is a rotating planar screen whose surface of emission consists of an array of optoelectronic devices (see Section 5.5.3). These elements are referred to as *voxel generation centers*.

The resulting classification scheme is illustrated in Figure 3.1. From this diagram it may be seen that we define two classes of voxel activation technique, these distinguishing between systems that are beam addressed and those that are addressed directly. The case referred to above (under "passive surface of emission"), in which image sequences are projected on a reflective surface, is considered to represent a special case of the beam addressed technique.

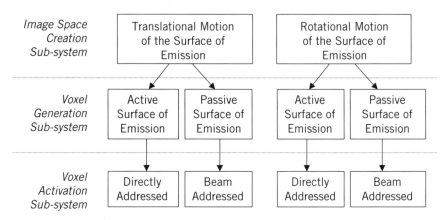

FIGURE 3.1 Classification scheme for swept-volume display units based on the underlying technologies employed in the image space creation: voxel generation and voxel activation subsystems. Note that systems which employ a reflective surface of emission upon which images are cast are considered as being included within the beam-addressed, passive surface class.

This final level of classification is not, in fact, essential, as it is determined by the chosen voxel generation technique. That is, a passive surface of emission will by its nature be addressed by a beam(s) and an active surface of emission will be addressed directly. It is, however, included in the diagram for clarity.

3.4   IMAGE SPACE CREATION BY ROTATIONAL MOTION

Display units in which the image space is created by rotational motion most commonly employ a surface of emission that has either helical or planar geometry.* Voxels may be activated as the surface of emission passes through the appropriate positions within the image space. Clearly, this necessitates careful synchronization between the output of voxel information and the constantly changing position of the surface of emission.

3.4.1   Image Space Update

The rate of rotation of the surface of emission places constraints on the image update frequency and the voxel activation capacity of the display unit.

---

* As mentioned previously, other screen shapes are possible (see, e.g., Sections 5.5.1 and 5.6.1).

Furthermore, the rate of rotation determines the perceived level of image flicker. The *maximum* frequency at which an image scene may be updated (f_r) is twice the value of the frequency of motion of the rotating surface of emission (f_s).* That is,

$$f_r^{\max} = 2f_s \tag{3.1}$$

Often, however, this maximum image update rate cannot be attained since it requires that the surface of emission extends symmetrically on both sides of the axis of rotation and therefore passes through every possible voxel location *twice* per revolution. If each voxel is refreshed only once per revolution, the update frequency is equal to the frequency of screen rotation. In practice, f_r must exceed the flicker fusion frequency. In the case of systems that permit each voxel to be updated twice per revolution, the screen rotation frequency may, in principle, be half the flicker fusion frequency.

Screens that are planar or helical in geometrical shape may take several forms, as illustrated in Figure 3.2. As may be seen from this illustration, the single-bladed helical surface of emission would have a maximum image update rate

Planar surface

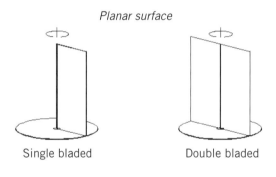

Single bladed   Double bladed

Helical surface

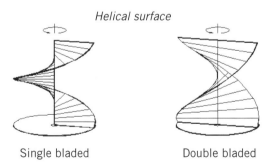

Single bladed   Double bladed

FIGURE 3.2 Single- and double-bladed forms of planar and helical surfaces of emission.

---

* This assumes that no interlacing technique is employed.

equal to its frequency of rotation. The use of a double-bladed helical or planar surface of emission could, in principle, result in the maximum image update rate indicated by Eq. (3.1). This is, however, not an inherent property of a double-bladed surface of emission as a further requirement is that each of the two blades must be able to generate voxels with identical characteristics. This requirement can be illustrated by considering the display unit outlined in Section 5.4.1, in which electron beams are used to address a planar phosphor surface of emission. In this configuration, to permit the depiction of multicolor images (see Figure 5.10), the two blades of the surface of emission are composed of different types of phosphor (Ketchpel 1963, Blundell et al. 1993b). Therefore, as the two blades are not able to generate identical voxels, the maximum image update rate remains equal to the frequency of rotation of the surface of emission.

Perhaps the most significant underlying difference between swept and certain types of static volume display units concerns restrictions on the voxel activation period. This difference can be illustrated by considering a voxel positioned at an arbitrary location within an image space. In the case of a static volume display unit such as the elementary system model (see Section 1.3), a voxel can be activated and remain in its active state until such a time as it is no longer required to form a part of an image scene. Such a display unit is, in principle, able to generate flicker-free images. Other static volume implementations may provide transient voxel activity in response to the activation stimulus.

However, in the case of a swept-volume display unit, a voxel may only be activated for a short period as extensive activation results in discernible elongation of the voxel due to the motion of the surface of emission during the activation period (see, e.g., Plate vii). Following the activation period, no further reactivation (refresh) is possible until the surface of emission has completed a further cycle of motion or, in the case of the double-bladed configuration described above, turned through an angle of 180°. Voxels generated within swept-volume systems do not, and cannot, act as continual light sources and rely on the persistence of vision to *appear* continually present in the image space (see Section 6.3). As we will see in Section 10.2.3, this precludes the use of swept-volume display units for the generation of opaque (solid) images — irrespective of the technique used for voxel generation.

By combining Eqs. (2.2) and (3.1) we obtain a maximum value for the voxel activation capacity (N_a^{max}):

$$N_a^{max} = \frac{P}{2Tf_s} \qquad (3.2)$$

This confirms our intuitive notion that as the rotational frequency of the screen is increased, the voxel activation capacity of the display unit must be diminished. There is therefore a conflict between the need to increase the rotational frequency of the screen so as to reduce image flicker and the desirability of attaining a high voxel activation capacity so as to achieve a high voxel density.* The majority

---

* As described in Section 2.8, the term *voxel density* is used loosely and we neglect any anisotropic image space characteristics.

of swept-volume systems implemented to date are operated with a value of f_s such that the level of image flicker is tolerable during brief viewing periods. This permits display systems to be operated at their greatest voxel activation capacity.*

3.4.2 Mechanical Considerations

The rotational motion of the surface of emission and supporting structure is, when considered from a mechanical perspective, considerably more easily implemented than the alternative translational motion. The use of a helical screen geometry is best suited to the production of an image space that has a small height to width ratio. This is a direct consequence of the need to span the complete height of the image space by means of a helix that possesses no more than a single turn. The reasons for this requirement are as follows:

- In order that any viewing angle restrictions are minimized and that image clarity remains invariant with viewing direction, it is undesirable to use a helical geometrical form having more than a single turn (see Figure 3.3). Should a helical structure having, for example, two turns be employed, portions of the screen may obstruct or distort an observer's view of an image scene.
- For display units that employ a passive surface of emission, the use of a helical screen with multiple turns is likely to result in shadowing dead zones

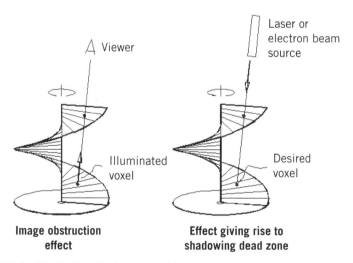

FIGURE 3.3 Shadowing dead zone and image obstruction with a helical surface of emission. In general, these effects become worse as the number of turns in the helical screen is increased.

---

* The effect of flicker in volumetric displays during prolonged periods of viewing has not yet (to the authors' knowledge) been researched in depth.

(see Figure 3.3 and Section 4.4.2). These are regions within the image space that cannot be addressed, due to another part of the screen presenting a physical obstruction to the voxel activation beam(s).

For these reasons it is highly desirable to span the height of the image space with a helical structure that has, at most, a single turn. Therefore, to increase the height of the image space, it is necessary to increase the pitch of the helix, and this results in an increase in its surface gradient. For display units that employ a passive surface of emission and beam source(s) positioned in line with the axis of rotation, this has the effect of increasing the radius of the cylindrical distortional dead zone* centered about the rotational axis. In the case of a helical screen employing an active surface of emission, this problem is likely to be less acute.

The helical form of supporting structure offers greater mechanical rigidity than its planar equivalent. In the latter case, rigidity may be enhanced by increasing the thickness of the supporting structure. However, this may have undesirable consequences for the image quality. Alternatively, added support may be achieved by the inclusion of a frame around the screen (see, e.g., Plate xii). Unfortunately, such a frame is likely to increase the severity of the visual dead zone (see Section 4.2 and Plate viii).

3.4.3   Uniformity of the Image Space

An image space that is created by means of rotational motion has associated shortcomings. These are caused primarily by the nature of the motion and by the presence of the supporting structure, which must be of a sufficient thickness to prevent flexing of the surface of emission as it sweeps out the volume.

Voxel Density Variations

A fundamental consequence of the rotation of the surface of emission is the variation in the maximum voxel density with radial distance from the axis of rotation. That is, the maximum achievable density of voxels decreases with increasing distance from the rotational axis. This is illustrated below.

Let us suppose that the activation of each voxel occupies a time T and that during this period, the surface of emission moves through an angle $\delta\theta = 2\pi f_s T$. If P voxels may be activated simultaneously, the activation of n voxels may occupy an angle:

$$\Delta\theta = \frac{n}{p}\delta\theta \qquad (3.3)$$

If these voxels are activated at a distance r from the rotation axis, they will be distributed across a distance Δd along the direction of motion of the surface:

---

* As discussed in Sections 4.2.1 and 4.3, a distortional dead zone is a region within image space in which voxel elongation is unacceptably high. Such dead zones arise when the angle between the incident beam and surface of emission is acute.

$$\Delta d = r\Delta\theta = \frac{2\pi n f_s T}{P} r \qquad (3.4)$$

This confirms our intuitive notion that voxels may be clustered more closely together toward the rotation axis.

This restriction arises as a direct consequence of the nature of the motion, and affects the achievable density of voxels *in the direction of the screen's motion*. Restrictions in the achievable voxel density on the surface of the screen itself (i.e., in the vertical and radial directions) are determined by the voxel activation and generation mechanisms. In combination, these factors lead to an anisotropic voxel distribution, and therefore any reference to voxel density must be interpreted with care (see Section 9.6). A related effect may be visualized by considering a line segment lying along an instantaneous screen position (e.g., in the case of a planar screen consider a vertical line parallel to the axis of rotation). If the activation of voxels is sequential, the lines will be distorted in the direction of screen motion.*

Should we wish to provide a uniform image space, it is undesirable to take advantage of the increasing voxel density toward the axis of rotation. The density specified for the display unit should, ideally, be achievable everywhere and hence will be limited by the worst-case performance — that at the periphery. This is an inherent problem for display units employing rotational motion and has conse-quences for the maximum achievable image space radius. We revisit these issues in the context of active surface of emission systems in Section 3.6.

Effect of the Supporting Structure

The presence of the supporting structure affects the properties of an image space in two important respects. First, let us consider the region of image space close to the axis of rotation. As discussed in Section 3.2.1, the supporting structure forms a rigid base on which the surface of emission exists. Suppose that a planar supporting structure is positioned symmetrically about the rotation axis. In such a configuration there will be a cylindrical region of the image space that is contin-ually occupied by this structure and in which voxels cannot be activated. If the screen has a thickness k and rotates about an axis that is located equidistant from each face of the screen, this cylinder will have a radius $k/2$ and will occupy the full height of the image space. Any region of image space in which voxels cannot *possibly* exist, whatever the arrangement of the voxel activation mechanisms, will be termed *dead image space* (see Section 4.2).

To minimize this effect and any optical distortion caused by the structure (see below), it should be fabricated so as to be as thin as possible. However, in the case of anything other than a small image space (perhaps of the order of 300 mm in diameter), it may be necessary to provide added support to a thin (less than 1 mm) supporting structure by the provision of a center shaft. The physical presence of such a shaft will itself create a region within which voxels cannot exist.

---

* See also Section 3.5.4 for the case of translational motion.

Although a helical screen is inherently more rigid than is its planar equivalent, in the case of a large image space, a center shaft may still be required. As with the planar screen configuration, this will give rise to dead image space. Furthermore, a center shaft will also cause the visual occlusion of a portion of the image space and consequently will affect an image when it is viewed from certain locations [i.e., when the image is viewed through this central region (see Section 4.4.2)].

As we have mentioned previously, a supporting structure interferes with light emitted in certain directions from the surface of emission. Light that emerges from a transparent supporting structure at an angle that is normal to its surface will be transmitted without significant diminution. However, light emerging at other angles will be subjected to refraction, and in the worst case, total internal reflection may cause a reduction in image brightness (see Figure 3.4). Consequently,

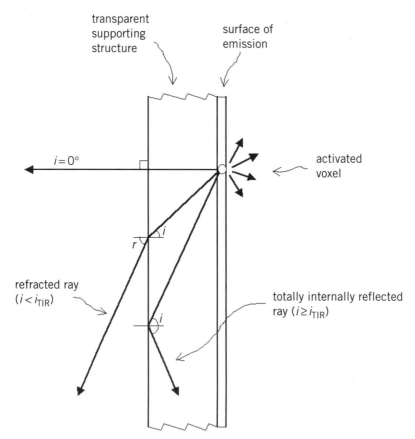

FIGURE 3.4 Refraction of light emitted from the surface of emission and passing through a largely transparent supporting structure. The light progresses undistorted only for the special case of emission normal to the surface. Other rays are refracted or may undergo total internal reflection.

light may not propagate rectilinearly and isotropically in all directions from the surface of emission to the exterior of the image space and so to the observer. In general, these effects become more severe as the surface of the screen lies at more acute angles with respect to an observer. This can result in a perceived darkening of an image across a narrow region as illustrated in Plate viii. This region, which depends on the position of the viewer, is termed a *visual dead zone* (see Section 4.2).

Fortunately, the effect of binocular vision somewhat ameliorates this defect; while a part of an image may be degraded in intensity from the viewpoint of one eye, the diminution seen from the other is often less severe. The net effect is a smaller reduction in image brightness within the visual dead-zone region than would otherwise be the case. For this reason, a planar screen rotating about a horizontal axis would suffer more severely from the visual dead zone and should therefore be avoided. As the dimensions of an image space are reduced, the extent of the visual dead-zone will remain unaltered.* Consequently, the perceived impact of a visual dead zone may be greater for image spaces of smaller diameter.†

Should the screen have a nonplanar geometry, the visual dead zones will take a different form and position. For example, consider a helical screen rotating about a vertical axis. When the image space is viewed from the side, the visual dead zone will appear as a darkened region extending diagonally across a portion of the image space, corresponding to the supporting structure lying at an acute angle to the viewing direction. Should the supporting structure be translucent, voxels may appear to be less intense or less clearly defined when viewed through the supporting structure, compared to their appearance when the surface of emission is viewed directly. This is due to the emitted light being partially scattered or absorbed by the material comprising the supporting structure. If the supporting structure is opaque, regions of an image space will be occluded and this will introduce severe visual dead zones.

3.5   IMAGE SPACE CREATION BY TRANSLATIONAL MOTION

Display units that employ translational motion for image space creation generally utilize a planar screen that moves back and forth along an axis perpendicular to its plane. The amplitude of the motion determines the maximum depth of the image space, its other dimensions corresponding to those of the surface of emission. As the screen moves through each cycle of motion, voxels may be activated as it passes through the appropriate locations. In the case of display units that employ a passive surface of emission, the image space is generally divided into a number of slices, as illustrated in Figure 3.5. Each voxel descriptor (see Section 8.4)

---

* Assuming that the thickness of the supporting structure remains constant.

† Some time ago, the authors developed a display unit (see Plate ix) without paying proper regard to this problem. This device employed the rotational motion of a small square planar screen (sides of length 20 cm). The axis of rotation was located horizontally. Consequently, the visual dead zone was very obtrusive, and so the perceived image quality was poor.

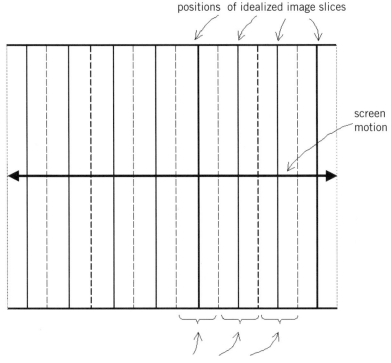

portions of image space containing
data assigned to the slices indicated

FIGURE 3.5   The division of an image space into slices provides a convenient method of ordering data that are to be output to a display unit employing a passive surface of emission. The extent of each slice determines the maximum permissible error in voxel positioning along the direction of motion of the surface.

within the image data set is then mapped into the most appropriate image space slice. As we will see in Section 8.5, this provides a simple and highly effective method of ordering the data and in principle may ensure that the magnitude of any error in voxel positioning cannot exceed a defined threshold.

3.5.1   Image Space Update

As with display units employing a rotational motion, the cyclic frequency imposes constraints on both the image update frequency and the voxel activation capacity of the display unit. This frequency also determines the perceived level of image flicker. In general, systems employing translational motion permit the activation of voxels throughout the complete cycle of motion, as the screen moves first in

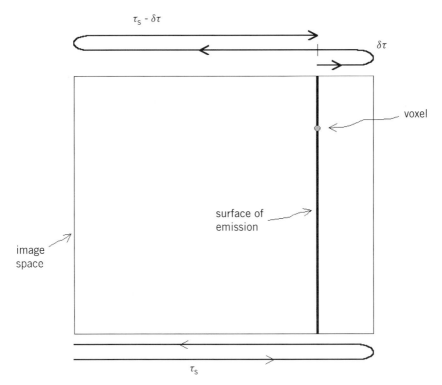

FIGURE 3.6 Voxels in a display unit employing translational motion may be updated twice per period. However, only voxels located at the midpoint of the surface's motion will be updated at a frequency equal to twice that of the motion. The time between successive updates for other voxel positions oscillates between $(\tau_s - \delta\tau)$ and $\delta\tau$. The refresh frequency should therefore equal the frequency of motion of the screen. However, by activating different voxels during each half cycle of motion, it is possible to maximize the voxel activation capacity of the display unit.

one direction and then as it reverses it path. However, the image refresh frequency is *not* twice the frequency of motion. This may be illustrated as follows. Let us consider a voxel that is activated at a location that is close to one extremity of the screen's range of motion (Figure 3.6). Three consecutive activations of this voxel will not be equally separated in time. Two of the voxel updates will be separated by the time $(\delta\tau)$ required for the screen to move to the nearest extremity of its motion and return, and the other interval will correspond to the time $\tau_s - \delta\tau$ required for the surface to move to the farthest extremity and return. As we consider voxels closer to the extremities of the motion, $\delta\tau \to 0$, so the largest interval between successive updates approaches the screen period τ_s. Therefore, if voxels are activated throughout the range of the screen's motion, the worst-case time delay between successive voxel activations is τ_s. Consequently, the image update frequency must be equal to the frequency of screen motion.

3.5.2   Uniformity of the Image Space

The essential difficulty in the implementation of a display employing translational motion concerns the development of the mechanical system responsible for effecting movement of the screen. Since it is desirable to maximize the depth of the image space, it is important to ensure that the screen's motion is optimally utilized for this purpose. For example, consider an image space that is to be created by a surface moving with constant velocity (thereby sweeping out equal volumes in equal times), as illustrated in Figure 3.7. In this case, so as to maximize the depth of the image space, the linear portion of the velocity profile should be as great as possible, and therefore the extent of the regions of acceleration and deceleration minimized. As a consequence, the screen must be able to withstand large forces, and so to minimize both the physical size of the mechanical drive and the power consumption, its mass must be as low as possible. Reductions in the mass of the screen will restrict its dimensions, and furthermore the large forces to which it will be exposed are likely to affect the life expectancy of the system.

Let us now turn our attention to the implementation of a display unit in which the screen moves with a sinusoidal velocity profile. Implicitly, the screen will not sweep out equal volumes in equal times. Consequently, to compensate for

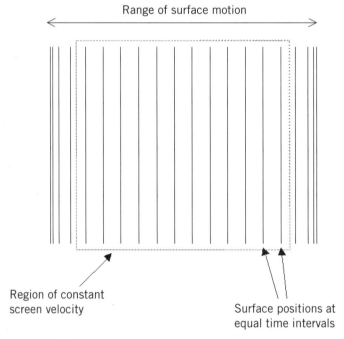

FIGURE 3.7 Screen undergoing translational motion such that a constant velocity is provided during the central portion of the displacement. The necessary accelerations and decelerations are restricted to the extremities of the motion beyond this section.

the nonlinear motion and so ensure that images will not be subjected to any associated distortion, the timing of data output to the display unit may be varied in accordance with the screen's position. Alternatively, it is possible to restrict the image space to a subvolume of the region swept out by the screen. For example, we could make a linear approximation to a screen's sinusoidal velocity profile. The proportion of the total volume swept out by the screen that is employed as an image space would then be determined by the maximum permissible error in voxel positioning.

Both of these scenarios will now be examined in greater depth. We begin the discussion by assuming that an approximate ordering technique has been applied to the image data so as to map it into ξ equally spaced slices (see Figure 3.5 and Section 8.5). Each of these slices will be output to the surface of emission as it passes through the appropriate region of the image space.

Compensation for a Sinusoidal Velocity Profile

If the rate at which image slices are output to the surface of emission is varied to compensate for nonuniform velocity, the amplitude of the screen's motion may equate to the depth of the image space. Consequently, the extent of the image space is maximized. If we ascribe to the image space a rectangular coordinate system (x,y,z) in which the z-axis lies parallel to the direction of the screen's motion, the position of the surface as a function of time may be described by

$$z(t) = \frac{A}{2} \sin \omega_s t \tag{3.5}$$

where A represents the peak-to-peak amplitude of the motion and ω_s is the angular frequency of the motion such that $\omega_s = 2\pi f_s$. If the ξ slices are spread throughout the full range A of the motion, the distance between the start of each slice is

$$\Delta z = \frac{A}{\xi} \tag{3.6}$$

The z position of the jth slice may then be expressed as:

$$z_j = j\Delta z = \frac{jA}{\xi} \tag{3.7}$$

Using this result and reexpressing Eq. (3.5), the time t_j at which the jth slice is activated may be written

$$t_j = \frac{1}{\omega_s} \arcsin \frac{2j}{\xi} \tag{3.8}$$

and therefore the time between consecutive slices is given by

$$\Delta t = t_j - t_{j-1} = \frac{1}{\omega_s} \left[\arcsin \frac{2j}{\xi} - \arcsin \frac{2(j-1)}{\xi} \right] \tag{3.9}$$

TABLE 3.1   Numerical Example Data: Translational Motion[a]

Parameter values used in passive surface example, Sections 3.5.2 and 3.5.3:

- Number of image slices: $\xi = 256$
- Mass of screen structure: $m = 0.4$ kg

Parameter values used in active surface example, Section 3.6.2:

- Voxel diameter: $V_d = 1$ mm
- Adjacent voxel spacing (between voxel center): $V_s = 1$ mm

Parameter values common to both examples:

- Frequency of motion: $f_s = 25$ Hz (hence $\omega_s = 50\pi$ rad $\cdot$ s^{-1})
- Area of surface of emission: $L_x \times L_y = 40$ cm $\times$ 40 cm $= 1600$ cm^2
- Amplitude of motion: $A = L_z = 30$ cm.

[a]Partial specification for display units employing a translational motion. The first set of parameter values is used in the passive surface example expounded in Sections 3.5.2 and 3.5.3, the second set applies to the active surface example discussed in Section 3.6.2, and the third set contains values common to both.

Let us illustrate this numerically by means of the example data provided in Table 3.1. The smallest time between consecutive slices occurs at the midpoint of the motion, where the velocity of the screen is at its greatest. Taking $j = 1$, we obtain from Eq. (3.9) the value $\Delta t \approx 50$ μs. The largest time between slices occurs at the extremities of the motion, where the screen slows down and changes direction. Using $j = \xi/2 = 128$ in Eq. (3.9), we obtain $\Delta t \approx 800$ μs.

Unfortunately, the variation in the temporal duration of each slice has undesirable consequences for image space uniformity (in terms of voxel placement) and/or for the voxel activation capacity of the display. Clearly, to preserve uniformity it is desirable to ensure that each slice is able to accommodate an identical maximum number of voxels. This may be achieved by determining the maximum number of voxels that may be activated within the slice that has the smallest time duration. The maximum number of voxels that may be activated in each of the other slices would then be set to this value (despite their ability to contain a greater number of voxels). As a consequence, the voxel activation capacity would then be smaller than it would otherwise be if we permitted each slice to be populated to its maximum extent. In this case, uniformity has been attained by making a reduction in the maximum attainable voxel activation capacity.

Linear Approximation to a Sinusoidal Velocity Profile

Should the graphics engine output image slices at a constant rate to a display unit in which the screen moves with a sinusoidal velocity profile, most of the slices will be subjected to a positioning error in the direction of screen motion. The constant slice output rate corresponds to a linear approximation to the screen's

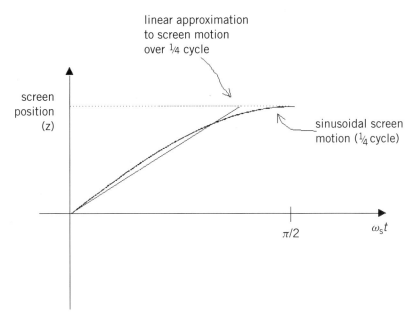

FIGURE 3.8 Linear approximation $z(t) = (\alpha A\omega_s/\pi)t$ to the sinusoidal screen motion $z(t) = (A/2)\sin\omega_s t$, over a quarter-period $(0 \leq \omega_s t \leq \pi/2)$. Errors in voxel positions arise due to this approximation underestimating or overestimating the displacement of the surface of emission. The depicted line has a slope of approximately 1.14, this being the optimum for the case where the surface is addressed throughout its entire range of motion (see the text).

motion, and it is of interest to determine the magnitude of the errors incurred in this approximation and the time between slices that minimizes these errors.

Let us consider the motion during a quarter-period $(0 \leq \omega_s t \leq \pi/2)$, as illustrated in Figure 3.8. The straight line shown in this diagram represents the linear approximation to the sinusoid described by Eq. (3.5). The equation of the linear approximation is

$$z^{\text{approx}}(t) = \alpha \frac{A\omega_s}{\pi} t \qquad (3.10)$$

where the parameter α enables the slope of the line to be varied. When $\alpha = 1$, the line defined by Eq. (3.10) will intersect with the sine curve at its apex $(\omega_s t = \pi/2, z = A/2)$. When $1 < \alpha < \pi/A\omega_s$, the line intersects the sine curve at some value $\omega_s t < \pi/2$. In the case of $\alpha > \pi/A\omega_s$, the line lies above the sine curve throughout this range, and when $\alpha < 1$ the linear approximation value is lower than the sine curve for all values of $\omega_s t$ in this quarter-period.

Consider the case where the ξ slices are spread throughout the full range of the motion and the slices are output throughout the entire quarter period. The time at which the jth slice is activated is $t_j = j\Delta t$, where the time between the activation of consecutive slices is $\Delta t = \pi/\omega_s\xi$. Thus

$$t_j = \frac{j\pi}{\omega_s\xi} \tag{3.11}$$

The positional error δz_j of the jth slice is thus the difference between the sine curve [Eq. (3.5)] and the linear approximation [Eq. (3.10)] at the time t_j:

$$\delta z_j = A\left(\frac{1}{2}\sin\left(\frac{j\pi}{\xi}\right) - \frac{\alpha j}{\xi}\right) \tag{3.12}$$

The value of α that minimizes the maximum positioning error is in this case $\alpha = 1.14$. The value of $\omega_s t$ at which the curves intersect is obtained by solving $\partial(\delta z_j)/\partial(\omega_s t) = 0$, which yields

$$(\omega_s t)^{\text{intersect}} = \arccos\left(\frac{2\alpha}{\pi}\right) \tag{3.13}$$

Thus, for $\alpha = 1.14$, the inter Section point occurs at $\omega_s t = 0.76$ and $z = 0.28A$. It can be seen from Figure 3.8 that this slope balances the positioning errors [given by Eq. (3.12)] toward the extremity of the motion range, where the linear

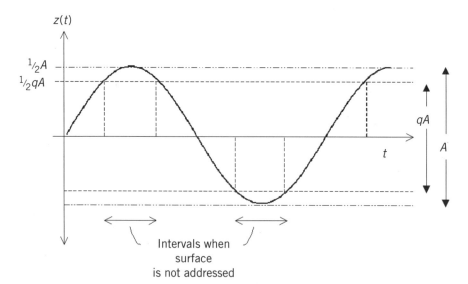

FIGURE 3.9 The worst slice positioning error in the linear approximation to a sinusoidal motion may be reduced by addressing the screen over a fraction q of it motion range A. The optimal slope α increases with decreasing q. The fraction of time during which the surface is addressed is lower than the spatial fraction q, due to the slower screen speeds near the extremities of its motion as it decelerates and reverses direction.

approximation overestimates the screen position, and earlier in the cycle where the linear approximation underestimates the screen's position. As the slope is increased, the former error increases while the latter decreases, and vice versa. The slope of 1.14 represents a balance between these trends, with a positioning error of magnitude of about $0.07A$ at each of these points. This is important if the positioning of each slice is to be within, say, half the slice separation of its ideal depth. With the error noted above, only *seven* slices could be depicted within the image space and conform to this positioning accuracy!

If a constant data activation rate is to be employed in addressing a sinusoidal motion, the linear approximation can be restricted so that the screen is addressed only during part of its motion. That is, the image space now comprises only part of the volume swept out by the screen, and the linear approximation is then applied to the central portion of the sinusoid. Let q represent the fraction of the amplitude of motion during which the screen is addressed (Figure 3.9). As q is decreased, the optimal slope of the linear approximation increases and the worst-case positioning error is reduced, as illustrated in Table 3.2. However, if the volumetric image is to consist of 256 slices and these are to be positioned to an accuracy of within half the slice separation of their ideal location (for the desired case of evenly spaced slices), q is required to be less than approximately 0.3. (This can be seen by comparing the separation of evenly spaced slices over a depth qA: namely, qA/ξ, with the worst-case error from Table 3.2.) Under these conditions, the image space occupies less than one-third of the range of motion. To enlarge the useable subvolume, either the number of slices can be reduced or the positional accuracy condition relaxed. However, the latter option will increase the distortion of data within the image space.

TABLE 3.2   **Optimal Slope (α) of Linear Approximation to a Fraction q of the Full Range of Sinusoidal Motion of a Translating Surface, and Worst-Case Positioning Error (normalized to range of motion, A) of Data Slice in Each Case**

q	Optimal Slope, α	Worst-Case Error/A
0.1	1.57	0.06×10^{-3}
0.2	1.56	0.27×10^{-3}
0.3	1.55	0.72×10^{-3}
0.4	1.54	1.7×10^{-3}
0.5	1.52	3.2×10^{-3}
0.6	1.49	5.5×10^{-3}
0.7	1.46	9.8×10^{-3}
0.8	1.41	15.5×10^{-3}
0.9	1.34	26.7×10^{-3}
1.0	1.14	69.7×10^{-3}

We conclude from these results that for a display unit employing sinusoidal reciprocating motion, compensation for the nonlinear motion is preferable to a linear approximation (constant data rate). There are a variety of alternative approaches that may be adopted to compensate for nonlinear motion. However, these are not discussed here.

3.5.3   Mechanical Considerations

The mechanical aspects of any display unit that is to employ translational motion for image space creation must be given great consideration. For example, due attention must be given to the following:

- The need for a highly rigid supporting structure
- The need to minimize the mass of the screen
- The energy needed to obtain the required acceleration
- General reliability, life expectancy, and acoustic noise

Let us consider briefly a screen moving with a sinusoidal motion as described by Eq. (3.5). From this equation the velocity of the screen is given by

$$v = \frac{\omega_s A}{2} \cos \omega_s t \tag{3.14}$$

Using the numerical values given in Table 3.1, the maximum screen velocity would be $\omega_s A/2 = 15\pi/2 = 23.6$ m $\cdot$ s^{-1}. The acceleration a of the screen may be expressed as

$$a = -\frac{\omega_s^2 A}{2} \sin \omega_s t \tag{3.15}$$

In the case of the numerical example given above, this would result in the magnitude of the peak acceleration being $\omega_s^2 A/2 = 375\pi^2 = 3701$ m $\cdot$ s^{-2}.

It is instructive to determine the approximate power that must be applied to the screen drive mechanism so as to achieve the proper motion of the surface of emission and associated supporting structure. For the sake of simplicity we consider that the screen structure moves within an evacuated chamber and so air resistance may be neglected. The screen is instantaneously at rest when at the extremities of its motion and is accelerated in one-quarter of its motion period to its maximum velocity, $\omega_s A/2$, at the midpoint of its movement. The instantaneous kinetic energy supplied to the screen is given by $E_k(t) = \frac{1}{2}mv(t)^2$, where $v(t)$ is given by Eq. (3.14). Over a quarter-period of the motion, this integrates to a total applied kinetic energy of

$$E_k^{\text{tot}} = \frac{\pi m \omega_s A^2}{32} \tag{3.16}$$

Therefore, the power required to impart this energy during one quarter-period $(\pi/2\omega_s)$ is given by

$$P = \frac{m\omega_s^2 A^2}{16} \tag{3.17}$$

Inserting into this equation the numerical data presented in Table 3.1 indicates a required power of approximately 56 W.

3.5.4 Effect of Sequential Voxel Activation on a Line Segment

Consider the depiction of the line segment illustrated in Figure 3.10, which is to be drawn within an image space created by either the translational or rotational motion of a screen. This line segment is intended to have no component in the direction of the screen's motion and consequently, must be drawn instantaneously.

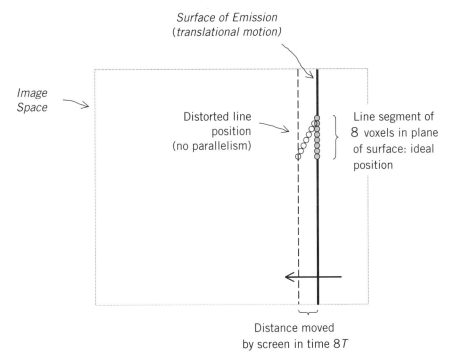

FIGURE 3.10 A line segment oriented so that it has no component in the direction of the screen's motion may being displayed erroneously if the parallelism supported by the display unit does not permit all the voxels that comprise the line segment to be activated simultaneously. The eight-voxel line segment illustrated here is tilted as a consequence of the screen's motion. This illustration assumes that the voxels are activated in order from top to bottom; if the activation order is otherwise, the line could appear discontinuous. Although illustrated here for the case of translational motion, this effect applies equally to rotational motion systems.

However, should voxel activation take place sequentially, the finite time required for the production of each voxel will result in the line being tilted in the direction of the screen's motion [by an amount given by Δd in Eq. (3.4) for the case of rotational motion].*

In the case of translational motion the positioning error e_i for the ith voxel in such a line segment is given by

$$e_i = 2iL_z f_s T \qquad (3.18)$$

where T is the total voxel time. Should other image data descriptors exist having z coordinate values between z_0 and $z_0 + 2iL_z f_s T$, it is apparent that their positioning will also be subject to error. The result is the possible presence of a compounded distortion along the z axis.

3.6 ACTIVE SURFACE OF EMISSION

Swept-volume display units that are equipped with an array of discrete elements[†](voxel generation centers) and in which these elements can be addressed individually by, for example, direct application of an electrical signal are classified in this book as having an active surface of emission (see Section 3.3). When instructed to do so, each element is able to undergo a visible optical change (and *ideally*, remain in a transparent state otherwise)[‡]. The surface of emission may move in either a rotational or translational manner, and hence each voxel generation center is responsible for the production of multiple voxels along the path that it traverses during each cycle of motion. For example, in the case of systems employing a rotational motion, each element on the surface of emission is responsible for the generation of voxels along a circular track, and in the case of translational motion for a series of voxels located on a linear path (Figure 3.11).

3.6.1 Rotational Motion

The use of rotational motion gives rise to spatial variations in the maximum achievable voxel density similar to those discussed in Section 3.4.3. We now revisit these considerations in the specific context of active surface display units.

---

* This applies in a straightforward manner when all voxels that constitute the line segment are generated sequentially ($P = 1$). However, should a degree of both parallel and sequential activation be involved in the production of the line segment, voxels (or groups of voxels) will be shifted from their proper location. To predict the nature of this shift, it is then necessary to examine the precise configuration of the voxel activation mechanisms. Should all the voxels constituting the line segment be activated in parallel, the line may be activated instantaneously and will therefore have no component in the direction of the screen's motion.

† Bonded to, or formed on, a supporting structure.

‡ Also, the refractive index of each element should match that of the supporting structure *and* that of the unoccupied image space. Since the two are generally different, this is somewhat problematic!

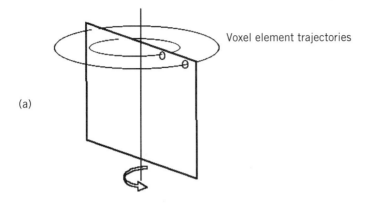

(a)

Voxel element trajectories

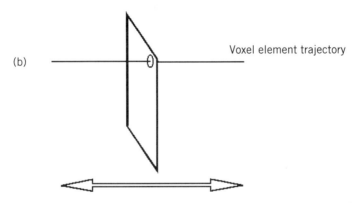

(b)

Voxel element trajectory

FIGURE 3.11   Voxel generation in display units employing an active surface of emission through the use of (*a*) rotational and (*b*) translational motion.

Consider a display unit in which an active surface of emission moves with a rotational motion (see, e.g., Sections 5.4.4 and 5.5.3). Elements that are located at a greater radius from the rotational axis clearly traverse a greater distance in the same period than do those that are closer. However, to produce an image space with a uniform voxel density, consecutive voxels (created by each element) should have the same spatial separation, *irrespective* of their radial distance from the rotational axis. Furthermore, and as a consequence, the workload (peak data throughput) is not shared equitably among the elements. In a single revolution, an element at radial distance r from the axis of rotation travels around a circle of circumference $2\pi r$, so that the peak modulation rate (voxels $\cdot$ s^{-1}) is directly proportional to an element's radial position.

This problem may be illustrated by reference to the display unit whose partial specification is presented in Table 3.3. Consider two voxel generation centers

TABLE 3.3   Partial Specification for a Display Unit

Numerical Example Data — Rotational Motion

- Square double bladed surface of emission with sides of length 40 cm.
- Rotational frequency: $f_s = 25$ Hz.
- Voxel diameter: $V_d = 1$ mm.
- Adjacent voxel spacing (between voxel centers): $V_s = 1$ mm

---

located at 5 and 20 cm from the rotational axis. During the course of a single revolution of the surface of emission, the elements travel a distance of approximately 31 and 125 cm, respectively. For an intervoxel spacing of 1 mm, these two elements are responsible for the generation of approximately 310 and 1250 voxels respectively. At a rotational frequency of 25 Hz, the element located at $r = 5$ cm must be capable of operation with an activation frequency of approximately 7.8 k voxels $\cdot$ s^{-1}. This contrasts with the peak voxel throughput of 31.3 k voxels $\cdot$ s^{-1} for the voxel generation center located at a radius of 20 cm.

As the diameter of the surface of emission is increased so as to provide for the creation of a larger image space, the required operating frequency of the elements at the extremities will rise in proportion to their distance from the rotational axis. Consider the case in which the workload of the voxel generation centers is not redistributed on the basis of radial position, but maintained at the same value for all elements. The intervoxel separation in the angular direction will then increase toward the axis of rotation, leading to a highly nonuniform image space. As a consequence, the various characteristics of an image component may be affected by the position at which it is located within the image space.

Continuing with the numerical example, from the specification given it is apparent that the surface of emission would be equipped with 160,000 elements, and these would have the capability (in principle) of providing a voxel activation capacity in excess of 50 million. This very large voxel activation capacity is made possible by the high degree of parallelism introduced into the display unit's voxel generation mechanism. In principle, this display architecture is able to accommodate a 100% fill factor (see Section 2.9). The passage of electrical signals to the rotating surface of emission and in turn to each appropriate voxel generation center clearly represents an added complication in the implementation of display units of this type. An optical link provides an obvious solution. An example of a parallel data transfer link suitable for use with this type of architecture is discussed in Section 12.3.1. In the case of the display unit outlined in Section 5.5.3, a serial optical link is employed to pass voxel data to processing hardware that corotates with the screen.

3.6.2   Translational Motion

Consider a display unit in which an active surface of emission moves with a translational motion, such as the volumetric device described in Section 5.6.7.

Unlike the configuration described above in which rotational motion is employed, the peak workload is equitably shared between all of the voxel generation centers. As the size of the surface of emission is increased, the workload for each element remains unaltered. However, as the depth of the image space is increased, each element must move at a greater velocity over a larger distance, and as a consequence the bandwidth which must be supported by each voxel generation center increases.

From the partial specification provided in Table 3.1, it is apparent that each of the 160,000 voxel generation centers will be responsible for the production of up to 300 voxels during each cycle of motion of the surface of emission, providing a voxel activation capacity of 48 million.* As with the configuration outlined in Section 3.6.1, this display unit also supports a high degree of parallelism in the voxel activation subsystem and, in principle, may permit a 100% fill factor. For this configuration each voxel generation center must be capable of operation with a peak throughput of approximately 7.5 k voxels $\cdot$ s^{-1}.

3.7 DISCUSSION

Although the graphics engine is able to compensate for a number of display unit deficiencies (see Section 10.5), it is most unlikely that it can be made to accommodate optical and mechanical weaknesses in the surface of emission and its associated supporting structure. The emphasis that we have placed on the design of the surface of emission and the structure within this chapter reflects the very considerable impact they have in determining the overall image quality. In general terms the characteristics associated with varifocal mirror technologies are similar to those of volumetric systems employing a passive surface of emission moving with translational motion. However, the former type of display employs a very simple drive system and avoids the mechanical complexities associated with the latter. It is therefore natural to question the continued relevance of translational motion as a means of creating an image space. However, in Section 12.4 we propose a hybrid system that employs simultaneously translational and rotational motion. This display unit offers a number of advantageous characteristics and so demonstrates benefits that may be associated with translational motion. In the case of this embodiment the mechanical drive problems referred to above are simplified since only a very small amplitude of motion is required.

---

* This assumes that the full amplitude of motion is utilized (see Section 3.5).

CHAPTER 4

BEAM-ADDRESSED SWEPT-VOLUME DISPLAY UNITS

Between the idea and the reality
Between the notion and the act
Falls the Shadow.*

4.1 INTRODUCTION

In this chapter we focus on swept-volume display units employing a passive surface of emission in conjunction with one or more stationary voxel activation mechanisms. Emphasis is placed on the complications that may arise as a consequence of the continual variation in the orientation of the voxel activation mechanism(s) with respect to the surface of emission. An important aspect of the design of any volumetric display unit relates to the evaluation of potential dead zones that may be associated with any proposed architecture. These are regions in which at least one image (or image space) characteristic falls below an acceptable level of tolerance. The term *dead zone* was first used in the context of volumetric display systems in 1991 (Blundell and King 1991), and has now become a generic term used to refer to a range of adverse image space characteristics.

Dead zone regions may be caused by a number of distinctly different effects. For clarity, we extend this title to distinguish between the underlying technical problems that give rise to the occurrence of unacceptable characteristics throughout one or more regions of an image space. In Chapter 3 we mentioned several types of dead zone. In the next section we discuss dead zones in greater depth and classify them according to their characteristics. Once the form and extent of the various dead zones have been considered, it is possible to devise

---

* T.S. Eliot, *The Hollow Men*, in *Complete Poems and Plays: 1909–1950*, Harcourt Brace, NY, 1952.

strategies to minimize both their impact and their extent with the intention of improving the overall image quality. This is illustrated in Section 4.3, where we present an analysis of the distortional dead zone for two swept-volume configurations.* Although our emphasis is directed to display units employing a passive surface of emission, we shall, where appropriate, refer to swept-volume systems using an array of voxel generation centers (i.e., an active surface of emission). By definition these systems are addressed directly (e.g., by the application of electrical signals to each element) and the dead zones that are caused by the continual variation in geometry between the beam sources and surface of emission are avoided. However, as will become evident during the course of this chapter, displays of this type are not free of all dead zones.

4.2 GENERAL DEAD ZONE CONSIDERATIONS

A volumetric display device ideally provides a uniform three-dimensional image space that may be viewed from practically any orientation. In practice, it has not been possible to achieve these objectives fully. For example, many volumetric techniques give rise to voxel characteristics that vary throughout an image space. This gives rise to the possible existence of regions of image space in which image fidelity, in some respect, falls below a desired criterion. Such regions are generically known as *dead zones*. Different dead zones arise depending on which image space parameter is affected, and the size, shape, and importance of each type of dead-zone may vary according to the display unit architecture. It is important to realize that the affected parameter(s) generally vary in a continuous fashion throughout the image space. Therefore, a dead-zone boundary is to some extent arbitrary and will often be determined by the performance demanded from the system.

The expression *dead zone* may be defined broadly as follows:

Dead zone: a region of an image space within which the desired performance cannot be achieved. The exact boundary of a dead zone is defined by the system performance required.

In the design of any display unit, potential dead zones must be identified and techniques sought that will circumvent, or at the very least minimize, their impact on the overall image quality. Although the uniformity of the image space may often be improved by altering the system design, this may be at the expense of viewing range. Furthermore, the mechanisms giving rise to dead zones often play an important part in determining the ultimate performance of a particular volumetric technique. Therefore, to reduce the severity of some types of dead zone, it may be necessary to make a reduction in one or more areas of system performance. Consequently, a proper understanding of the severity of the dead

---

* Dead zones may also affect static-volume display unit architectures (see Section 6.6).

zones that may be associated with particular display unit architectures allows a more realistic estimation of the performance of a system configuration to be gained prior to its implementation.

4.2.1   Classes of Dead Zone

Dead zones may arise as a consequence of an underlying weakness in one partic-ular display unit subsystem or may be caused as a result of the interaction between the subsystems employed in a particular display unit architecture. When a dead zone is attributed to the problems arising from the interaction between two subsys-tems, it is natural to question the desirability of employing these subsystems in combination. Unfortunately, it would appear that the dead zones associated with various display unit architectures have sometimes been identified *after* construc-tion of the system. In this case, their severity and extent may possibly have been disguised by careful scaling and positioning of any images prior to their depiction within the image space. In this way, dead zone regions are left void and would, therefore, not be visible. Clearly, this is most undesirable.

Dead zones may be divided into a number of classes:

- *Distortional dead zone* (affecting *voxel size/shape*): a region in which the size and/or shape of the individual voxels deviates too greatly from the ideal. As a consequence, voxel brightness may also be affected.
- *Voxel placement dead zone* (affecting *precision of voxel placement*): regions in which the nearest-neighbor voxel separation is unacceptably high and/or in which spatial accuracy at which voxels can be positioned falls below a desired value.
- *Shadowing dead zone*: a region in which voxels cannot be placed due to a part of the screen presenting a physical obstruction to the propagation of the activation beam(s).

In the case of some volumetric architectures, voxels appear to vary in their intensity when they are viewed from different positions relative to the image space. Regions of an image space that suffer from a diminution of intensity will be referred to as *visual dead zones*. In the case of swept-volume systems employing a rotational motion, they generally arise in those portions of the image space corresponding to the surface of emission lying close to the line of sight of the observer. Such visual dead zones thus "follow" the viewing position of the observer. They are of particular concern in the case of display units employing the rotational motion of a screen, irrespective of the passive or active nature of the surface of emission. A visual dead zone is illustrated in Plate viii.

Regions of an image space in which voxels cannot *possibly* exist will be referred to as *dead image space*. For example, the presence of a center shaft passing through the image space or the spatial separation of voxel generation centers on an active surface of emission will both give rise to dead image space. The latter applies to both static volume display configurations (such as the elementary model introduced in Section 1.3) and swept-volume systems employing an active surface of emission. In Chapter 9 we introduce the concept of subspace. Within this context, dead image space corresponds to regions of an image space in which subspace does not exist.

A particular volumetric display technique may exhibit all, some, or none of these types of dead zone. Unfortunately, the relevance of a particular dead zone to a certain display unit configuration may not be readily apparent during the early stages of the design process. For example, it may be considered that display units employing the rotational motion of an active surface of emission would be free from the voxel placement dead zone. However, if proper account is not taken of the variation in peak workload as a function of the radial distance of the voxel generation centers from the axis of rotation, a voxel placement dead zone may ensue. This would correspond to a portion of the image space in which one or more voxel generation centers cannot be activated at the required frequency.*

In practice, dead zones represent a more serious problem in the case of passive surface volumetric techniques, in which voxels are activated by means of a radiation beam impinging on a rotating surface of emission. With this in mind, we present in the next section a mathematical analysis of the distortional dead zone. This analysis will assume the use of a rotating passive surface of emission; both planar (Ketchpel 1963, Blundell and King 1991, Bains 1993) and helical (Williams and Donohoo 1991, Clifton and Wefer 1993a, Soltan et al. 1994) geometries will be considered.

4.3 DISTORTIONAL DEAD ZONES

Consider a display unit architecture employing a passive surface of emission moving with a rotational motion and in which the voxel activation mechanism takes the form of a single stationary† beam. In this case, increasing voxel elongation will occur as the angle between the incident beam and the local tangent to the surface becomes more acute. The distortional dead zone is defined by the region in which this voxel elongation is unacceptable. In the following pages, we explore this type of dead zone in some detail, and obtain solutions for both the helical and planar screen geometries.

The angle of incidence (ϕ) of a beam onto the target surface is given by

$$\phi = \arccos(\hat{\mathbf{b}} \cdot \hat{\mathbf{s}}) \tag{4.1}$$

where $\hat{\mathbf{b}}$ represents the beam direction and $\hat{\mathbf{s}}$ the inward normal to the surface at the point at which the beam impinges (Figure 4.1). As the surface normal moves away from the beam direction, ϕ approaches $90°$ and $\hat{\mathbf{b}} \cdot \hat{\mathbf{s}}$ approaches zero. If

---

* This is of greatest consequence for voxel generation centers located at the greatest distance from the axis of rotation.
† We preclude systems in which the voxel activation mechanism(s) corotate with the surface of emission.

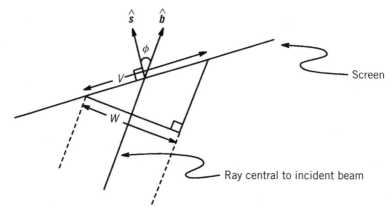

FIGURE 4.1 Voxel elongation in a display system employing a passive surface addressed by a radiation beam. When a beam of width W impinges on the surface of emission at an angle ϕ, it gives rise to a voxel elongated to length V in one direction.

we assume an ideal beam of circular crosssection and diameter W, the elongated voxel length V is given by

$$V = \frac{W}{\hat{\mathbf{b}} \cdot \hat{\mathbf{s}}} \tag{4.2}$$

For example, if the incidence angle is $60°$, $\hat{\mathbf{b}} \cdot \hat{\mathbf{s}} = \frac{1}{2}$, so the voxel will be twice as long as it is wide (in the plane of the surface). Alternatively, the elliptical shape of the voxel along the surface may be quantified by its eccentricity:

$$e = \sqrt{1 - (\hat{\mathbf{b}} \cdot \hat{\mathbf{s}})^2} \tag{4.3}$$

The precise extent of this dead zone depends entirely on the criterion chosen for acceptable/unacceptable voxel elongation.

Specific forms of this dead zone for particular surface geometries can be determined by making use of the image space frame, a Cartesian coordinate system (x, y, z) whose origin will be assumed to coincide with the center of the image space (Figure 4.2). The coordinates of the beam source in this frame are denoted (x_s, y_s, z_s). If a voxel is to be created at any point (x, y, z) in the image space, the unit vector $\hat{\mathbf{b}}$ describing the beam direction toward this point is given by

$$\hat{\mathbf{b}} = \frac{1}{B}(x - x_s, y - y_s, z - z_s)^{\mathrm{T}} \tag{4.4}$$

where the normalization factor (B), given by

$$B = \sqrt{(x - x_s)^2 + (y - y_s)^2 + (z - z_s)^2} \tag{4.5}$$

represents the distance between the beam source and the voxel.

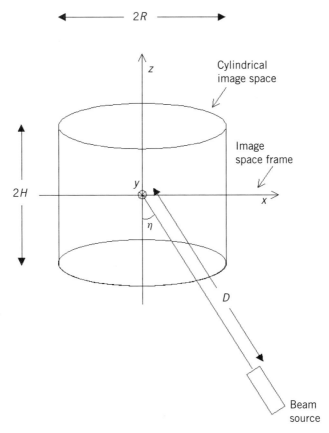

FIGURE 4.2 Coordinate systems for the dead zone analysis in Section 4.3. A right-handed Cartesian coordinate frame $(x, y, z)^\mathrm{T}$ is associated with the image space. It is oriented such that its z-axis lies along the axis of rotation and has its origin at the center of the image space. In this coordinate system, the position at which the beam deflection occurs (for simplicity, both horizontal and vertical beam deflections are assumed to occur at the same point in space) is denoted $(x_s, y_s, z_s)^\mathrm{T}$.

We will assume that the surface rotates about the z axis of the image space frame, and without loss of generality the beam source is considered to lie in the xz plane (i.e., $y_s = 0$). The image space created by the motion of the surface is assumed to be cylindrical, of height $2H$ and of diameter $2R$. The distance from the beam source to the center of the image space is

$$D = \sqrt{x_s^2 + z_s^2} \tag{4.6}$$

and its angle from the z axis is denoted η (Figure 4.2). With the foregoing assumptions, the beam source coordinates may be expressed explicitly as

$$x_s = D \sin \eta \qquad y_s = 0 \qquad z_s = H - D \cos \eta \tag{4.7}$$

In the following sections we turn our attention to analysis of the distortional dead zone in two specific cases: planar and helical surface forms.

4.3.1 Planar Surface Geometry

If the surface of emission takes the form of a plane rotating about the z axis, the position of the surface may be parameterized by a rotation angle β in the xy plane. In this case the Cartesian components intersecting the instantaneous surface position are

$$(x = r\cos\beta, \quad y = r\sin\beta, \quad z = z) \tag{4.8}$$

(Figure 4.3). The normal vector to the surface is therefore

$$\hat{\mathbf{s}} = (-\sin\beta, \cos\beta, 0)^{\mathrm{T}} \tag{4.9}$$

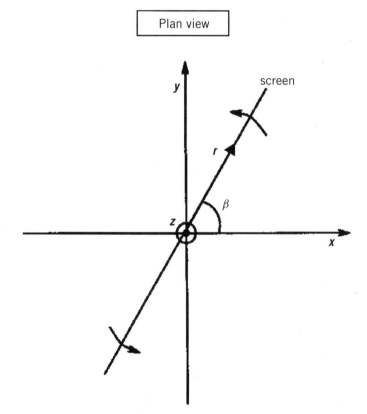

FIGURE 4.3   Polar coordinate description (r, z, β) of position within a cylindrical image space swept out by a rotating planar surface. The x and y axes of the Cartesian image space frame are also shown for reference.

Reexpressing Eqs. (4.4) and (4.5) in cylindrical coordinates gives

$$\hat{\mathbf{b}} = \frac{1}{B}(r\cos\beta - D\sin\eta, \, r\sin\beta, \, z - H + D\cos\eta)^{\mathrm{T}} \qquad (4.10)$$

$$B = \sqrt{r^2 + D(D\sin\eta - 2r\cos\beta)\sin\eta + (z - H + D\cos\eta)^2} \qquad (4.11)$$

The dot product $\hat{\mathbf{b}}\cdot\hat{\mathbf{s}}$ has the form

$$\hat{\mathbf{b}}\cdot\hat{\mathbf{s}} = \frac{D\sin\beta\sin\eta}{\sqrt{r^2 + z^2 + D^2 + D(2z\cos\eta - 2r\sin\eta\cos\beta - 2H\cos\eta) + 2H[(H/2) - z]}} \qquad (4.12)$$

This represents a function of the voxel position within the image space and therefore, in association with Eq. (4.2) and a defined criterion for the acceptable voxel elongation (V/W), permits those locations falling within the distortional dead zone region to be mapped.

Resultant Dead Zones

The distortional dead zone has a reflection symmetry through the xz plane.[*] In horizontal cross-section, the dead zone has a characteristic fan shape (Figure 4.4). The angular extent of the dead zone in a given horizontal plane increases with the distance of this plane from that containing the beam source. Thus, for an equatorial beam source ($\eta = 90°$) the dead zone is most severe at the top and bottom of the image space.

To reduce visual obstruction of the image space caused by the presence of the beam deflection apparatus, it is often desirable to position the beam sources below the equatorial plane ($\eta < 90°$) (see Plate x). As the beam source is lowered (η decreased), the extent of the dead zone at the top of the image space becomes larger, and for values of η such that $D\cos\eta > H$, the dead zone extent is increased in any horizontal plane through the image space.

The actual extent of the dead zone depends on some defined criterion relating to the acceptability of a certain degree of voxel elongation. The incidence angle ϕ approaches $90°$ as the surface approaches the xz plane, so the elongated voxel length (V) becomes infinite. Therefore, even if we tolerate a considerable amount of elongation, the distortional dead zone will continue to be present. See also Section 4.4.2 in connection with the shadowing dead zone.

One means of circumventing this dead zone[†] involves the use of several beam sources. These are positioned around the image space in such a manner that each is responsible for voxel activation when the orientation of the surface of emission is favorable.[‡] This approach is illustrated Figure 4.5. Unfortunately, this

---

[*] This assumes that the beam deflection apparatus is perfectly aligned with respect to the image space coordinate system. Misalignment may result in an asymmetric dead zone.

[†] And also the voxel placement dead zone.

[‡] When multiple beams are employed in this way, it is likely that only a single beam will be active at any one time. Consequently, the display unit does not necessarily support parallelism in the voxel activation process.

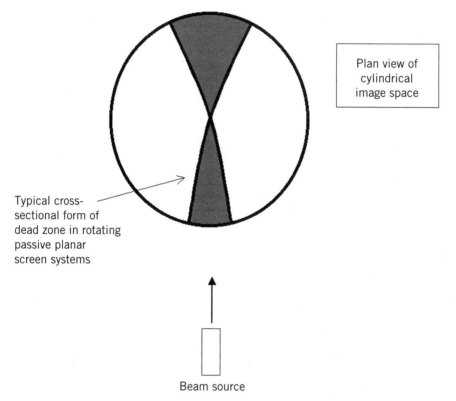

Plan view of cylindrical image space

Typical cross-sectional form of dead zone in rotating passive planar screen systems

Beam source

FIGURE 4.4 Characteristic fan-shaped cross-section of dead zone arising in passive-screen systems employing a rotating planar surface of emission. This figure depicts a horizontal cross-section through the cylindrical image space, with the beam source positioned down the page. The precise extent of the dead zone depends on the geometry and operating parameters of the display unit, and on the desired image space characteristics.

introduces other problems, including the following:

- Each set of beam deflection apparatus must be accurately aligned with respect to the image space coordinate system.
- This precise alignment must be maintained throughout the operational life of the display unit.
- The gray scale of each beam source must be calibrated. This is of particular consequence when electron beams are used for voxel activation. In this case, the beam energy and beam current associated with each electron gun must be carefully matched. Furthermore, dynamic focusing may be required.
- The visual impact associated with the presence of each voxel activation mechanism must be minimized so as to place the smallest possible restriction on the viewing freedom.

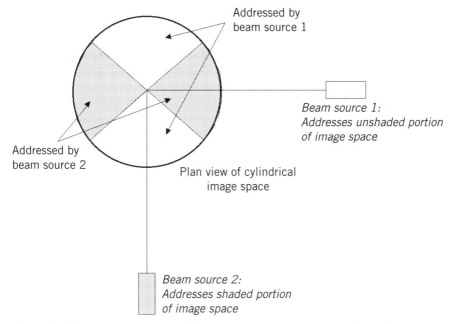

FIGURE 4.5 In passive, rotating planar-screen systems, the worst effects of dead zones can be avoided by employing several beam sources positioned around the image space. Each of these beams addresses the screen for only a portion of the image update period.

An alternative approach is outlined in Section 5.3.3. This mechanically complex display unit employs a planar phosphor-coated surface of emission addressed by two electron beams. The electron guns corotate with the screen, and therefore a constant geometry is maintained between the two. Both distortional and voxel placement dead zones are thereby avoided.

4.3.2 Helical Surface Geometry

If the surface of emission takes the form of a helix with a pitch $2H$, the set of points intersected by the surface and parameterized by the rotation angle β is

$$\left(x = r \cos \left(\beta + \frac{\pi z}{H} \right), \quad y = r \sin \left(\beta + \frac{\pi z}{H} \right), \quad z = z \right) \tag{4.13}$$

where $0 < r < R$ for a helix extending one side of the axis only, and $-R < z < R$ for a double-sided helix. The normal vector to the surface in Cartesian coordinates is given by

$$\hat{\mathbf{s}} = \frac{1}{\sqrt{H^2 + \pi^2 r^2}} (H \sin(\beta + \psi), -H \cos(\beta + \psi), \pi r)^{\mathrm{T}} \tag{4.14}$$

where $\psi = \pi z/H$ is the angle of the surface at height z relative to that at $z = 0$. From Eqs. (4.10) and (4.14), $\hat{\mathbf{b}} \cdot \hat{\mathbf{s}}$ has the form

$$\hat{\mathbf{b}} \cdot \hat{\mathbf{s}} = \frac{\pi r(z - H + D\cos\eta) - 2HD\sin\eta\sin(\beta + \pi z/H)}{B\sqrt{H^2 + \pi^2 r^2}} \tag{4.15}$$

where

$$B = \sqrt{r^2 + D[D\sin\eta - 2r\cos(\beta + \pi z/H)]\sin\eta + (z - H + D\cos\eta)^2} \tag{4.16}$$

In the special case of an axial beam source ($\eta = 0°$) these expressions simplify to

$$\hat{\mathbf{b}} \cdot \hat{\mathbf{s}}_{\text{axial}} = \frac{\pi r(z - H + D)}{B\sqrt{H^2 + \pi^2 r^2}} \tag{4.17}$$

$$B_{\text{axial}} = \sqrt{r^2 + (z - H + D)^2} \tag{4.18}$$

As with the planar screen, the dot product is a function of spatial position within the image space, and given a chosen tolerance in voxel distortion, the extent of the dead zone may be elucidated.

Resultant Dead Zones

The form and position of the distortional dead zones arising in connection with a helical screen geometry are strongly influenced by the location of the beam source with respect to the image space. If the beam source is positioned axially, the beam incidence angle increases toward the axis of the helix, and the resulting dead zone has an approximately cylindrical form (centered about the axis of rotation). Beams originating from off-axis positions ($\eta > 0°$) may be able to address portions of this cylindrical region with reduced voxel distortion. However, additional distortional dead-zone regions will be created, and furthermore, shadowing dead zones may be introduced. These arise due to part of the screen presenting a physical obstruction to the beam such that it cannot propagate into certain portions of an image space (see Figure 4.6 and also Section 4.4.2). As with the planar screen geometry discussed previously, the impact of both the distortional and voxel placement dead zones may be ameliorated through the use of multiple beam sources placed at the appropriate locations around the image space. Unfortunately, this then introduces the various problems that have been summarized in Section 4.3.1.

4.4   DEAD ZONES: FURTHER CONSIDERATIONS

As we have seen, in the case of systems employing a passive surface of emission, voxel activation is generally achieved through the use of one or more beams. For example, laser beams (operating in the visible region of the spectrum) may be

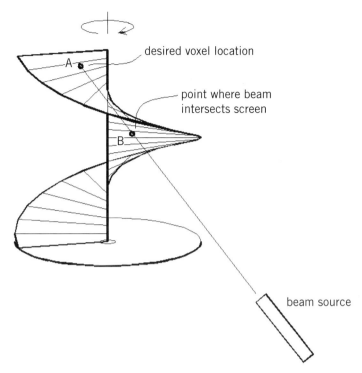

FIGURE 4.6   Shadowing dead zone effect for helical screen systems. A shadowing dead zone arises as a portion of the screen prevents the passage of the beam to a particular location on the screen's surface. For example, as depicted here, the beam is aimed toward the desired voxel location at position A, but intersects the surface instead at position B.

used in conjunction with a surface of emission that is able to scatter the incident light. Provided that both the beam deflection and its modulation are suitably controlled and synchronized to the position of the surface of emission, visible voxels may be produced. Images comprising more than one color may be created by using several laser beams of different wavelengths.* Should electron beams be used, the surface of emission would take the form of a thin layer of phosphor, enabling conversion of the kinetic energy of the electrons into visible light (for examples, see Sections 5.4.1 and 5.6.3).

Unlike an active surface containing an array of individual voxel generation centers, a passive surface will, in principle, not impose a constraint on the placement of voxels thereon. In practice, however, the distribution of voxels on the surface is generally determined by the accuracy of the beam deflection apparatus and associated controlling hardware. Within the following subsections we review the various dead zones that may be associated with swept-volume systems

---

* This introduces stringent requirements for the alignment of the sets of beam deflection apparatus.

employing a passive surface of emission moving in either a translational or rotational manner.

4.4.1 Translational Motion

By ensuring that the beam deflection apparatus is maintained at a sufficient distance from the surface of emission, it is possible to avoid — without recourse to multiple beams* — both distortional and voxel placement dead zones. The complexities associated with the use of multiple beams may therefore be avoided and the construction of the display unit is simplified. However, as discussed in Section 2.6 and elsewhere in this book, a sequential voxel activation process will ultimately limit the voxel capacity of this, and any other, display unit. Several examples of systems of this type are provided in Chapter 5.

4.4.2 Rotational Motion

As we have seen, should a single stationary beam source be used to address either a planar or helical surface of emission, regions within an image space will exist in which voxels cannot be placed to a sufficient degree of accuracy. Furthermore, there will also exist regions in which the incidence angle between the beam and target surface is such that voxels become unacceptably elongated. If the beam source is stationary, dead zones arise primarily in those parts of the image space corresponding to large beam incidence angles (see Figure 4.1). For example, in the case of a planar rotating surface, the dead zones occur when the surface of emission approaches a position end-on to the beam source. Within this region the image quality is affected by distortions in voxel shape and size (leading to distortional dead zones), by a reduction in the spatial accuracy of voxel activation, and by an increase in the separation of nearest-neighbor voxels (corresponding to voxel placement dead zones).

In Section 4.3, the form and position of distortional dead zones were analyzed for both planar and helical screen geometries. In the remainder of this section we discuss the manner in which other types of dead zone arise, and their consequences for system design.

Reduction in Beam Positioning Accuracy (Voxel Placement Dead Zone)
The voxel activation beam(s) may, in general, be deflected to a limited number of discrete angular positions. The minimum increment in the deflection angle is likely to be determined by the characteristics of the deflection apparatus in combination with the precision of the information held in the relevant portion of the voxel descriptor (see Section 8.4). In principle, a passive surface of emission differs from its active equivalent in that it does not restrict voxel activation to a set of discrete locations. However, the rotational motion in combination with the quantized range of deflection angles leads to variations in the nearest-neighbor intervoxel separation throughout the image space. Furthermore, as the

---

* Multiple beams would be required for the generation of multicolor images.

angle between the surface of emission and the incident beam becomes more acute, uncertainties in beam deflection will be magnified.

The uncertainty in the voxel position depends on any intrinsic uncertainty in the beam deflection angle, the distance of the voxel from the beam deflection apparatus, and the incidence angle of the beam onto the surface. As the beam incidence angle increases, small uncertainties in the direction of the beam become magnified into increasingly large voxel positioning inaccuracies. This gives rise to a voxel placement dead zone in which uncertainties in voxel positioning and the separation of nearest-neighbor voxels exceed some defined criterion.

So as to minimize the extent of the distortional and voxel placement dead zones, the distance between the beam deflection apparatus and the screen must be considered with great care. Unfortunately, any desire to increase this distance so as to reduce the extent of the dead zones is likely to conflict with a wish to produce a display unit that is compact and readily portable.

Anisotropic Voxel Intensity (Visual Dead Zone)

In the case of a swept-volume display unit employing a passive surface of emission, the radiation intensity from each voxel is often anisotropic. This may be caused by a number of factors and can be illustrated by considering Lambertian diffuse reflection (Figure 4.7). In this case when a given voxel is viewed from a perspective normal to the surface, it is seen at a maximum intensity. However, when the viewpoint lies closer to the plane of the surface, the perceived intensity of the voxel decreases.

The supporting structure also has a major impact on the propagation of light from each voxel activated. The visibility of voxels viewed through this structure may be reduced by partial absorption or scattering, and at certain angles by total internal reflection. In the case of display units employing an active surface of emission, the voxel generation centers may also hamper the propagation of light through a small range of angles about the plane of the surface.

Individually, and in combination, these effects give rise to a visual dead zone (see Plate viii). Each must be considered with care during the design of any display unit employing the rotational motion of a surface of emission, irrespective of the passive or active nature of the voxel generation subsystem. Should a screen moving with rotational motion lack rigidity and therefore flex, the severity of the visual dead zone will be increased.* Any attempt to increase the rigidity of the screen by the inclusion of a center shaft (passing through the image space) should be avoided. Not only will a center shaft give rise to a visual dead zone (by obstructing the passage of light emitted in certain directions), but it will also result in dead image space (see below).

Absence of Voxels (Shadowing Dead Zones and Dead Image Space)

The surface of emission is supported by a rigid substrate (supporting structure). As discussed in Section 3.4.3, in the case of a planar target surface, the finite

---

* Flexing will also affect the voxel placement dead zone.

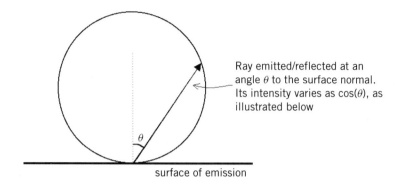

Ray emitted/reflected at an
angle θ to the surface normal.
Its intensity varies as $\cos(\theta)$, as
illustrated below

surface of emission

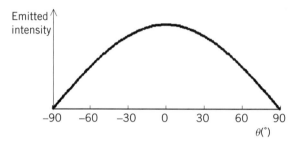

FIGURE 4.7 Lambertian (diffuse) angular dependence of emitted intensity, $I = I_0 \cos \theta$. The intensity of light emitted decreases as the direction of view moves away from the normal.

thickness of the screen gives rise to a cylindrical region (centered about the axis of rotation) in which voxels cannot be placed. The extent of this region of dead image space will be increased by the presence of a center shaft. If the surface of emission has a nonplanar geometry, shadowing dead zones may arise due to one part of the screen physically obstructing the passage of the beam to a location at which a voxel is to be activated (see Figure 4.6). In general, such regions can be avoided by using a number of appropriately positioned beam sources. However, this will result in a number of additional problems (see Section 4.3.1).

When the edge of the supporting structure passes across the beam axis, it is not possible for the beam to enter the image space. However, a fraction of the volume that lies in the shadow of the screen edge when it lies end-on to the undeflected beam may be addressed when the screen occupies other angular positions. The remaining portion of this volume cannot be addressed by this beam, irrespective of the screen position, as illustrated in Figure 4.8. The dashed lines in this diagram indicate the beam paths that mark the left and right-hand dead image space boundaries. (The portion of the boundary closest to the beam source is due to the central cylindrical region of dead image space.) These beam paths are those for which the beam, when deflected so as to lie parallel to the face of the screen, becomes coincident with it. Thus, the finite thickness of the

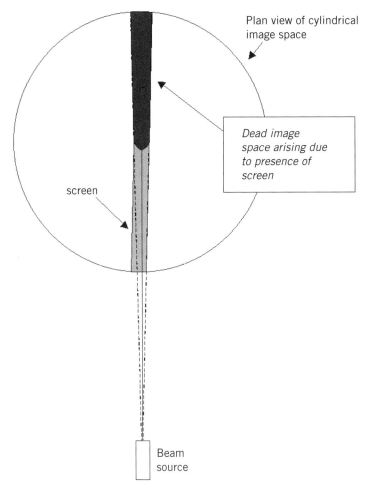

FIGURE 4.8 A region of dead image space arises in a rotating planar screen system as a consequence of the physical presence of the supporting structure. Voxels cannot be activated by this beam source within the dark region indicated toward the back of the image space, irrespective of the angular position occupied by the screen.

screen results in a narrow region of dead image space. This region is generally far smaller in extent than the distortional dead zone region within which it is located. If, however, a frame is employed around the periphery of the screen and this is wider than the thickness of the supporting structure (see Plate xii), the shadowing effect may be more severe.

4.5 BEAM REGISTRATION CONSIDERATIONS

Swept-volume display units employing the rotational motion of a passive surface of emission and in which voxels are activated by means of a radiation beam directed into the image space are prey to distortional and voxel placement dead zones (also possibly shadowing dead zones). As discussed previously, one approach to this problem involves the use of a number of spatially separate beam sources.* Each of these beams is made responsible for voxel activation during only a fraction of each image update period (corresponding to the surface of emission being at a favorable orientation with respect to the beam source). The beams therefore operate sequentially and a plurality of beam sources may be responsible for the creation of a single image component. In this situation, misalignment between the beam sources is likely to be readily apparent and might perhaps take the form of discontinuities or distortions of the image. It is therefore necessary to ensure that proper registration is maintained between each beam source coordinate system and the coordinate system assigned to the image space.

Consider a display unit employing a planar surface of emission and having two beams positioned at an angle of 90° with respect to each other, these being located in the equatorial configuration ($\eta = 90°$). So as to overcome the dead-zone problem, each beam is responsible for voxel activation within a certain region (probably corresponding to one half) of the image space. At the boundaries of these regions a smooth transition in voxel placement must occur to avoid the introduction of visible image discontinuities. This necessitates a high degree of registration (alignment) between the beams.

There are two possible approaches to the implementation of these boundary regions:

- *Sharp boundary.* In this case, as the surface of emission passes through the defined boundary location angle, the beam source that was active is made inactive, and that which was inactive becomes responsible for the activation of subsequent voxels. The boundary region is therefore sharply defined and occurs at a precise location within the image space.
- *Gradual boundary.* In this case, as the surface of emission approaches the defined boundary location angle, the inactive beam source becomes increasingly responsible for voxel activation, and the beam source that had been active gradually less responsible. This process continues across the boundary until the changeover has been effected completely. Thus at angles close to the boundary location, both beam sources are responsible for voxel activation. In this manner it is possible to avoid a sharp transition between beam sources and so make the boundary region less visible.

The use of multiple beam sources introduces an additional problem in ensuring that the gray scale associated with each source is properly calibrated with respect to each of the others. Failure to effect a suitable calibration procedure would result in inconsistent voxel intensities. In the case of multiple beam sources

---

* Multiple beam sources may be required for three distinctly different purposes: (1) to ameliorate the distortional, voxel placement, and shadowing dead zones; (2) to permit the production of multicolor images; and (3) to permit parallelism in the voxel activation process.

being employed for the creation of full color images or parallel voxel activation, a very high degree of registration between the beams is required *throughout* the image space. This is discussed further in Section 9.8.

4.6 VOXEL ORDERING CONSIDERATIONS

The voxels comprising the image are ideally activated as the surface of emission passes through their exact location within the image space. For display units employing a passive surface of emission and exhibiting a low degree of parallelism, the voxel activation capacity N_a is typically lower than the number of available voxel locations N_l (see Section 2.9), and only the visible voxels comprising the image are activated, in a sequential fashion for each beam source. Such a random-scan technique is referred to here as a dot-graphics technique. A simple and approximate means of ordering the voxel data for activation is to divide the image space into a sequence of thin subvolumes, each corresponding to a small range of screen positions. In the case of a rotating screen, this corresponds to dividing a cylindrical image space into S thin sectors (Figure 4.9). The image data are then divided into slices containing the voxels in each subvolume, and each slice is output as the surface of emission passes through the corresponding subvolume. The positions of the voxels in the direction of screen motion (e.g., in the case of a rotating screen, the angular coordinates) are then discarded. Each

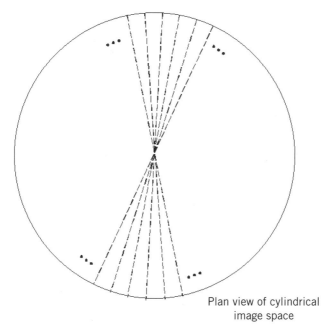

Plan view of cylindrical
image space

FIGURE 4.9 Plan view of an image space generated by the rotational motion of a surface of emission. An image slice is depicted on the surface of emission as it moves through each of S sector positions.

voxel therefore has a positional uncertainty and a worst-case positioning error in the direction of motion equal to the extent of the sector or slice. Clearly, in the case of rotational motion, the tangential positioning error has a maximum value at the periphery of the image space.

This ordering scheme provides a simple solution to what is in fact a difficult 3-D optimization problem, which in all but very sparse data sets involves a trade-off between the voxel positioning accuracy in the direction of screen motion and the efficiency of the beam deflection path. Once the voxels have been redistributed into slices, the total path traveled by the beam in activating the voxels within each slice may be shortened by treating each slice as a 2-D ordering problem. This matter is discussed further in Section 8.5.4.

In general, the need for voxel ordering arises in swept-volume systems that exhibit a low degree of parallelism in the voxel activation process. Such systems are typically passive-screen display units, where one or very few beams address the surface of emission concurrently. Display units employing an active surface of emission generally demonstrate a higher parallelism (in principle, any number of the voxel generation centers may be activated simultaneously). In this case, any bottleneck in the voxel throughput is likely to arise elsewhere in the graphics engine (see Chapter 9). Reflective surfaces may (Harris et al. 1986, Yamanaka et al. 1988) or may not (Fuchs et al. 1982) have a high degree of parallelism — in the latter case, voxel ordering procedures may be applied.

4.7   USE OF A REFLECTIVE SCREEN

One approach to the implementation of a swept-volume display unit involves the use of a reflective surface of emission on which images are cast (e.g., see Section 5.6.4). A conventional form of 2-D display device such as a high-performance cathode ray tube (CRT)* may provide the image source. In this case, the output of frames to the CRT is synchronized with the position of the surface of emission, so the volumetric image effectively comprises a sequence of slices (Harris et al. 1986, Kennedy and Nelson 1987, Yamanaka et al. 1988). Naturally, it is necessary during the computation of each frame to take into account any variations in the geometry between the CRT screen and the surface of emission. Should this type of projection technique be used in combination with a rotating planar screen distortional and voxel placement dead zones will occur. As with the beam addressed display units discussed previously, this problem may be resolved by employing two projection systems. An alternative approach described in Section 5.3.2 involves the rotation of a CRT about its axis.

The generation of image data on the CRT screen is a physically sequential process and the system bandwidth (and hence the voxel activation capacity) is ultimately determined by the performance of the electron gun (maximum rate of beam modulation, maximum rate of beam deflection, and image intensity). Volumetric display units of this type take advantage of a mature (and therefore

---

* Equipped with a short-persistence phosphor.

(a)   (b)

Plate i Two views of an animated stick figure depicted within the Cathode Ray Sphere. The green figure runs on a red track in a natural manner. The limb positions in each frame of the animation sequence were obtained from stroboscopic photographs of a person running.

Plate ii Pair of green wireframe spheres each containing a smaller red sphere and depicted within the Cathode Ray Sphere. The screen drive shaft may be seen at the top of the photograph.

Plate iii Wireframe AutoCAD drawing of the *Starship Enterprise* depicted within the Cathode Ray Sphere. The two electron guns responsible for voxel activation may be seen together with the screen drive shaft (above) and the inlet to the vacuum pumping system (below).

Plate iv Molecular structure depicted within the Cathode Ray Sphere. The atoms are depicted in red and the bonds in green. These data represent the unit cell of the compound calcium copper tetra (trichloroacetate) tetra aqua, obtained directly from x-ray crystallography analysis. The two electron guns responsible for voxel activation may be seen. Below the left-hand gun the amplifier module (responsible for the amplification of the deflection and blanking signals) is visible. Voxel dropout of the red (toward the top) and green (toward the bottom) components of this image may be seen. The early algorithms employed to depict this image did not optimally distribute the available voxels.

Plate v   Prototype HL3D display unit employing a helical screen and developed by Rüdiger Hartwig (see Section 5.5.2). The solid rotational axis may be seen clearly. (Courtesy of R. Hartwig).

Plate vi   Stack of squares depicted within the Cathode Ray Sphere. This image is representative of a number of test patterns used to align the coordinate systems of the two electron guns with the image space reference frame. The squares are depicted in yellow, a product of color mixing (red and green).

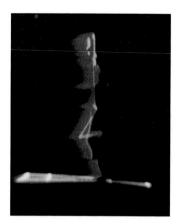

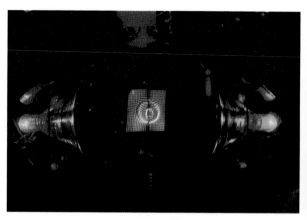

Plate vii Animated stick figure (see also Plate i) depicted within the Cathode Ray Sphere. On this occasion a long-persistence red phosphor was used. The motion of the surface of emission compounded with the excessive phosphor persistence results in extensive (and highly undesirable) voxel elongation. Within the region of image space in which the figure is depicted, the screen's motion is from right to left.

Plate viii Wireframe model of a math function depicted within the Cathode Ray Sphere. Clearly visible is the visual dead zone, a narrow region of diminished intensity passing vertically through the image. Due to binocular vision, the visual dead zone is less apparent when images are viewed directly.

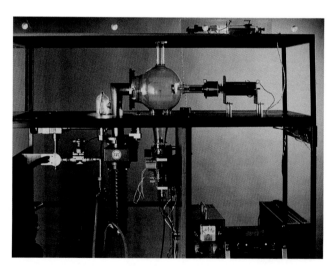

Plate ix MkIV Cathode Ray Sphere prototype. The display unit was developed to evaluate the characteristics of various phosphor materials employed in the voxel generation subsystem. The electron gun may be seen below the sphere (the electron beam propagating upward). The screen rotated about a horizontal axis (the motor may be seen on the right-hand side of the sphere). As a consequence of the horizontal axis of rotation, the visual dead zone was not ameliorated by binocular vision, and was particularly apparent.

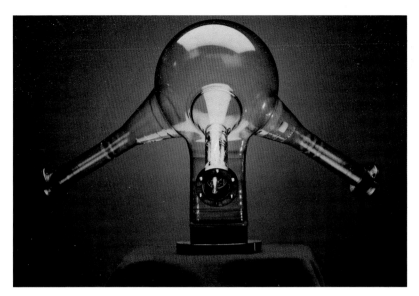

Plate x   Glass vacuum vessel constructed for the MkIII Cathode Ray Sphere. The three ports are oriented to enable the electron guns to be positioned below the equator of the sphere. The display provided excellent image space visibility. The large opening below the sphere permitted the insertion (and assembly) of the screen, contained the motor drive, and connected the sphere to the vacuum pumping system.

Plate xi   First Cathode Ray Sphere prototype (MkI). This display employed a custom-built stainless steel vacuum vessel. The screen drive may be seen at the top of a long drive shaft tube. The bulbous chamber was fitted with three viewing ports and three flanges to which electron guns were connected (one gun may be seen on the right-hand side). The screen may be seen within the chamber. This prototype operated in New Zealand in late 1989.

Plate xii A 3-D Lissajous figure depicted within an experimental prototype (ca. 1960) by R. Ketchpel. The supporting structure located around the screen may be seen. (Courtesy of R. Ketchpel).

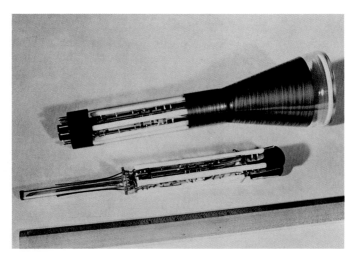

Plate xiii An electron gun employing electrostatic deflection. The oscilloscope tube (type 3WP1M) is similar to (although smaller than) the 5BP1 tube referred to in the text. The neck refers to the cylindrical portion of the glass envelope, which houses the electron gun. (Courtesy of Thomas Electronics, Wayne, New Jersey).

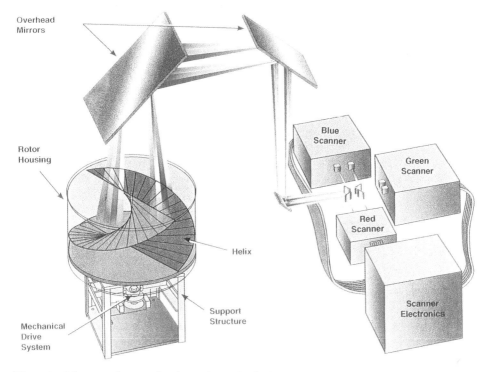

Plate xiv The second-generation 3-D volumetric display system (36-inch laboratory model) developed by the U.S. Navy. (Courtesy of P. Soltan).

Plate xv A two-color Lissajous figure depicted within the MkIII Cathode Ray Sphere prototype display system.

Plate xvi   The 3-D structure of glucose depicted within the Cathode Ray Sphere. The carbon, hydrogen, and oxygen atoms are represented as spheres of different radii.

Plate xvii   The Lorenz attractor depicted within the Cathode Ray Sphere. (Blundell and Schwarz 1995b).

highly refined) CRT display technology. Furthermore, they are able to make use of techniques that may otherwise be problematic. For example, the postdeflection acceleration (PDA) of the electron beams has been refined for use within various types of CRT (Sherr 1992). The implementation of this technique in a display unit employing the direct interaction of electron beams with a rotating phosphor-coated screen (e.g., see Section 5.6.3) is likely to require considerable effort.

4.8 DISCUSSION

The design of any volumetric display unit is a complex process in which many interrelated issues must be considered. The performance of each subsystem must be evaluated in isolation and the problems that may arise as a consequence of their interaction must also be anticipated. Oversights may have serious repercussions and may restrict image depiction to a limited portion of an image space. The majority of display unit architectures exhibit at least one of the various types of dead zone that have been discussed in this chapter. In the case of swept-volume display units employing one or more stationary beam sources, the continually varying geometry between the image space creation and voxel activation subsystems is particularly problematic and gives rise to distortional voxel placement and possibly shadowing dead zones. Attempts to eradicate dead zones (or limit their impact) are likely to increase the overall complexity of the display unit and possibly restrict aspects of its performance.

In the next chapter we review some of the swept-volume display unit technologies that have been proposed over the course of the last 50 years. A number of these systems employ the rotational motion of a passive surface of emission, and it is instructive to examine the techniques that have been used to produce an image space free of undesirable dead zone regions. Despite their mechanical complexity, the display units described in Sections 5.3.2 and 5.3.3 are considered to be of particular interest.

CHAPTER 5

SWEPT-VOLUME DISPLAY UNIT DEVELOPMENT

Ah, but a man's reach should exceed his grasp,
Or what's a heaven for?*

5.1 INTRODUCTION

Swept-volume volumetric display system techniques have been the subject of research for at least 50 years. During this period, numerous systems have been discussed in scientific literature and documented in patent applications. From this published material it is apparent that many working in this area have demonstrated considerable ingenuity in both the electronic and mechanical domains.

In preparing this chapter, it has been necessary to review many publications relating to swept-volume systems, and unfortunately, it has sometimes been difficult to differentiate between systems that were simply conceptualized and those that were actually constructed. The impetus for the work that has been carried out together with levels of success achieved must therefore on occasion remain the subject of conjecture. Furthermore, it is unfortunate that the reasons for the discontinuation of research programs in this area remain undocumented.

During the perusal of both this chapter and Chapter 7, it is interesting to reflect on the limitations and deficiencies of the various volumetric system architectures that have been developed over the years and so judge their potential for 3-D visualization. It will become evident that although the majority (if not all) of the systems discussed in this chapter are, in principle, able to display volumetric images, their image space characteristics are often far from ideal. Furthermore,

---

* Robert Browning, *Andrea del Sarto*, in *The Complete Works of Robert Browning: With Variant Readings and Annotations*, Vol. 16, Ohio University Press, Athens, OH, 1998.

a number of the display units employ complicated mechanical devices intended to overcome deficiencies in the voxel activation subsystems. As a consequence, it is natural to question (in terms of long-term stability and general reliability) their suitability for production.

Most image space generation techniques and voxel activation mechanisms employed within swept-volume display units create anisotropic and nonhomogeneous image spaces. Performing suitable manipulations on an image data set before its passage to the display unit may provide a method for artificially improving these attributes (see Section 10.5). This does, however, significantly increase the burden placed on the graphics engine. Ongoing improvements in computer performance have made it possible to upgrade the quality of volumetric images by compensating for some display unit deficiencies in this way. However, although image data manipulations within the graphics engine can improve the homogeneity and isotropy of an image space, such techniques can have no effect on any perceived variation of image quality with viewing direction. This is caused by both the image space creation and voxel generation mechanisms and the effect these have on the propagation of light from each activated voxel through the image space, and ultimately to the observer.

The systems described in this chapter are presented in approximately chronological order. We have taken the dates of various developments from the available literature, so the actual work is likely to have been carried out some time beforehand. This chapter is not intended to provide a definitive account of the history of swept-volume volumetric display unit development. It does, however, represent a first attempt to record briefly some past and current research activity within this discipline.

5.2 PIONEERING WORK: 1940–1950

In 1948 one of the most notable scientific papers concerning volumetric display systems was published (Parker and Wallis 1948). The authors of this extensive work demonstrated considerable insight into the potential benefits that could be derived from volumetric systems and attempted to place these systems in context with various other 3-D visualization techniques available at that time. Their work related to the application of swept-volume volumetric systems (referred to within their article as "truly three-dimensional displays") for the visualization of radar information,* so a portion of the article naturally concerns the acquisition of such information. The remainder relates to visualization techniques, and despite the passage of 50 years since its publication, this article continues to be of considerable relevance.

The paper, when presented to a meeting of the Radio Section of the IEE on March 24, 1948, gave rise to a most interesting discussion (Parker and Wallis

---

* This work was connected with the development of the *Sea Slug* missile.

1949). Some of the comments made at this meeting remain highly relevant. This may be seen from the extracts that follow:

R. A. Smith:

The paper gave me the impression that we are suffering from an *embarras de richesse*; anyone coming into this field finds so many possible displays that he may be in doubt about which line to investigate. ...

C. W. Earp:

In view of the number of ways in which a position in space is appreciated by the human brain, it seems that, despite certain difficulties which are particularly associated with the truly three-dimensional display, only displays of this general class could ever be completely satisfactory. However, being electronically rather than mechanically minded, I am disappointed to find that all of the "solid" displays involve the use of mechanical scanner.[*] Can the authors give any indication as to whether we are likely to be able to avoid the use of moving parts?

C. W. Thomas:

I think that it is fair to say of all these pictorial methods, which include the truly three-dimensional and the perspective displays, that their volume of coverage is inherently small and that it is not convenient to obtain quantitative information from them, or, if this were really to be done, the operational significance of the display would disappear because it would be no longer quicker to use.[†] This probably means that such a display would be used not quantitatively but qualitatively, accompanied by some system for extracting the value of the coordinates of the signals. ...

W. Ross:

It seems to me that perhaps one useful application of these displays not mentioned by the authors is as an instrument for teaching what is essentially a difficult subject, namely solid geometry.

The following extracts are taken from the response made by E. Parker and P.R. Wallis to the matters raised by the audience:

We are in complete agreement with Mr. Earp concerning the desirability of electronic rather than mechanical methods. ... It is conceivable, however, that some form of gas-filled tube throughout which a small discharge glow could be scanned may be developed. ... A number of speakers have commented that

---

[*] Referring to the image space creation technique.

[†] The speaker was considering the application of volumetric display systems to the visualization of radar data. Within this context volumetric systems offer to increase the rate at which information may be interpreted.

these displays tend to favor qualitative rather than quantitative use. Most three-dimensional observation in actual life tends to be qualitative. When it becomes quantitative, it usually reduces to a coincidence of two points, as in catching a ball or touching an object, which is analogous to our use of electronic markers in the displays. It is difficult to treat the problem theoretically, and detailed comparisons between different displays for different applications must await actual experience. We have no doubts about the fundamental possibility of usefully presenting three dimensions, but we are still sounding the problem before setting the course.

It is interesting to reflect on the concluding sentence and consider its continued applicability after the passage of 50 years.

During the course of 1948 and early 1949, E. Parker and P. R. Wallis presented their work (and demonstrated the experimental prototype displays) at a number of venues in the United Kingdom. This gave rise to further well-informed discussion (Parker and Wallis, 1949). Several extracts are presented below.

W. Macrae*:

I feel that three-dimensional displays must be regarded as giving a general picture only, and cannot be used to give accurate quantitative data. However, there are many applications where a general picture would be extremely useful along with normal measuring systems, for example in ground control of interception fighter aircraft and in airport control systems. . . .

I. N. Vaughan-Jones[†]:

I should like to ask whether there are not severe mechanical difficulties attached to the reciprocating-mirror system, since a mirror of at least 6 inches square is required to move at speeds of the order of 20 c/s. It would be expected, at least, to give rise to a great deal of noise. Is this system therefore a practical one or is it described to indicate a possible approach to the problem?

The response of E. Parker and P. R. Wallis to the latter question:

Mr. Vaughan-Jones's doubts about the moving-mirror display are unduly pessimistic, incidentally, as our demonstration equipment includes a 14 inch square mirror driven at 20 c/s with little noise or difficulty. . . .

Let us now turn our attention to the publication that gave rise to the discussion above. As mentioned previously, this paper is extensive and relates particularly to the visualization of radar data. Since electronic digital computer systems were in their infancy when this work was undertaken, information processing was carried out by analog computation. The display systems described by the authors were

---

* North-Eastern Radio and Measurements Group of the IEE in Newcastle-on-Tyne, December 1949.
[†] North-Western Radio Group of the IEE in Manchester, January 1949.

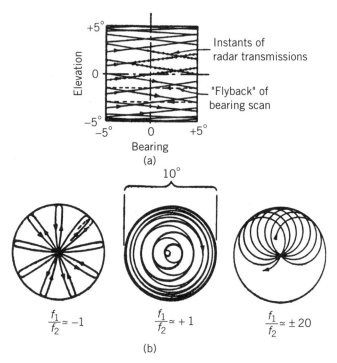

FIGURE 5.1   The rectangular (*a*) and petal (*b*) scanning patterns illustrated in the article by Parker and Wallis (1948). (Reproduced by permission; © 1948 IEE.)

intended to operate in real time, depicting the incoming data stream without intermediate storage facility. The radar installation acted as the source of volumetric information providing the range, elevation, and bearing of airborne objects.

The authors outlined several scanning patterns that may be traversed by a radar beam to provide coverage of portions of the surrounding sky. Two such patterns are illustrated in Figure 5.1. In designing the architecture of their volumetric display unit, the researchers sought to achieve a *direct* correspondence between the aerial radar scan pattern and the image space scan. By arranging the appropriate synchronization between the two, the mapping of radar echoes into the image space was greatly facilitated and the need for computation minimized.

The authors described two types of volumetric display units, each being suited to the depiction of information derived by means of a particular radar scanning pattern. These two display unit architectures and associated scanning patterns are outlined below.

5.2.1 Rectangular Scan

Figure 5.1*a* illustrates the form of a rectangular aerial radar scan. Within their work the authors consider the scan to operate with a pulse repetition frequency of 8 kHz, a linear bearing scan of 400 Hz, and a sinusoidal elevation scan having

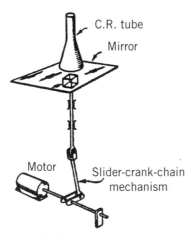

FIGURE 5.2 Moving mirror display described by Parker and Wallis (1948). In this embodiment data depicted on a CRT is cast on a mirrored screen moving with a translational motion. (Reproduced by permission; © 1948 IEE.)

a repetition frequency of 20 Hz. The volumetric display unit constructed for the visualization of this radar information employed the translational motion of a mirror upon which cathode ray tube (CRT) images were projected as illustrated in Figure 5.2 (see also Section 4.7). The motion of the mirror represents the elevation scan, and it therefore moved with a sinusoidal velocity profile having a frequency of oscillation of 20 Hz. The amplitude of motion is reported as being 3 in.

The bearing scan was depicted on one axis of the CRT screen and the range scan on the other. Target echoes were placed within the image space by the modulation of the CRT electron beam. Since the radar scanner and image space scan were synchronized, any target echo signal could be used directly to turn on the CRT's electron beam and a "blip" would automatically be placed at the correct location within the image space. The exhaustive scanning of the image space was made possible by the low scan resolution and small image space dimensions (a cube with sides of length 3 in.). This would have been sufficient for a prototype display given the nature of the information to be depicted therein.

The authors comment on an alternative configuration in which the CRT itself moved along its axis with a translational motion. In this embodiment, the CRT screen would sweep out the image space, avoiding the need for the projection of the CRT image onto a mirror. The authors considered this to be undesirable, for obvious mechanical reasons.

5.2.2 Petal or Spiral Scan

Three-dimensional Lissajous patterns may be displayed with great clarity on volumetric display systems. In the case of swept-volume beam-addressed displays, they may be generated with ease and without the need for any form of graphics engine. Their depiction may be achieved by the direct application of time-varying waveforms produced by, for example, function generators to the beam deflection apparatus.

In Figure 5.1*b*, a Lissajous pattern representing the petal (or spiral) scan discussed by Parker and Wallis is illustrated. As with the rectangular scan illustrated in Figure 5.1*a*, the diagram shows the pattern followed (in terms of elevation and bearing) by the radar scanner — the third dimension (into the page) corresponding to the range scan. In an appendix to their paper, the authors discussed these scanning techniques. It is sufficient for our purposes to consider that they may be attained by the vector addition of two sinusoidal frequencies.

As with the rectangular scanning pattern, the authors sought to reproduce the track followed by the radar scanner within the image space. This they achieved by the projection of a CRT image onto a screen moving with rotational motion as illustrated in Figure 5.3. The operation of this system may best be understood by considering the application of a vertical sinusoidal sweep to CRT 1, and assuming that the screen is stationary, a vertical line normal to the axis of rotation will be projected onto it. The amplitude of this line will be affected by the position of the screen, being at a maximum when the surface of the screen is parallel to the CRT screen and falling to zero when the screen's surface is normal to that of the CRT. If the surface is now rotated, a two-dimensional Lissajous figure will be generated, its form being determined by the relative motions of the sinusoidal sweep mapped onto the CRT and the speed of rotation of the screen. By making suitable adjustments and effecting the appropriate synchronization, the petal and spiral patterns may be generated within the image space. The third dimension (along the axis of rotation) was used by the authors for depiction of the range scan.

It is interesting to note the proposed incorporation of a second CRT (Figure 5.3), whose function is described by the authors as follows:

> A second projection system at right angles, with a bearing/range raster, can be added to avoid any loss of resolution when the screen is very oblique to the direction of projection.

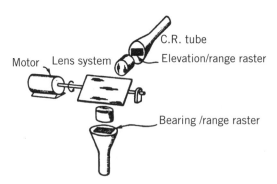

FIGURE 5.3 Moving screen display, described by Parker and Wallis (1948). In this embodiment data depicted on two CRTs are cast on a screen moving with rotational motion. (Reproduced by permission; © 1948 IEE.)

Although the reference to the second CRT projecting a bearing/range raster rather than an elevation/range raster may seem confusing, it is clear that the authors had found an appropriate solution to the voxel placement dead zone discussed in Chapter 4. Unfortunately, the authors do not discuss the way in which registration between the images projected by the CRTs was to be achieved.

In their paper the authors provide an excellent review of the various types of 3-D display techniques available at that time. They also discuss metrics that may be used to describe 3-D display characteristics. Finally, in a section entitled "The General Theory of Three Dimensional Displays," they discuss aspects of 3-D display systems.

It is appropriate to conclude this section with a further brief extract from the paper:

> The displays have a limitation, in that they appear always as a "transparency." Light is radiated from each point in the display without regard to the radiation from other points "between" it and the observer. This would make its direct application to a television system awkward, even supposing "range" information could be made available from the television cameras. ... The application of three-dimensional displays to X-ray work for medical and other purposes would be assisted, however, by this "transparency."*

This is an issue to which we return in the context of image space characteristics in Chapter 10.

5.3 PIONEERING WORK: 1950–1960

During this period, interest in vacuum-tube technologies was at its strongest and a considerable infrastructure existed for the design and mass production of these components. As a consequence, it is not surprising that during the 1950s and early 1960s we encounter articles relating to the implementation of volumetric display units based on novel types of CRT. Clearly, the large number of manufacturers able to mass-produce such devices would have facilitated the low-cost production of relatively high-performance displays. Had our present computational systems (and associated visualization requirements) been available during this period, it is likely that volumetric display systems would have gained widespread acceptance.

5.3.1 Peritron

The *Peritron* (Withey 1958) constituted a novel type of CRT able to display volumetric images. From Figure 5.4 it may be seen that the peritron differs from a conventional CRT by the inclusion of a mechanical drive system responsible for the motion of the phosphor-coated (P11 phosphor) CRT screen. The creation

---

* Although this may be intuitively correct, the validity of this statement can be ascertained only when volumetric systems are properly used within these contexts (e.g. medical applications).

NEW MODEL

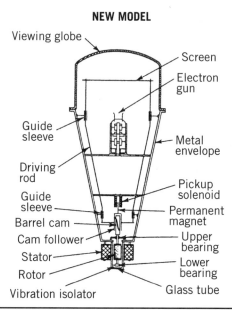

Improvements include: induction motor with stator outside vacuum system; barrel-cam screen-drive for true harmonic screen motion; larger screen moving through a greater amplitude; glass dome; and sound isolators

FIGURE 5.4   The Peritron, a device that employs the translational motion of a passive surface of emission. Voxel activation may be achieved by means of a single electron beam. (From Withey, 1958.)

of the image space was thus achieved by the translational motion of a passive phosphor-coated surface of emission, and voxel activation by means of a single electron beam.

The author describes the display unit as generating a cylindrical image space measuring 18 cm in diameter and 3 cm in depth. The frequency of motion and therefore the maximum image update frequency was 30 Hz. All the mechanical components were contained within the vacuum vessel. The consequent reduction in air resistance would considerably reduce the power required to affect the motion of the surface of emission and its supporting structure (see Section 3.5.3). The Peritron appears to have been intended for use in air traffic control applications.

5.3.2   Generescope

Early in 1958, Max Hirsch filed a detailed patent (U.S. patent 2,967,905) that describes two volumetric display system architectures. The first of these

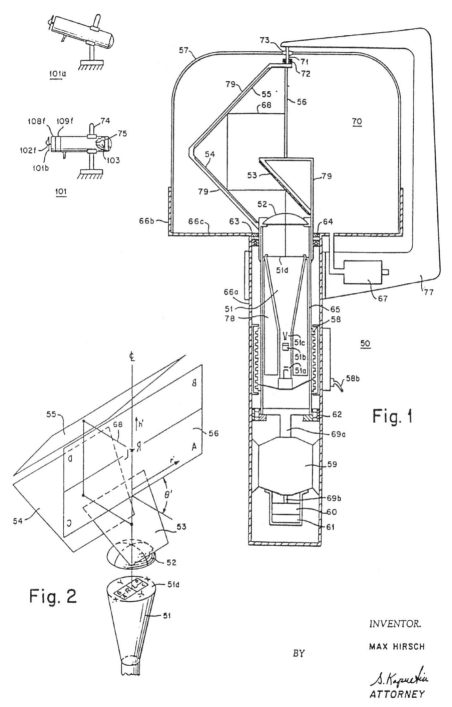

Fig. 1

Fig. 2

INVENTOR.

MAX HIRSCH

BY

S. Kapueketic

ATTORNEY

FIGURE 5.5   The Generescope. The screen, mirror assembly, and CRT rotate so as to avoid various dead zones. (From U.S. patent 2,967,905.)

was referred to as the *Generescope*. Immediately apparent from the cross-sectional diagram reproduced in Figure 5.5 is the level of mechanical complexity involved in the implementation of the device. The system employed a standard CRT — images generated therein being reflected by a series of mirrors (labeled 53, 54, 55 in the diagram) onto a rotating projection surface (56). The image space was created by the rotation of this surface about its vertical axis. The corotation of the CRT, mirror assembly, and surface ensured that a constant angle was maintained between them and so distortional and voxel placement dead zones were avoided.

The rotation of the CRT clearly introduces mechanical complexity and necessitates the use of a commutator assembly able to pass the required static voltages (together with the beam gating and beam deflection information) to the CRT. The mirror assembly and projection screen are described as being located within a transparent enclosure through which the image space could be viewed. A small pump was provided to reduce the air pressure within this enclosed region and hence lower the air resistance to the motion of the projection screen and other rotating components.

It is interesting to consider the impact of the mirror assembly (54, 55) and supporting structure (79) on the visual characteristics of the image space. They form a nontransparent structure that rotates around the image space, and this causes a limitation in the viewing angle. In the case of the Generescope, two preferential viewing directions were defined. The graphics engine described by the inventor was able to output an image scene twice during each revolution of the screen. This allowed for the creation of two viewing zones: one at the front and the other at the rear of the image space. Within these zones observers would have considerable freedom in viewing position. The occlusion caused by the presence of the mirror assembly ensured that only one image scene was visible to each group of observers.*

Within this extremely comprehensive patent the inventor also describes a pointing device for interacting with the image space and discusses the potential range of applications for volumetric display systems. These include both air traffic control (including training) and the visualization of computer-generated information.

5.3.3 A Volumetric CRT

In the patent introduced in Section 5.3.2, Max Hirsch also describes a second type of volumetric display. As in the case of the Generescope, this system was mechanically complex. The display unit employed two electron guns, the beams originating from which directly addressed a phosphor-coated surface of emission. This approach eliminated the need for the optical projection of images, which was necessary in the case of the Generescope. From the plan view of the display unit (Figure 5.6) it may be seen that each gun is made responsible for addressing one half of the surface of emission. The screen rotates around its vertical axis

---

* This is perhaps not obvious and may require some consideration.

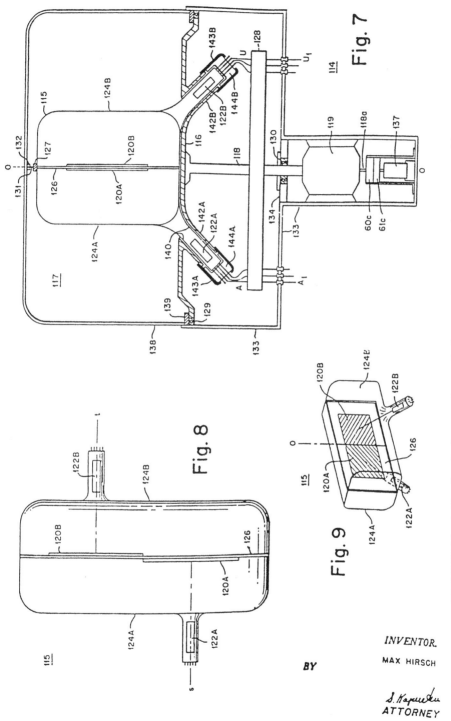

FIGURE 5.6 A Novel cathode ray tube employing two electron guns that corotate with a passive surface of emission. (From U.S. patent 2,967,905.)

INVENTOR.

MAX HIRSCH

BY

S. Kapuuten
ATTORNEY

103

and so sweeps out an image space. The electron guns corotate with the screen, so a constant geometry is maintained between the two subsystems. Distortional and voxel placement dead zones were therefore avoided. As with the Generescope, a commutator was employed for the passage of static voltages, deflection signals, and beam blanking information to the electron guns.

As may be seen from Figure 5.6, the electron guns are positioned below the image space, and any image occlusion that may be caused by the presence of this voxel activation apparatus is therefore minimized. The system described would, in principle, have been capable of depicting images composed of two different colors. This could have been achieved by the deposition of two different phosphors on each face of a supporting structure, each capable of generating light of a different wavelength when activated. Possible resonant vibration of the electrodes within the electron guns as a consequence of their rotation may have resulted in difficulties in registration between the two beams and therefore could have made it difficult to achieve color mixing. The generation of multicolor images within a volumetric display unit was a major aspect of the work undertaken by Ketchpel and is outlined in the next section.

Within his patent Hirsch demonstrated a deep understanding of both the theoretical and practical aspects of volumetric display system architectures. As a consequence of their mechanical complexity it is doubtful that the systems proposed could have operated reliably for extended periods, and the vibration of electrodes within the electron gun as a consequence of their motion may ultimately have been problematic. However, this in no way detracts from his outstanding work and the contribution that he made to the field. It would be most interesting to see the performance of either one of these display units when used in conjunction with current processing and control technologies.

5.4   PIONEERING WORK: 1960–1970

During this period there was considerable activity in the development of both swept and static-volume display systems. Some of the systems then proposed continue to be the subject of ongoing research, including the Cathode Ray Sphere (CRS), which has been the subject of the authors' work for a number of years (see Section 5.6.3). This display builds on the original work undertaken by Ketchpel, which is described below.

5.4.1   Volumetric CRT for Multicolor Images

In June 1960, Richard D. Ketchpel filed a patent at the U.S. Patent Office which concerned a novel cathode ray tube able to depict 3-D images. It would appear that this system was developed within research laboratories at the Hughes Aircraft Company in California. The patent was granted in 1964 (U.S. patent 3,140,415), and in the interim period several publications concerning this work appeared in scientific journals (Ketchpel 1962, 1963).

Like Hirsch, Ketchpel sought to produce a 3-D display device by constructing a new type of CRT based on well-known technologies of the era. However, instead of projecting the images generated on a conventional CRT screen onto a moving surface, Ketchpel eliminated the projection process and employed the direct interaction between an electron beam and a rotating phosphor-coated screen. In this way Ketchpel claimed to have developed a more compact display unit that was able to depict images of greater intensity. The display described by Ketchpel in his patent is reproduced in Figures 5.7 and 5.8.

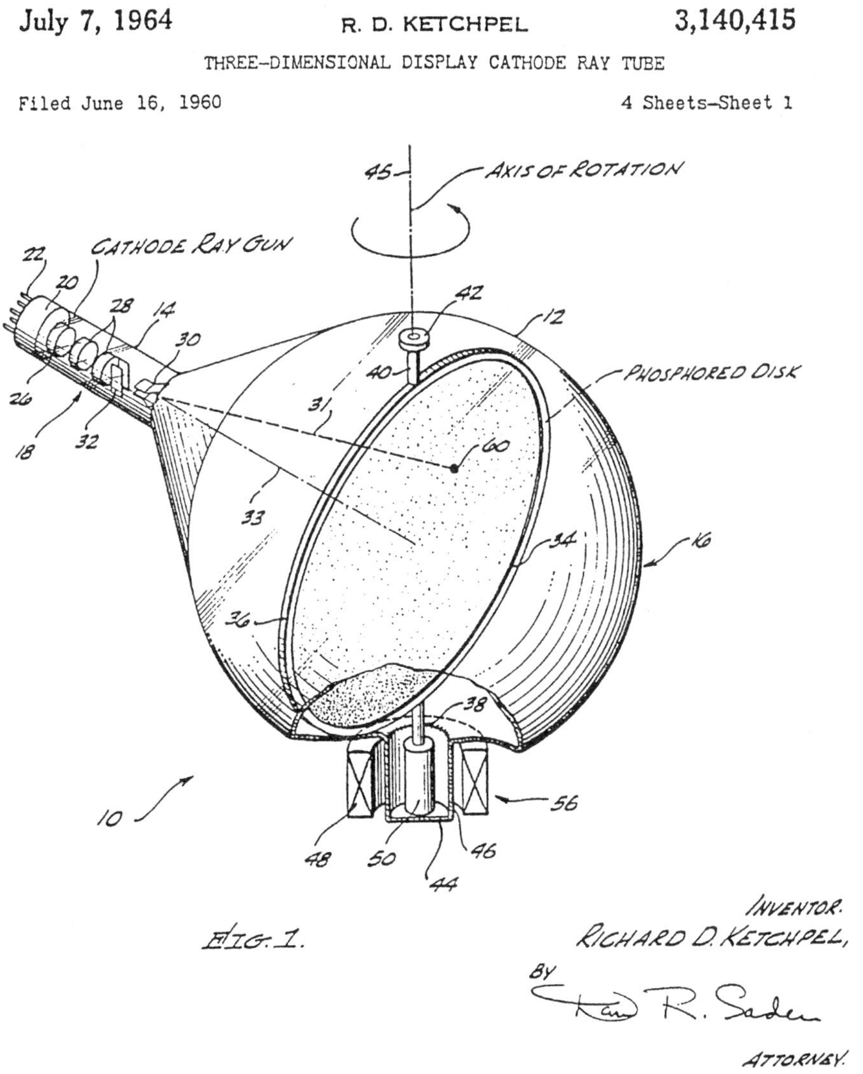
July 7, 1964
R. D. KETCHPEL
3,140,415
THREE-DIMENSIONAL DISPLAY CATHODE RAY TUBE
Filed June 16, 1960
4 Sheets-Sheet 1
AXIS OF ROTATION
CATHODE RAY GUN
PHOSPHORED DISK
FIG. 1.
INVENTOR.
RICHARD D. KETCHPEL,
BY
ATTORNEY.

FIGURE 5.7 Ketchpel's display, shown here with a single electron gun and a rotating surface of emission (34). (From U.S. patent 3,140,415.)

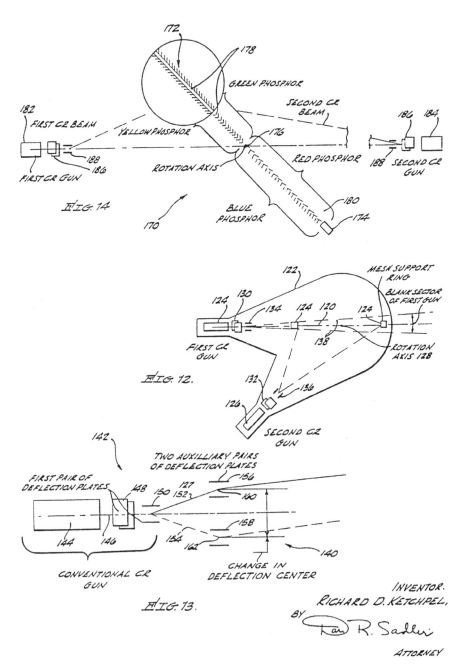

FIGURE 5.8 One of Ketchpel's prototype display units employing two electron guns. (From U.S. patent 3,140,415.)

In his publications Ketchpel describes the need for two electron guns to be positioned 45° apart to overcome *frame shadow*, described as being caused by the finite thickness of a thin supporting frame located around the periphery of the rotating screen (see Plate xii). When the surface of the screen lies parallel to the electron beam axis, the supporting structure gives rise to the frame shadow. Ketchpel further explains that "the sole function of the second gun would be to illuminate the blank sector of the display" (Ketchpel 1963).

The graphics engine described by Ketchpel is illustrated in Figure 5.9. Of academic interest is the method of synchronizing image data flow with the position of the surface of emission. Electron beam deflection angles are stored on a magnetic storage drum in digital form. The position of these data on the drum's surface provides the temporal coordinates. Synchronizing the rotation of the storage drum with that of the surface of emission then permits the synchronized flow of image data.

Ketchpel examined the feasibility of employing a raster scanning technique to exhaustively sweep out an entire image space with a diameter of 8 in. which was divided into 180 sectors. He assumed a raster scan resolution and image update period equal to that of a conventional but special-purpose CRT then available (and with a screen diameter of 21 in.). Ketchpel recognized that the 3-D display device would possess an effective screen area that would be approximately 26 times greater than that of the conventional tube. He concludes that an extremely high scan frequency would be required, and this would have greatly diminished the maximum attainable intensity of any images depicted within the display unit. Ketchpel therefore quite rightly sought to employ a more practical approach that employed a discontinuous deflection scan.

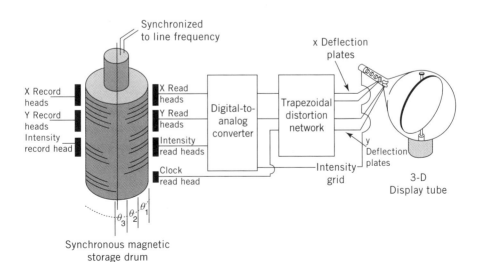

FIGURE 5.9 Simple graphics engine described by Ketchpel (1963) employing a magnetic storage drum. (Reproduced by permission; © 1963 IEEE.)

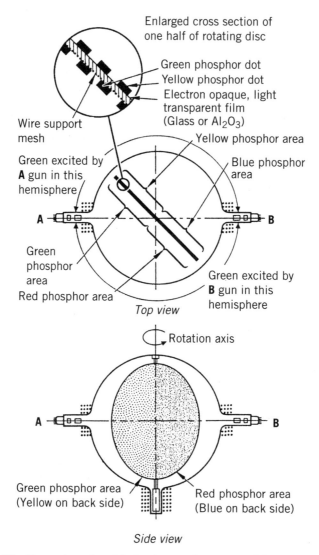

FIGURE 5.10 Generation of multicolor images by means of four types of phosphor deposited on a planar supporting structure and described by Ketchpel (1963). (Reproduced by permission; © 1963 IEEE.)

Also described is a technique for the generation of multicolor images in which the screen was divided into four regions (see Figures 5.8 and 5.10). Each of these regions was coated with a different type of phosphor which when excited by an electron beam produced light of a different color (Ketchpel suggests red, green, blue, and yellow). During the course of a single revolution of the screen, each phosphor coating sweeps out the entire image space, so any of the four colors may be excited during each image refresh period* (Figure 5.10). In an attempt to preserve the transparency of the surface of emission, Ketchpel proposed that the phosphor be deposited as a matrix of dots, the dots on either side of the supporting structure being offset from each other. However, due to the constantly varying geometry between the screen and observer(s), the benefits that could be derived from such an implementation are questionable.

5.4.2 Projection System Employing Rotational Motion

In October 1960 a brief article appeared in *Aviation Week* describing a volumetric display system developed in prototype form by ITT Laboratories (*Aviation Week* 1960). The system appears to have been intended for use in air traffic control and military applications, with particular emphasis on the depiction of point form information such as radar blips. There is no indication within the article that it was intended for the display of more complex images constructed from voxels.

The display unit employed a high-intensity CRT and rotating screen. The images depicted on the CRT were cast upon the rotating screen via two corotating mirrors (Figure 5.11), the output of information to the CRT being synchronized to the screen position. The determination of the location of image points (e.g., radar blips) within image space was governed by their position on the face of the CRT and the time at which they were illuminated. The relative motion between the CRT and rotating screen clearly demanded the use of a CRT with a short-persistence phosphor so that elongation of image points could be avoided.

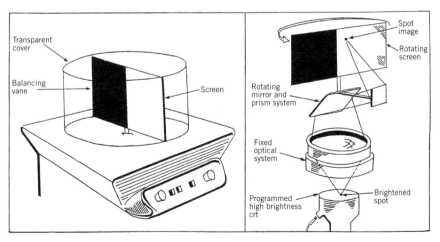

FIGURE 5.11 Volumetric display unit prototype in which image data depicted on the surface of a CRT is cast into an image space. [From *Aviation Week* (1960).]

---

* To permit voxels of each color to be generated at any location within an image space created by the rotational motion of a planar surface of emission, two or three electron guns are required.

The article describes the construction of a prototype display unit having an image space with a diameter of 12 in. and a height of 6 in. In the case of this prototype system, image point generation is reported as having been achieved through the use of small lamps rather than by means of a CRT. As a consequence, it is doubtful that image points were sharply defined. The black vane illustrated in Figure 5.11 is intended to provide a counterbalance to the rotation of the screen and is equipped with white markers which when illuminated with a stroboscope locked to the motion of the screen, produce static guidelines within the image space.

5.4.3   Volumetric CRT with a Curved Surface of Emission

A.M. Skellett filed a patent in July 1962 (U.S. patent 3,204,238) providing details of a volumetric display unit suitable for use in the visualization of aircraft radar information. This system is described as employing a single electron gun that directly addresses a phosphor-coated surface of emission. From information available to the authors this appears to be the first occasion on which a nonplanar surface of emission was used for image space creation. As may be seen from the illustration provided in Figure 5.12, the curved surface of emission and its supporting structure rotate around the electron guns axis. Skellett describes the motion as follows:

> The transparent screen rotates around the tube axis, and, as it rotates, passes through every portion of the volume within the hemisphere to describe or sweep out the volume in the hemisphere.

The shape of the screen was described in detail by Skellett and may best be understood by the following extract taken from his patent (the appropriate diagrams are reproduced in Figure 5.13):

> The screen tilts upward at a constant rate with change in position around the axis of the tube, or axis of rotation, going from perpendicular to the axis of rotation to parallel with the axis of rotation in 324 degrees around the axis. That is, in every 36 degrees around the axis of the face of the screen will increase its tilt by 10 degrees so that along line OA the screen will be perpendicular to the axis, along line OB, displaced 36 degrees clockwise from OA, the screen will be inclined 80 degrees to the axis of the tube, and so on until along line OJ, displaced 324 degrees clockwise from OA, the screen will be parallel to the axis of the tube.

Skellett recognized that the screen should be transparent. In the case of a surface of emission employing the deposition of phosphor, this can be difficult to achieve without a loss of efficiency in the conversion of electron beam energy into visible light (resulting in a diminution of image intensity). Often, such screens tend to be only semitransparent. The screen geometry suggested by Skellett is such that light emitted from an image point *may* pass through another region of the

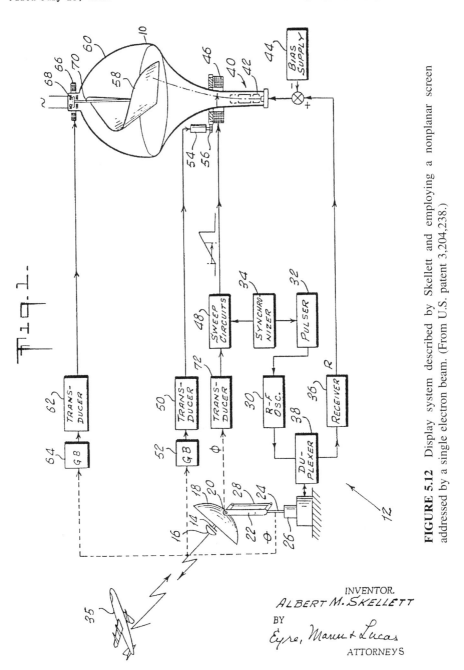

FIGURE 5.12 Display system described by Skellett and employing a nonplanar screen addressed by a single electron beam. (From U.S. patent 3,204,238.)

INVENTOR.
ALBERT M. SKELLETT

BY

Eyre, Mann & Lucas
ATTORNEYS

111

Aug. 31, 1965   A. M. SKELLETT   3,204,238

CATHODE RAY TUBE FOR THREE-DIMENSIONAL PRESENTATIONS

Filed July 13, 1962   2 Sheets—Sheet 2

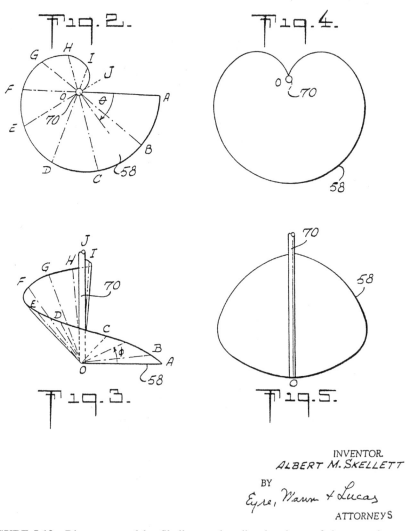

INVENTOR.
ALBERT M. SKELLETT

BY
Eyre, Mann & Lucas

ATTORNEYS

FIGURE 5.13 Diagrams used by Skellett to describe the shape of the curved screen employed within his display unit (see the text for details). (From U.S. patent 3,204,238.)

screen before reaching the observer. The semitransparent nature of the screen, coupled with refraction effects within the supporting structure, could have had a negative impact on image quality by introducing visual dead-zone regions (see Section 4.2).

5.4.4   Display Unit Employing an Electroluminescent Panel

Also in 1962, an article was published describing the development of a swept-volume display employing the rotational motion of an active surface of emission (Coddington and Schipper 1962). In this work the authors consider various attributes of volumetric systems and discuss a number of possible areas of application. The display unit described therein employed a planar electroluminescent panel that swept out a volume measuring 10 in. in diameter and 11 in. in height. The rotation frequency is reported to have been 20 Hz, and the voxel activation signals were passed to the array by means of a 100-ring commutator. This approach is similar to the display unit architecture discussed in greater depth in Section 5.5.3.

5.5   PIONEERING WORK: 1970–1980

5.5.1   Display Developed for the Reconstruction of Ultrasonograms

In the early 1970s a prototype display system was developed as an aid to the visualization of ultrasound scans (Szilard 1974). The very simple and effective display unit produced for this purpose employed a series of rectangular screens attached to the periphery of a cylinder. This cylinder rotated about its central axis and, as may be seen from Figures 5.14 and 5.15 the screens were each displaced

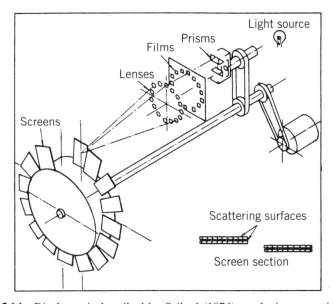

FIGURE 5.14 Display unit described by Szilard (1974) employing a rotating cylinder attached to the periphery of which are a number of screens. The extent of the image space is considerably less than the dimensions of the volume swept out by the screens. (© 1974 Ultrasonics.)

FIGURE 5.15   A photograph of the display unit described by Szilard (1974) (© 1974 Ultrasonics.)

from their neighbors so as to follow a helical path. The developers of this system described its operation in the following manner:

> By rotating the cylinder (or disc) the screens will pass, one after the other, through the same angular position but in a different 'depth.' If we project the image of a section of the body onto the screen of the corresponding 'depth' while it passes through the appropriate position and if this is done at a sufficiently high frame rate then we will see a real three-dimensional image.

As a consequence, the height and width of the image space were determined by the dimensions of screen employed, and its depth by the extent of the helical form mapped out by them. The projection of the ultrasonic scans (which were stored on photographic film) was achieved by ensuring that the illumination of each photograph occurred when the appropriate screen was passing through the viewing region.

It is interesting to note that this approach gives rise to an image space whose dimensions are considerably less than the volume swept out by the rotating screens. This applies to several other swept-volume image space creation techniques and is generally undesirable, as it may result in cumbersome mechanical components. However, despite the physical size of the rotating components, the display unit developed by Szilard would appear to have been capable of producing a uniform image space with excellent optical characteristics. Furthermore, the projection method employed to cast each depth plane into the image space would have facilitated the production of high-quality images therein.

5.5.2   Helix Laser 3-D Display

In May 1976, Rüdiger Hartwig filed a patent (DE 26 22 802 C2) describing a volumetric display unit incorporating a helical screen and which used visible laser radiation for voxel activation. The patent was disclosed in 1977 and granted in 1984. This display unit is illustrated in Plate v. In this approach, the laser beam may be directed to any part of the helical surface, and by synchronizing the modulation of the beam to the rotation of the screen, voxels may be activated. In order that each voxel is visible from practically any orientation, two essential conditions must be fulfilled:

- The helix must be semitransparent.
- The helix must be single bladed.

Fundamental to this invention is the combination of a rotating helical surface (as we discussed previously, this form of screen has a number of associated advantages) with a collimated laser beam (alleviating the need for dynamic focusing). Therefore, this class of volumetric 3-D display has generally been referred to as the *helix laser* 3-*D display* (HL3D). The double-bladed configuration (subsequently used by both Texas Instruments and the U.S. Navy; see Section 5.6.6) enables voxels to be refreshed twice during each rotation of the helix. Unfortunately, transparency may then be lost.*

As with so many of the volumetric display systems that have been developed in prototype form, the story of the initial research is fascinating. In Hartwig's own words:

> After the disclosure discussions took place with colleagues at the University of Stuttgart [where Hartwig is an honorary professor] and also with colleagues at the University of Heidelberg [where Hartwig at that time worked in the IBM Scientific Center as a chief advisor for Computer Graphics applications]. Doubts arose that such a simple principle could ever give rise to 3-D images positioned in space. Finally the Director of the Institute for Applied Physics of the University of Heidelberg, Prof. Dr. J. Bille, offered to fund the construction of a prototype model. Under my advice an enthusiastic group of young scientists and skilled workers built the very first helix-laser-3D-display during the period 1978 to 1979.

This display unit prototype had a cylindrical image space measuring 60 cm in diameter and 40 cm in height. It employed a semitransparent single-bladed helix that was constructed from white plastic and rotated at a frequency of 20 Hz. The system was able to depict Lissajous figures as well as computer-controlled geometrical curved images.

In 1982, Hartwig presented the helix laser 3-D display to European scientific journalists. This denoted the first public demonstration of this architecture

---

* In the case of any swept volume system employing rotational motion, the transparency of an image space can be easily assessed by attempting to read a newspaper through the volume. The visual dead zone that may be associated with the presence of a center shaft will then be readily apparent.

(Cross 1982, Hartwig 1982, Brinkmann 1983). Hartwig's pioneering work formed the basis for subsequent research and development activity in a number of institutions, and by means of current laser technologies, enormous increases in performance have been made (see also Section 5.6.6 and U.S. patent 3,300,779).

5.5.3   Display Unit Employing an Active Surface of Emission

It will be recalled from Section 2.6, that major increases in the voxel activation capacity are best achieved by the introduction of parallelism in the voxel generation and activation processes. Many of the display unit architectures described in this chapter support only sequential voxel activation, and their voxel activation capacities are therefore determined by the time required to create each voxel, and the image refresh period.

In 1977, E.P. Berlin, Jr. (MIT) filed a patent detailing a volumetric display system able to support a high degree of parallelism in the voxel generation process and so allow a large voxel activation capacity to be achieved (U.S. patent 4,160,973) (see also Jansson et al. (1979)). The display unit detailed in this patent is illustrated in Figure 5.16. This system employs an active surface of emission consisting of a 2-D array of opto-electronic devices such as LEDs mounted on a planar supporting structure. This rotates around a vertical axis, so each LED is responsible for the production of voxels along the circular track through which it moves. Since, in principle, it is possible to address all the opto-electronic elements in parallel, it follows that the maximum degree of parallelism that might be exhibited by this volumetric display unit architecture is equal to the number of elements within the array.*

Berlin employed a serial optical link to pass voxel descriptor information from stationary electronics connected to the host computer through to electronic hardware that corotated with the screen. This was responsible for addressing the LED matrix and was powered via a slip ring commutator. In an alternative scenario, Berlin used a surface of emission comprising a plasma array. In this case the array dimensions are described as being 128 by 256 (with a rotational frequency of 30 Hz).

The technology described in this patent does not aim to solve the visual dead zone problem associated with a planar surface of emission employing rotational motion (see Sections 4.2 and 4.4.2). However, it does provide a technique for achieving a very high voxel activation capacity. It is also interesting to consider the display unit introduced in Section 5.4.4, which was developed more than 10 years earlier and which also offered a high degree of parallelism in the voxel generation and activation processes.

More recent work employing this approach is discussed by Solomon (1993).

---

* As described in Section 6.4, the parallelism exhibited by such a display unit is ultimately determined by any row and column addressing technique employed in the interconnection of the voxel generation centers with the data pathways at the output of the graphics engine.

U.S. Patent   Jul. 10, 1979   Sheet 1 of 6   **4,160,973**

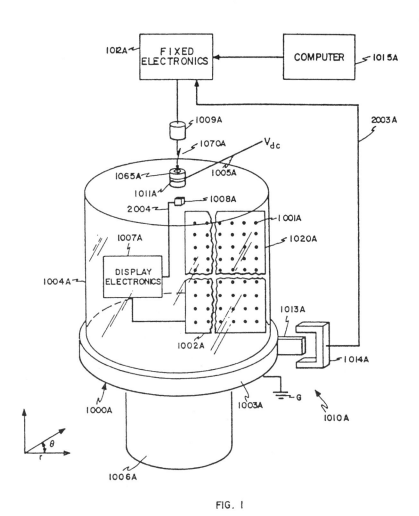

FIG. I

FIGURE 5.16   Display unit employing the rotational motion of a planar active surface of emission. (From U.S. patent 4,160,973.)

5.5.4   TOMAX System

TOMAX was the name given to a volumetric display system intended for use in medical imaging. The system was designed for the 3-D visualization of, for example, computed tomography (CT) scans that generate image data sets in the form of a series of 2-D slices. In principle, data of this type may easily be reconstructed within the image space of a suitable volumetric display system. Unfortunately, such medical scans generate very large data sets, and only a fraction of these data can be depicted on existing volumetric technologies during each image refresh period. The compression and/or the filtering of these data are somewhat problematic, and any arbitrary omission must not be permitted (see Chapter 10).

The TOMAX system was described briefly in a publication (Mark and Hull 1977), and a diagram of their display unit is reproduced in Figure 5.17. It may be seen that the display unit employed a conventional CRT upon which a series of image frames (slices) were displayed sequentially and repetitively. These images were projected on two mirrors whose motion was synchronized to the output of the image slices on the CRT and thus provided image depth. The motional frequency of the mirrors, reported as 30 Hz, corresponded to the image sequence cycle frequency of the CRT.

The motion of the mirrors employed by this display is somewhat unusual, and as a consequence, this display does not fit properly into the classification scheme provided in Section 3.3. Furthermore, according to the definition of a volumetric display system presented in Section 1.2, any volumetric system should be equipped with a physical volume within which images may be depicted and viewed. Since the TOMAX display unit has no such physical volume but employs a virtual image space, it may be considered as not representing a truly volumetric system. However, the mirror motion makes the system particularly interesting, and since its operation resembles that of a swept-volume system employing translational motion, it is included in this book.

From Figure 5.17 it may be seen that the two mirrors are each pivoted and move synchronously in such a way that they remain parallel to each other at all times. This motion is illustrated in the diagram by positions A and B, which

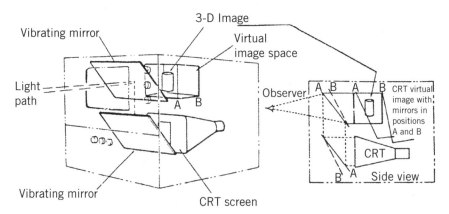

FIGURE 5.17 TOMAX display system described by Mark and Hull (1977). Of particular interest is the motion of the mirrors, which gives rise to a virtual image space. (Reproduced by permission; © 1977 SPIE.)

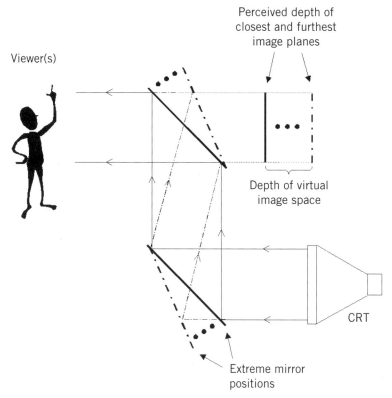

FIGURE 5.18   Ray diagram illustrating the generation of a virtual image space by means of the TOMAX system. The two mirrors alter their aspect with respect to the CRT and viewer in a synchronized manner, as shown. Sequential image slices are thus perceived as lying at different depths along the viewer's line of sight.

correspond to the extremities of travel of the mirrors. The CRT images are cast on mirror 1 and consequently reflected onto mirror 2. The virtual image generated is observed through this second mirror (Figure 5.18). Unfortunately, no information is provided about the manner in which the mirror motion was achieved.

The TOMAX system is reported to have operated in two configurations. The first of these allowed the depiction of seven image slices, each consisting of 256 by 256 voxels (each voxel having an associated 16 gray-scale levels). The second configuration permitted the depiction of 12 image slices, each slice being composed of up to 128 by 128 voxels. The rectangular virtual image space is reported to have had dimensions of 4 in. by 5 in. by 4 in. (deep).

5.5.5   Display Unit Employing a Rotating Mirror

Also in 1977, an article was published describing a swept-volume display unit that employed the rotational motion of a mirrored surface on which images depicted

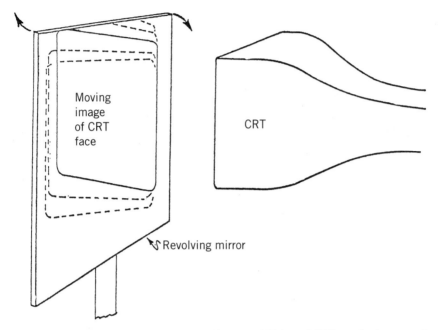

FIGURE 5.19   Display unit described by Simon and Walters (1977) employing a rotating mirror on which CRT images are projected. The graphics engine must ensure that the generation and timing of image frames produced on the CRT place voxels at the required location within the image space. (Reproduced by permission; © 1977 SPIE.)

by a CRT were cast (Simon and Walters 1977). This display unit architecture was intended to overcome the visual dead zones that had previously been experienced by the researchers (Simon 1969). In this earlier work, CRT-generated images were cast on a rotating mirror (Figure 5.19). The authors describe the limitations of the earlier system as follows:

> This method, while producing a three-dimensional image, has the disadvantages of a narrow field of view obstructed by the oscilloscope* and the fact that the observer sees the edge of the mirror passing through his field of view twice during each rotation.

Their modified display unit employed a mirrored surface mounted on the end of a diagonally sliced cylinder. The cylinder rotated about its axis, and the preferred viewing direction was along this axis. As a consequence, when observed from this orientation, the edge of the mirrored surface never passed through the field of view, so the visual dead zone (see Section 4.2) was eliminated. Unfortunately, this configuration introduces a weakness—the mirrored surface fails to sweep

---

* CRT projection tube.

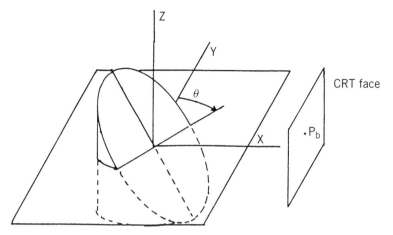

FIGURE 5.20 Mirror located on the end of a diagonally sliced cylinder as described by Simon and Walters (1977). The cylinder rotates about the z axis, so the preferred direction of viewing would be from above. Unfortunately, this configuration gives rise to two conical regions of dead image space. (Reproduced by permission; © 1977 SPIE.)

out two conical regions located above and below the "equator" of the image space (Figure 5.20). The authors have encountered similar proposals for image space generation in other publications, indicating that perhaps in the first instance this deficiency has sometimes been overlooked.

5.6 PIONEERING WORK: 1980–PRESENT

Various volumetric display system development projects initiated during the 1980s are still the subject of active research. Within this section we have therefore amalgamated projects undertaken within both the previous and current decades. During this time there has been considerable activity in volumetric display research, and a number of very promising technologies could give rise to commercial products in the not too distant future.

5.6.1 Display Unit Employing an Archimedes Spiral

In 1984 researchers in Japan described work undertaken in producing an experimental volumetric display system employing an unusual image space creation technique (Yamada et al. 1984). Although this system is interesting from an academic point of view and its development is a considerable achievement on the part of those who carried out the work, the image space characteristics do not appear to be favorable.

To generate their image space, the researchers employed two Archimedes' spirals rotating around a fixed axis (Figure 5.21). The curvature of each spiral is

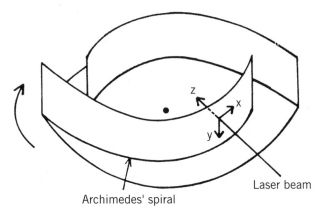

Archimedes' spiral

FIGURE 5.21 Image space created by means of two Archimedes' spirals and as described by Yamada et al. (1984). (Reproduced by permission; © 1984 ICALEO.)

such that its radius (r) is a linear function of angle (θ). Each spiral may therefore be represented in polar coordinates by the expression

$$r = -a\theta + b \qquad (5.1)$$

where b represents the radius at $\theta = 0$. Since the first term is negative, b corresponds to the maximum radius (r_{max}). The extent of each spiral used for the image space generation is reported as being π radians, and at this angle the spiral is at its closest to the axis (r_{min}).

The workers describe each of the spirals as being formed from thick paper coated with an orange-yellow organic daylight fluorescent pigment (intended to increase the luminance of the screen and to reduce speckle when addressed by laser radiation). As each spiral rotates, it maps out a volume that has a radial width $r_{max} - r_{min}$. A scanned laser beam (10 mW argon laser) was used to sweep out an approximately cubic region of image space using a raster scan geometry. Each 2-D frame of the scan, while being synchronized to the rotation of the spiral screens, was performed at a considerably higher frequency than the screen's frequency of rotation, and in extent, each frame occupied only a small portion of the spiral curve. Therefore, as each spiral rotates it is possible to generate multiple frames at differing depth. The difference between r_{max} and r_{min} determines the extent of the third dimension of the scanned region.

In this display unit embodiment, only a small region of the volume swept out by the screen's motion was employed for image generation (see also Section 5.5.1). The authors determined the extent of the scanned region by ensuring that the nonlinear behavior of the image space creation technique was sufficiently small across the dimensions of the scanned region so as to be negligible.

A source of nonuniformity arises when the laser beam is deflected along the surface of the spiral. In Figure 5.22 this problem is exaggerated somewhat for the

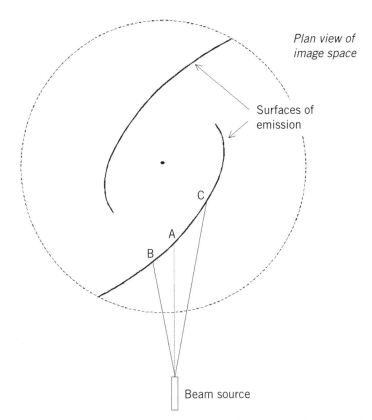

FIGURE 5.22   Should a linear scan be employed for beam deflection, it is readily apparent that this will give rise to a nonlinear scan on the surface of the spiral.

sake of clarity. The undeflected laser beam intersects the screen at point A. Points B and C indicate the points of intersection of the laser beam when deflected at equal and opposite angles relative to A. It is apparent that the distances AB and AC along the surface of the spiral are unequal. A linear scan of the laser beam will therefore generate a nonlinear scan along the surface of the spiral. Since the gradient of the spiral changes continually, the magnitude of this nonlinearity varies as the screen rotates. Consequently, successive frames generated on its surface will each be distorted by a varying amount.

The various distortions caused by the spiral's surface on the linear frame scanning technique employed by the researchers are analyzed in the article referred to above. This analysis leads to the intuitive result that a/b should be made as small as possible. This is achieved by making r_{max} as large as possible and $r_{max} - r_{min}$ as small as possible. Unfortunately, reducing the latter also reduces the depth of the image space. The researchers therefore tried to obtain a satisfactory compromise between the size of the spiral and the dimensions of the image space. Since the image space is scanned exhaustively, its size is further limited by the performance of the scanning system and the desired scanning resolution.

The research team reports the use of a rotating screen equipped with two spiral surfaces each having maximum and minimum radii of 10 and 7 cm. Technical problems are said to have limited the rotation frequency to 4.5 Hz, giving an image refresh frequency of twice this value. The horizontal scan frequency (generated using a polygon mirror scanner) for a 3-cm scan was 25,560 Hz. That of the vertical scan (employing a moving core type of scanner) was 182.6 Hz, permitting each frame to consist of 140 lines, 12 of which corresponded to a blanking period. The sequence of frames within each image refresh period was generated across the length of a spiral, that is, within approximately 0.1 second. This therefore permitted the depiction of 18 frames.

In summary, the image space described by the research team measured approximately 3 cm in each direction. This region was scanned exhaustively and contained 18 frames, each consisting of 140 horizontal lines. Unfortunately, the low image refresh frequency would have caused considerable image flicker. The exhaustive scanning technique coupled with low scan speeds and lack of a computational system responsible for proper matching of the scan with the nonlinear image space characteristics restricted the size of the image space. In fact, the extent of the scanned image space represented less than 6% of the volume swept out by the spiral screens. Finally, the form of the image space creation apparatus would have imposed considerable viewing angle restriction.

This volumetric system implementation serves to highlight some of the problems encountered when attempts are made to employ screens other than those with planar or helical geometries for image space generation. The problems associated with scanning an image space exhaustively by means of a sequential voxel activation process are also clearly illustrated. These comments do not, however, detract from the excellent achievement of constructing this display system prototype.

5.6.2 Volumetric Display System for Medical Imaging

In 1988 an article was published in the *IEEE Transactions on Medical Imaging* describing work undertaken by a team in Japan on the development of a prototype volumetric display system (Yamanaka et al. 1988). The authors describe the evaluation of this system by applying it to the depiction of ultrasonic echogram images.

The display unit, which is illustrated schematically in Figure 5.23, operates by means of the projection of CRT images onto a screen moving with translational motion. Although such a configuration is by no means novel, the system was implemented using high-performance components within the graphics engine and a random-scan 23-in. CRT-based display monitor as the image source within the display unit. The screen is reported to have consisted of a plastic mirror set into an aluminum frame weighing a total of 300 g and moving with a maximum velocity of $1 \text{ m} \cdot \text{s}^{-1}$.

The image space was cubic in shape, having sides of length 8 cm within which it was intended to depict 128 image slices, each comprising up to 128 by 128 voxels (and with a refresh frequency of 12 Hz). However, the system throughput

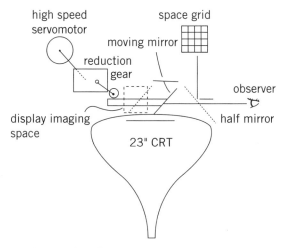

FIGURE 5.23 Display unit described by Yamanaka et al. (1988) employing a projection CRT and a screen moving with translational motion. (Reproduced by permission; © 1988 IEEE.)

is reported to have restricted the number of image slices that could be depicted on the display unit's projection system to 1075 frames per second instead of the 1536 frames per second required to generate 128 slices. The number of image space slices was therefore limited to 89. The researchers report that for the foregoing performance restrictions the graphics engine "thins out a block of 2-D cross sections" and further that it "deleted part of the cross sections at random without losing the smoothness of 3-D pictures." So although the source data consisted of 128 slices, the graphics engine was responsible for reducing this number to match the display unit's characteristics. The voxel activation capacity of the display unit appears to have been very high (approximately 1.5 million). This is an outstanding achievement when we consider that that the voxel activation mechanism was sequential. Unfortunately, the low image refresh frequency would have caused significant flicker problems.

5.6.3 Cathode Ray Sphere

In 1988, one of the authors (B.G.B.) initiated a research program in New Zealand in relation to volumetric systems and developed various prototype displays. For a number of years, development work centered on a swept-volume approach which employed a number of electron guns able to activate voxels on a rotating phosphor-coated surface of emission. The general form of a typical prototype is illustrated in Figure 5.24. The spherical shape of display vessel coupled with the use of electron guns for voxel activation led to the adoption of the name Cathode Ray Sphere (CRS) for this type of system (Blundell and King 1991, Blundell et al. 1993b, Blundell et al. 1994, Schwarz and Blundell 1997, Schwarz and Blundell 1994a, b). See Plates i–iv, vi–xi, and xvi–xvii.

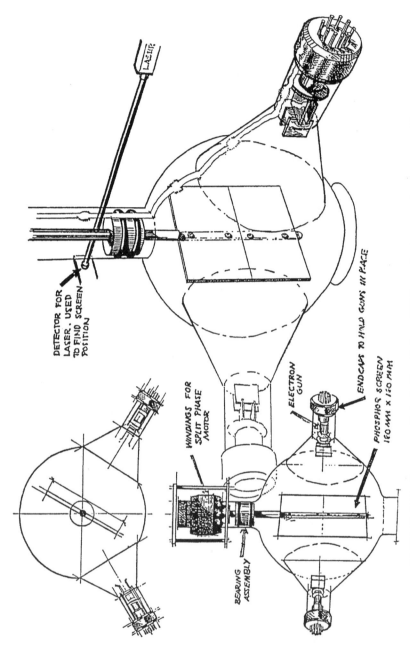

FIGURE 5.24 Prototype Cathode Ray Sphere (CRS) employing two electron guns. The use of two or more types of phosphor permits the generation of multicolor images. Proper alignment between the beam sources permits color mixing. (Drawing by W. King.)

126

The origins of this display can be traced back nearly 30 years to the ideas and proposals of Ketchpel (see Section 5.4.1 and Plate xii). However, the availability of appropriate computational systems and control hardware now permits the proper registration to be achieved between several beam sources. Consequently, significant progress has been made in minimizing the impact of both the distortional and voxel positioning dead zones. The first prototype display unit is illustrated in Plate xi. Blundell and King (1991) first reported on this system, and aspects of this work are discussed in more detail in Chapter 11. See also U.S. Patent 5,703,606, P.C.T./N.Z./96/00028.

5.6.4 Display Unit Employing a Helical Screen and CRT Projection System

A patent filed by R. Morton in 1989 details the architecture of a volumetric display system employing a rotating helical surface onto which images depicted on a CRT screen are cast (U.S. patent 4,922,336). The form of this display unit is illustrated in Figure 5.25. Of particular interest is the projection system employed to cast images onto the rotating surface. Since the flow of information to the CRT is essentially sequential, this display does not permit parallelism in the voxel activation process.*

5.6.5 Image Space Viewable from Within

Texas Instruments applied for a number of patents relating to volumetric display system technologies. In a European patent application, a display unit architecture is described that employs a rotating surface located at an angle of 45° with respect to the rotational axis (European patent application EPI 0 491 284 A1). Voxel activation within the image space is described as being achieved by means of laser beams in the usual manner.†

As we mentioned in Section 5.5.5, this orientation of the screen with respect to the axis of rotation gives rise to two conical regions located above and below the rotational center within which voxels cannot be generated (as the screen does not sweep through these regions). In their patent the researchers claim to use advantageously the lower conical void (i.e., located below the equator of the image space) by ensuring that it is of a sufficiently large extent so that an observer's head (or camera) may be located therein (Figure 5.26). Consequently, the image space may be viewed from the interior. Should the voxel activation technique be capable of attaining a high voxel activation capacity, some merit may be derived from this approach. However, since the display unit does not appear to support parallelism in the voxel activation process, it would appear that the system was capable of depicting only sparse data sets, and therefore the intended purpose of the display is most unclear. Furthermore, the advent

---

* We neglect the possibility of multiple electron guns within the CRT.

† Of related interest are U.S. patent 5,042,909 and European patent applications EPI 0 418 583 A2 and EPI 0 310 928 A2.

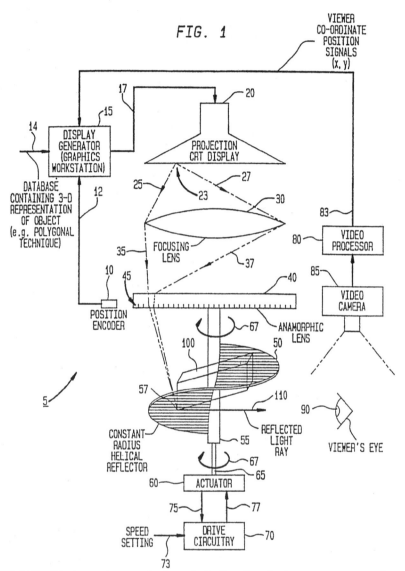

FIGURE 5.25   Display system employing a helical surface for image space generation. Images depicted on a CRT are cast into the image space and a tracking system is included to determine the position of the observer. (From U.S. patent 4,922,336.)

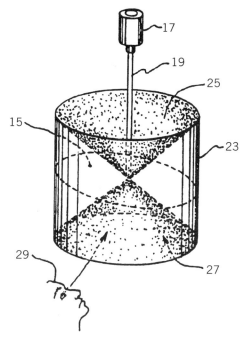

FIGURE 5.26 Image space created by means of a planar screen located at an angle of 45° with respect to the axis of rotation. Two regions of dead image space exist within the cylindrical volume. (From European patent application EP 0 491 284 A1.)

of high-quality headgear associated with virtual reality systems now provides a superior method of immersive image visualization.* This patent does, however, serve to highlight the form of the dead image space associated with this image space creation technique.

5.6.6 HL3D Systems Developed for the U.S. Navy

In 1992, following work at Texas Instruments (Williams and Garcia 1988, Williams and Donohoo 1991) Parviz Soltan demonstrated a U.S. Navy first-generation HL3D display system (Soltan et al. 1992) (see Section 5.5.2). This display unit employed a double-bladed helix, as illustrated in Figure 5.27. Subsequently, other versions of HL3D display have been developed, including a transportable system (Soltan et al. 1994, Lasher et al. 1996). These displays are equipped with a highly sophisticated laser deflection system. (See U.S. patent 5,854,613).

The transportable display shown in Figure 5.27 is based on a single-bladed helix similar to that originally patented by Hartwig (see Section 5.5.2). The laser

---

* However, this type of headgear does not provide the observer with the opportunity of a complimentary hair cut!

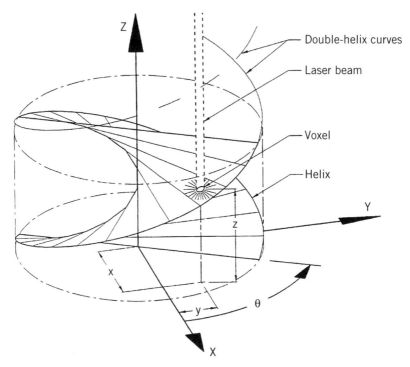

FIGURE 5.27 The double-bladed helical configuration employed by the U.S. Navy (courtesy of P. Soltan.)

beam is directed to the helix from below. The image space is reported to be 40 cm in diameter and 20 cm high. The helix rotates about a vertical axis at a frequency of 20 Hz. The image space is divided into four quadrants, and within each of these it is reported that 20,000 red, green, and blue voxels may be activated, corresponding to a voxel activation capacity of 80,000. This system was exhibited at the International Air Traffic Controller Association Conference in Nashville Tennessee in 1996, and an upgraded version was installed in 1997 onboard the U.S. Navy's largest aircraft carrier, USS *Stennis*. This work is discussed further in Section 11.5.

5.6.7 System Able to Project Volumetric Images into Free Space

Researchers in Japan published an article discussing the development of a prototype display system able to project volumetric images into free space. They describe this as being accomplished by means of a simple optical system employing two concave mirrors (Kameyama and Ohtomi 1993, Kameyama et al. 1993). Normally, an operator is physically excluded from an image space and is therefore unable to touch images generated therein. Depending on the type of application, this may be disadvantageous and may make direct image interaction less intuitive.* The projection of images into free space was intended to overcome this difficulty.

The researchers describe the development of a swept-volume display employing an active matrix panel moving with translational motion. The general specifications presented within their publication indicate that the physical image space has dimensions measuring 94 mm by 30 mm by 50 mm and a voxel activation capacity of 38,400. Voxel generation is described as having been accomplished by means of an LED array. The optical projection system described within their publication is reproduced in Figure 5.28. The generation of a real image within free space permits a wireless pointer to be used for direct image manipulation. Although there are clearly a number of potential difficulties in properly generating "free space" volumetric images in this manner, it is possible that this type of approach may ultimately permit images generated within a physical image space to be projected in such a way that they are magnified. Consequently, this approach may be particularly advantageous in the case of image spaces which are composed of a high density material or which are created by a translational motion.

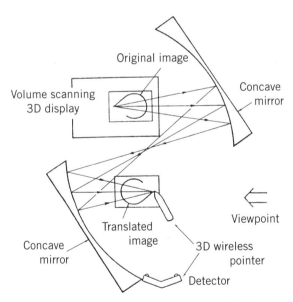

FIGURE 5.28 Display system described by Kameyama (1993) in which images generated within a physical image space are cast into free space by an optical system employing two converging mirrors. (Reproduced by permission; © 1993 SPIE.)

---

* In the case of systems possessing very large image spaces, this potential problem may become more severe.

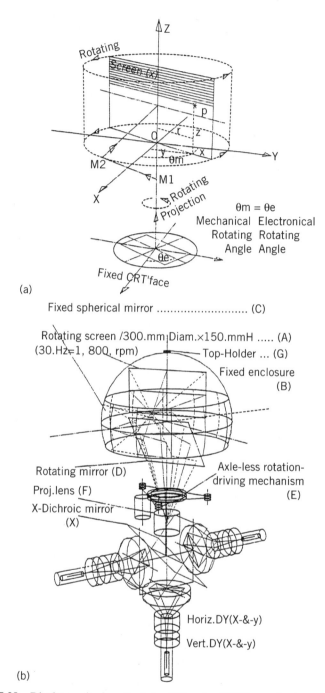

FIGURE 5.29 Display unit described by Shimada (1993) employing a CRT-based projection system and planar rotating screen. The optical projection system ensures that dead zones are eliminated. (Reproduced by permission; © 1993 SID.)

5.6.8 3-D Rotatron Display

Researchers in Japan associated with the Sony Corporation describe a swept-volume display system that they refer to as the *3D-Rotatron* (Shimada 1993). This display unit employs a planar rotating screen on which images depicted by a CRT projection system are cast (the CRT optical projection axis being coaxial with the axis of rotation of the screen). The major display unit components are shown in Figure 5.29. Images generated by the CRT projection system are reflected by a rotating mirror onto a spherical mirror that forms a ring about the cylindrical image space. A complete understanding of this optical arrangement and projection system is best obtained by reference to the original publication. It is sufficient for our purposes to note that the arrangement of the mirrors permits images to be projected on the rotating screen (from the stationary projection system) in such a way that distortional and voxel placement dead zones are avoided.

5.7 DISCUSSION

A comprehensive history of swept-volume display unit development could, in itself, occupy an entire book. In general, the systems described in the previous pages have been selected so as to highlight various issues relating to display unit implementation and to provide instances of principles discussed elsewhere in this work. At the very most, we estimate that within this chapter we have included 25 percent of the total number of swept-volume systems that have been reported in either patent applications or scientific publications.

During the extensive period that this type of research has been carried out, it is interesting to observe the cyclic nature of the activity. Ideas previously proposed have been resurrected at intervals and reworked within the contexts of new technologies and the ever-changing visualization requirements. Unfortunately, in the case of some architectures, essential weaknesses have not been identified in advance, indicating perhaps that development work may have been initiated before undertaking a proper review of published literature. Several swept-volume display unit architectures outlined in this chapter are discussed in greater depth later in the book.

CHAPTER 6

STATIC-VOLUME DISPLAY UNITS

Through the dense din, I say, we heard him shout
"I see your lights!" But ours had long died out.*

6.1 INTRODUCTION

In Chapters 3 through 5 we concerned ourselves with display units in which
the creation of a 3-D image space depends on the cyclic motion of a surface
of emission. In contrast, this and the following chapter describes display unit
architectures in which no reliance is placed on mechanical motion, the image
space being composed of a static-volume of material (or arrangement of materials)
within which voxels may be produced. For the purposes of this book, display
units of this type are classified as static-volume systems (see Section 3.1). The
elementary display introduced in Section 1.3 provides a simple illustration of a
system that employs a static-volume image space. Various other displays of this
type are described in both this chapter and the next.

The use of mechanical motion for image space creation may have consequences for a display unit's reliability, overall power consumption, portability,
and perhaps even give rise to unacceptable levels of noise. By removing the need
for mechanical components, static-volume systems offer a solution to these potential problems. However, a great deal of research remains to be done in the area
of static-volume system implementation. Consequently, display units employing
swept-volume architectures provide the most likely solution to the implementation of volumetric display systems in the short term. However, the advantages
offered by static-volume systems make it highly probable that they will ultimately

---

* Wilfred Owen, *The Sentry*, in *Collected Poems of Wilfred Owen*, WW Norton and Co., NY, 1965.

replace swept-volume displays and enable the production of high-quality images within a homogeneous and isotropic image space.

We begin this chapter with the presentation of a classification scheme for static-volume systems based on the nature of their three subsystems. Subsequently, we examine several types of display unit architecture. One particular approach employs the stepwise excitation of fluorescence. As with beam-addressed swept-volume display units employing the rotational motion of a surface of emission, this type of display may have associated distortional and voxel placement dead zones (see Section 4.2). To highlight various design considerations, we therefore provide in Section 6.6 an analysis of the distortional dead zone that is applicable to display units in which voxel activation occurs within the region of intersection of two directed beams.

It is interesting to note that the first attempt to develop a static-volume volumetric display unit appears to have been made in 1912 (see Section 7.2). It would seem that this predates the first swept-volume display implementation by approximately 30 years.* It is recommended that this chapter be perused in conjunction with Chapter 7 where we provide details on a number of static-volume display unit implementations.

6.2   CLASSIFICATION OF STATIC-VOLUME DISPLAY UNITS

In Section 3.3 we distinguished between swept and static-volume displays and introduced a general classification scheme for the former. This was based on the operation of the three display unit subsystems (dealing with image space creation, voxel generation, and voxel activation). It is appropriate to classify static-volume displays in a similar manner, and such a classification is presented in Figure 6.1.

Our classification of static-volume display units begins with an identification of the nature of the image space. In this respect an image space may be formed from one (or a uniformly dispersed mixture of more than one) material. The uniformity of such an image space will ensure the isotropic transmission of light — this is of vital importance. We will classify an image space of this type as *homogeneous*. By definition, such an image space must comprise components of a single density and refractive index. Alternatively, an image space may comprise an arrangement of materials assembled into a 3-D structure. Such an image space will be classified as *nonhomogeneous*.[†]

The second level of classification takes account of the voxel generation technique employed within the display unit. As with swept-volume systems, we are able to distinguish between display units that employ active and passive voxel generation processes. The elementary display unit introduced in Section 1.3

---

* Should any reader be aware of any earlier development work the authors would be very pleased to learn of it.
[†] The terms *homogeneous* and *nonhomogeneous* are also widely used in this book when referring to voxel form and voxel positioning within an image space. The meaning of these terms should therefore be interpreted from the context in which they are used.

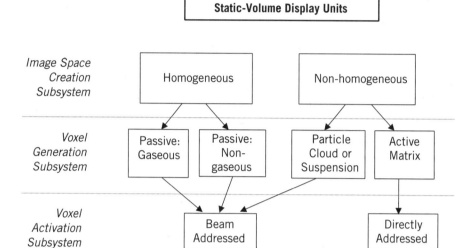

FIGURE 6.1   Classification scheme for static-volume display units based on the techniques employed within the three subsystems.

provides an example of the former (and clearly represents a nonhomogeneous image space). Possible voxel locations are specified by the positioning of a matrix of voxel generation centers during the assembly of the image space. The voxel activation process is achieved by the application of a stimulus to each element, and as a consequence we refer to such a configuration as a *directly addressed active matrix*. The elements may be embedded within a transparent solid material or may be supported by a framework (see, e.g., Section 7.4.1). As with directly addressed swept-volume displays, the activation mechanism may accommodate some form of multiplexing arrangement.

Alternatively, the image space may comprise a solid, liquid, gas, or particle cloud (for an example of the latter, see Section 7.6.2). The generation of voxels within such a system requires a physical process to occur in the material as a consequence of some external activation stimuli, for example, within the region of intersection of two nonvisible energy beams. This process results in the conversion of the incident energy into a visible voxel and may be achieved through, for example, the stepwise excitation of fluorescence. Other physical processes that have been proposed for the generation of voxels in beam-intersection static-volume display units are discussed in Chapter 7. Display unit architectures of this type will be referred to as *beam-addressed passive media systems*.

The use of a passive or active voxel generation technique determines the nature of the voxel activation subsystem. As with swept-volume systems, inclusion of the voxel activation technique within the classification scheme is not strictly necessary. In the classification scheme illustrated in Figure 6.1, we distinguish between gaseous and nongaseous voxel generation processes. This distinction is important, as it provides an indication of refraction effects that may occur as light crosses the image space and its boundary and also affects the physical density of the image space. In the case of display units in which the image space is formed from a solid material, considerable refraction may take place at the interface between the image space and the air. In such cases the shape of the image space is of considerable importance in determining its apparent dimensions [these are likely to differ from its physical size (see Section 10.4)]. Finally, we include in the classification scheme display units employing a cloud or suspension of particles. We consider that such image spaces are nonhomogeneous in their density. In some cases, the presence of the particles may cause the scattering of light as it passes through the image space.

6.3 VOXEL VISIBILITY LIFETIME

Before continuing with a discussion on the various types of static-volume display unit, we turn our attention to the issue of residual voxel visibility. Following removal of the voxel activation stimulus, a voxel location will, in the case of most voxel generation techniques, gradually return to its inactive (nonemissive and ideally nonvisible) state. We define the transition time required for a fully activated element to return to a state in which it is no longer visibly apparent to an observer* as the *voxel visibility lifetime*.

In the case of swept-volume display units, it is highly desirable to ensure that the voxel visibility lifetime is minimized (see Section 3.4). This requirement arises as a consequence of the motion of the voxel generation mechanism(s). When this is combined with any gradual and perceptible diminution of a voxel's intensity (following the removal of the activation stimulus), visible image trails are likely to ensue (see Plate vii). Since static-volume display units do not rely on such motion, image trails are not caused in this way. A similar effect may, however, occur in static-volume systems that are used to depict rapidly changing or animated image sequences. Unless the voxel visibility lifetime is sufficiently short, any afterglow of voxels that are changing from an activated to a deactivated state may reduce image contrast, or in the worst case, result in trails being left behind moving image components. An observable voxel visibility lifetime is therefore undesirable for both swept and static-volume display units.

6.4 ACTIVE MATRIX DISPLAY UNITS

Static-volume display units employing a 3-D array of voxel generation centers are classified as directly addressed active matrix systems. In this case there is a direct correspondence between each element within the array and each permitted voxel location. As a consequence, the voxel location capacity of the display unit will

---

* By this definition, the voxel visibility is affected by the ambient lighting conditions.

equate to the number of elements located within the image space. The elementary display unit introduced in Section 1.3 provides an example of a system of this type, and further examples are given in Chapter 7.

In general, there are three major difficulties associated with the implementation of a display unit of this type. These relate to the large number of elements that must be positioned within the image space, the considerable number of interconnections between these elements, and the graphics engine responsible for the processing and throughput of image data and the need to maintain a homogeneous image space.

This problem is best illustrated by means of a simple example. Let us consider an image space that takes the form of a cube with sides of length 10 cm (a very small image space). Should we wish the available voxel locations to take the form of a regular 3-D matrix having a nearest-neighbor voxel spacing of 1 mm, the number of elements required would be approximately 1 million. Clearly, the fabrication of such a large number of elements within the volume would represent an enormous undertaking and could only be achieved by automated assembly techniques and the use of a low-cost technology for the fabrication of each voxel generation center.

In accordance with the terminology introduced in Section 2.4, the voxel activation mechanism is responsible for the stimulation of the voxel generation subsystem so as to produce visible voxel(s). In the case of an active matrix display unit, the voxel activation subsystem takes the form of a number of pathways linking the output ports of the graphics engine to the individual elements within the matrix. Should there exist a one-to-one correspondence between the output channels of the graphics engine and the number of elements within the array, the display unit will exhibit a maximum degree of parallelism. In this case, since the voxel activation capacity may equal the voxel location capacity, the display unit may, in principle, exhibit a 100% fill factor.

The maximum rate at which image data must flow through each of the output ports of the graphics engine is determined by the image refresh frequency and the number of voxel generation centers that are connected to each port. To make maximum use of the bandwidth of each graphics engine output port and to minimize the number of ports, a row-and-column addressing technique may, for example, be employed (in a manner similar to that used on conventional LCD display panels).

Consider a cubic image space containing N elements within which monochrome images are to be generated. We can visualize this volume as consisting of $N^{1/3}$ 2-D planar arrays each containing $N^{2/3}$ elements. Each of these arrays may be addressed by $2N^{1/3}$ connections, as illustrated in Figure 6.2. We may now interconnect the arrays as shown in Figure 6.3 so that they form (from the point of view of their interconnection) a single square array that may be addressed by $2N^{1/2}$ connections.* For example, in the case of a display unit comprising 1 million elements, only 2000 connections are needed.

---

* This applies to configurations in which $\sqrt{N}$ is an integer.

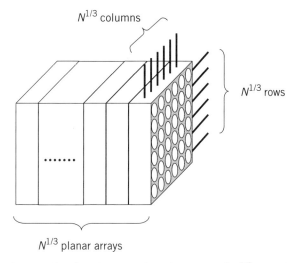

FIGURE 6.2 3-D matrix of N elements viewed as a set of $N^{1/3}$ 2-D planes, each plane being addressed by $2N^{1/3}$ data connections.

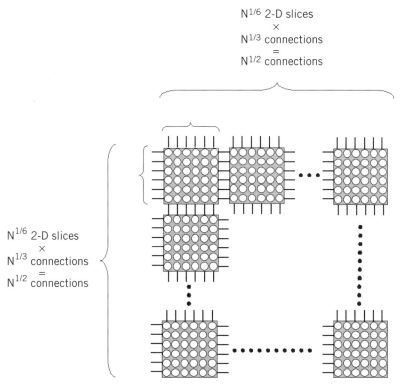

FIGURE 6.3 Interconnection of the 2-D element arrays illustrated in Figure 6.2, so that a total of $2N^{1/2}$ connections into the image space are required.

6.5 BEAM INTERSECTION APPROACH

A number of physical mechanisms by which a visible voxel may be generated at the intersection of two or more energy beams have been investigated (see, e.g., Section 7.5.1). One of the most popular techniques has employed the stepwise up-conversion of infrared radiation into visible light. Other proposals make use of electron beams directed into a volume of gas or photochromic changes in the optical properties of the image space medium (see Section 7.5.5). In this section we discuss the stepwise excitation of fluorescence and present a simple mathematical model that encapsulates some of the important aspects of this method. In Section 6.5.2 we consider the optimal relative timing of two radiation pulses applied so as to activate a voxel within the region of their intersection, and in Section 6.5.3 we discuss issues relating to voxel intensity.

6.5.1 Three-State Model for a Stepwise Excitation Process

A mechanism giving rise to a localized and visible optical change in an image space material at the intersection of two radiation beams may be thought of in terms of an idealized model of three distinct states in a medium (Figure 6.4). In some cases, such as the stepwise excitation of fluorescence, the mechanism is a quantum-mechanical process, and these three states correspond to quantized electronic energy levels. This process occurs in the molecules or atoms of certain gases under appropriate thermodynamic conditions (Fuchtbauer 1920,

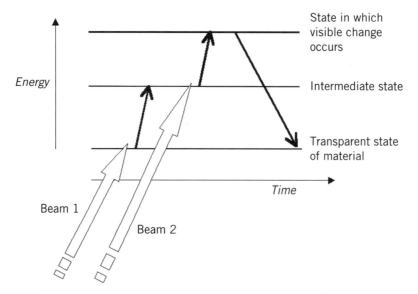

FIGURE 6.4 General three-state process for the conversion of the energy from two independent activation beams into visible radiation.

Wood 1924, Zito and Schraeder 1963, Kim et al 1996) or in rare earth ions doped into a transparent crystalline or glass medium. (Lewis et al 1971, Downing et al 1996). Each ion, atom, or molecule contributing to fluorescence output will be referred to as a *fluorescent center*, after Lewis et al. (1971).

In the stepwise excitation process, two beams of nonvisible electromagnetic radiation* excite the quantum system to a visible fluorescent energy state, as illustrated in Figure 6.5. Beam 1 has a frequency resonant with the transition $|1\rangle \rightarrow |2\rangle$ and excites the quantum system from the initial (usually the ground) state $|1\rangle$ to an intermediate excited state $|2\rangle$. The second beam (beam 2) is resonant with the transition $|2\rangle \rightarrow |3\rangle$ and excites the system to a higher state $|3\rangle$. The system subsequently returns by radiative decay to its ground state, $|3\rangle \rightarrow |1\rangle$, emitting visible light. The energy-level configuration shown in Figure 6.5 in which the energy difference $E_{fl} = h\nu$ between states $|3\rangle$ and $|1\rangle$ is greater than each of the two activation transitions, suggests that each of the activation beams has a frequency in the infrared (IR) region of the spectrum. Alternatively, for the energy-level configuration illustrated in Figure 6.6, ultraviolet radiation may be used for voxel activation. This relies on the presence of a fourth level, and the decay of the system proceeds via a radiative transition $|3\rangle \rightarrow |4\rangle$, and a nonvisible[†] transition $|4\rangle \rightarrow |1\rangle$ (or vice versa). If the nonvisible transition is rapid compared with the visible transition, the lifetime of the nonvisible transition may be neglected and the behavior modeled by the three-level system described below. In this case the energy state $|4\rangle$ is amalgamated with state $|1\rangle$.

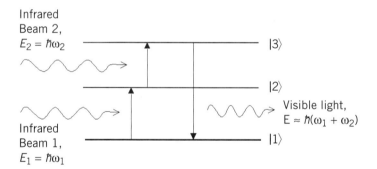

FIGURE 6.5 Three-level system encapsulating the stepwise excitation of fluorescence. The energy states $|1\rangle$, $|2\rangle$, and $|3\rangle$ are discrete states (typically, electronic) of the quantum system. Exciting transitions occur in the presence of radiation resonant with each transition. In the scheme illustrated here, each of the pumping radiation beams has an infrared frequency that is up-converted by this process into visible radiation.

---

* Laser radiation is generally employed for this purpose.

[†] This may be a nonradiative decay, which corresponds to a loss of energy from the fluorescent center to the host media that does not involve the emission of a photon. Often, this involves coupling with phonon resonances within the host material, and for this reason, special low-phonon glasses, which increase the probability of the desired radiative decay from state $|3\rangle$, have recently found application in displays based on this principle (Downing et al. 1996).

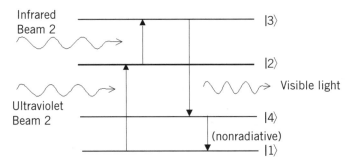

FIGURE 6.6   Four-level process in which one of the pumping beams is in the ultraviolet region of the spectrum. By amalgamating the energy states $|1\rangle$ and $|4\rangle$ (describing the sequential transition by one lifetime parameter) this situation may be described by the three-level system developed in Section 6.5.1

In reality, since more than three or four energy states are generally involved (see Figure 6.6), the energy-level scheme is more complicated than that assumed in this simple model (Zito 1963, Lewis et al. 1971, Downing et al. 1996, Soltan and Lasher 1996). However, the simple three-level scheme presented here serves to encapsulate the important characteristics of the stepwise excitation process.*

Rate Equations for the Three-Level System

When used in the context of a volumetric display unit, it is likely that the temporal profile of each beam will consist of a rapid sequence of pulses. Each pair of pulses will activate the fluorescent output for a particular voxel at the required location within the image space. A system of rate equations modeling the temporal behavior of a three-level quantum system may be constructed by associating a parameter with each of the transitions as illustrated in Figure 6.7. This allows for both absorption and stimulated emission due to each activation beam, and for radiative and nonradiative decay from each excited state. In this last respect, we include in the equations only a single relaxation transition from each of states $|2\rangle$ and $|3\rangle$. These are governed by a characteristic decay lifetime which is the average of all decays (thermal, spontaneous, radiative, and nonradiative) to the ground state. This simplifies the resulting rate equations and is permissible since the mechanism governing the $|2\rangle \rightarrow |1\rangle$ transition does not directly affect the population of $|3\rangle$, and hence the fluorescent output, for a single pulse combination. For state $|3\rangle$, the radiative decays from $|3\rangle \rightarrow |1\rangle$ that give rise to the fluorescence may be counted by considering the relevant fraction of all decays from $|3\rangle$.

---

* The system often occupies the other states transiently, and their effect can be incorporated into the transition parameters for one of the states in the three-level model.

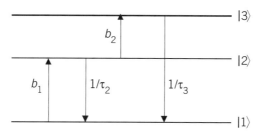

FIGURE 6.7 Transition parameters employed in the three-level stepwise excitation system model. The parameters b_1 and b_2 are, for each of the pumped transitions, the product of the pulse intensity and absorption cross section, divided by the photon energy. The lifetimes τ_2 and τ_3 govern the spontaneous decay of levels $|2\rangle$ and $|3\rangle$. Stimulated emission is accounted for in the rate equations via b_1 and b_2.

The absorption rates are governed by the population differences between the two states concerned, and by coefficients of the form

$$b_i = \frac{\sigma_i I_i}{\Delta E_i} \quad [\text{s}^{-1}]$$

where σ_i is the resonance absorption coefficient for the transition $|i\rangle \rightarrow |i+1\rangle$, I_i is the peak intensity of the activation radiation resonant with this transition* and ΔE_i is the energy difference between the two states $|i\rangle$ and $|i+1\rangle$. Decay from each of states $|2\rangle$ and $|3\rangle$ is governed by the time constants τ_2 and τ_3, which are assumed to represent an average of all decay transitions, as mentioned above. These can, where appropriate, include nonradiative transitions involving additional states. The fraction of fluorescent centers excited to level $|3\rangle$ which subsequently produce a fluorescent radiative decay (thus contributing to the visible voxel) is specified by the *fluorescence efficiency*, $0 \leq \eta \leq 1$.

If the population of each energy state is denoted $N_i (i = 1, 2, 3)$, the rate equations governing the model described above are

$$\frac{dN_1}{dt} = \frac{1}{\tau_2} N_2 + \frac{1}{\tau_3} N_3 + p_1(t) b_1 [N_2 - N_1]$$

$$\frac{dN_2}{dt} = -\frac{1}{\tau_2} N_2 - p_1(t) b_1 [N_2 - N_1] + p_2(t) b_2 [N_3 - N_2]$$

$$\frac{dN_3}{dt} = -\frac{1}{\tau_3} N_3 - p_2(t) b_2 [N_3 - N_2]$$

where the functions $p_j(t)$ (normalized to unity) represent the temporal profile of the activation pulses $(j = 1, 2)$, $N_i = N_i(t)$ and $N_1 + N_2 + N_3 = N$. The

---

* The activation radiation intensity and beam power are related by $I_j = P_j/a$, where a is the cross-sectional area of the beam at the point of intersection.

differences $(N_2 - N_1)$ and so on, allow for both stimulated excitation and stimulated emission in the presence of activation radiation. If the radiation is sufficiently strong that $N_2 > N_1$, for example, deexcitation of this transition is stimulated until $N_2 < N_1$. These equations may be expressed more concisely as the matrix equation

$$\frac{d\mathbf{N}}{dt} = \mathbf{A}(t)\mathbf{N}(t) \tag{6.1}$$

where $\mathbf{N}(t) = [N_1(t), N_2(t), N_3(t)]^T$, and the time-dependent coefficient matrix is

$$\mathbf{A}(t) = \begin{bmatrix} -p_1(t)b_1 & \dfrac{1}{\tau_2} + p_1(t)b_1 & \dfrac{1}{\tau_3} \\[2ex] p_1(t)b_1 & -\dfrac{1}{\tau_2} - p_1(t)b_1 - p_2(t)b_2 & p_2(t)b_2 \\[2ex] 0 & p_2(t)b_2 & -\dfrac{1}{\tau_3} - p_2(t)b_2 \end{bmatrix}$$

To maximize the voxel activation capacity, it is desirable to activate the voxels as rapidly as possible. The length of time for which the beams need to dwell at each voxel location to excite sufficient intensity is determined by the transition parameters outlined above, the intensity of the pumping beams, and the desired voxel brightness.* Therefore, it is of interest to determine the relative timing of the two excitation pulses that for a given pulse duration maximizes the fluorescent output, and hence the voxel brightness. This is considered in the next section.

6.5.2 Optimal Pulse Timing

We define the symbols t_1 and t_2 as representing the times at which the radiation of beams 1 and 2 respectively, arrive at the voxel location. If each beam pulse [with a profile described by $p(t)$] is assumed to have a characteristic duration T_P, we can define the normalized relative timing between the two pulses as

$$\Delta = \frac{t_2 - t_1}{T_p} \tag{6.2}$$

This is illustrated in Figure 6.8 for rectangular pulse profiles. We seek to determine the optimum value of this quantity, which we denote Δ_{opt}. This is defined as the value of Δ that maximizes the total population S_3 excited to the fluorescent state over the two-pulse sequence

$$S_3 = \int_{t_2}^{t_2 + T_p} [b_2 p_2(t)(N_2(t) - N_3(t))] \, dt \tag{6.3}$$

---

* The rate at which the beams can be deflected from one voxel position to the next will also affect the voxel activation capacity.

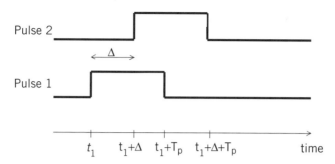

FIGURE 6.8 Schematic illustration of the relative timing (Δ) of pulses 1 and 2. Both of the pulses are assumed to have a rectangular temporal profile.

Since the number of fluorescent decays is directly proportional to S_3 via η, the voxel brightness is also maximized by Δ_{opt}. The integral is over the duration of the second pulse, with the population $N_2(t)$ of the intermediate state implicitly including the effect of the first pulse. As in Eq. (6.1), a population *difference* appears here to account for stimulated emission effects, which tend to equalize the populations if N_3 approaches N_2.

If the pulse intensities do not saturate the transitions, we can assume that the population of each excited state is small compared to that of the state immediately below. That is, $N \sim N_1 \gg N_2 \gg N_3$. If the intermediate state lifetime is also long compared with the pulse duration T_p, then $\Delta_{opt} \approx 1$ (Verber 1977). This can be appreciated as follows. During the application of pulse 1 alone, the population in state $|2\rangle$ increases linearly and thus reaches its maximum at the end of the pulse. Since $\tau_2 \gg T_p$, this population will decay only slowly back to the ground state. The number of centers excited by pulse 2 to the fluorescent state is, by Eq. (6.3) in the nonsaturating limit, proportional to N_2. This will therefore be maximized by the application of pulse 2 at the end of pulse 1, when the population of the intermediate state is at its highest.

Much recent work in static-volume displays employing this process has involved rare earth ions doped into low-phonon heavy metal fluoride glasses (see Section 7.7.2). In these systems, the excited-state lifetimes of relevant electronic energy levels in, for example, erbium ions is about 10^{-2} s (Downing et al. 1996, Honda et al. 1998). For pulse durations of the order of microseconds, the condition $\tau_2 \gg T_p$ is valid.

General Numerical Solution

The value of the optimal relative pulse timing may be found via a numerical solution of Eqs. (6.1) and (6.3) for a range of parameter values (Schwarz and Blundell 1993). It depends on both the temporal pulse intensity profile, the lifetime τ_2 of the intermediate state, and the radiation pulse energy $I_j T_p a$, where a is the cross-sectional area of the beam. In the limit of nonsaturating pumping radiation, the optimal relative timing (Δ_{opt}), depends strongly on the ratio τ_2/T_p (i.e., the ratio of the intermediate-state lifetime to the pulse duration). As discussed

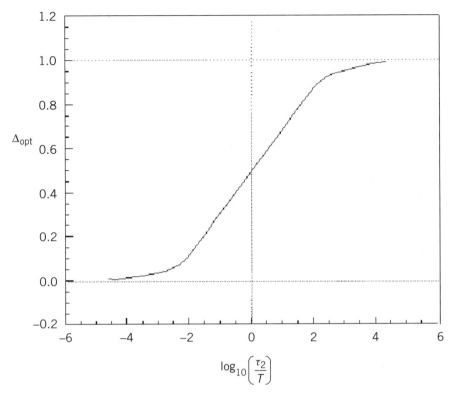

FIGURE 6.9 Dependence of Δ_{opt} on the intermediate-state lifetime τ_2 for nonsaturating activation radiation and rectangular temporal pulse profiles.

above, if $\tau_2 \gg T_p$, then $\Delta_{opt} \sim 1$. Voxel brightness is therefore maximized if beam 2 is applied immediately following beam 1. However, the optimum relative timing decreases with τ_2 until $\Delta_{opt} \sim 0$ for $\tau_2 \ll T_p$ (Figure 6.9).* In this case, fluorescent output is maximized by having both pulses impinge on the material simultaneously. The behavior is modified, however, if one or both of the pulses saturates the transition. The resultant stimulated emission in the presence of the activation radiation decreases the value of Δ_{opt} relative to the nonsaturating case. The intensity of beam 1 in particular then has a strong effect on Δ_{opt}. If both beams saturate their respective transitions, $\Delta_{opt} \approx 0$, independent of τ_2/T_p.

6.5.3 Voxel Brightness Considerations

In Section 6.5.2 we considered the general behavior of the three-level quantum system in the presence of transient activation radiation. In practice, the nonsaturating $\tau_2 \gg T_p$ approximation is often valid (Downing et al. 1996), and we now

---

* In this limit, the rate equation system describes the two-photon absorption process, as discussed in Section 7.4.4.

consider voxel brightness issues in relation to this case. For a given material, the total energy (E_V) output from a voxel due to a single two-pulse combination is proportional to the total number (S_3) of fluorescent centers excited to $|3\rangle$, the fluorescent efficiency (η), and the energy (E_{fl}) of each fluorescent photon. The energy output per voxel is given by

$$E_V = \eta E_{fl} S_3 \qquad [\text{J}] \qquad (6.4)$$

In the nonsaturating limit, $N \approx N_1 \gg N_2 \gg N_3$ and Eq. (6.3) takes the form

$$S_3 = b_1 b_2 N T_p^2 \qquad (6.5)$$

If the number of fluorescent centers (N) comprising a voxel is reexpressed as DV, where D is the density of fluorescent centers in the medium and V is the volume occupied by the voxel, Eq. (6.4) may be written as

$$E_v = \eta DV \frac{E_{fl}}{\Delta E_1 \Delta E_2} \sigma_1 \sigma_2 I_1 I_2 T_p^2 \qquad [\text{J}] \qquad (6.6)$$

So as to evaluate the power emitted from each voxel, it is necessary to include a factor that takes into account the time scale over which the visible light is emitted. If τ_3 is less than the time (T_p) over which the pulses are applied, T_p may be used to indicate the time scale of fluorescence; otherwise, the fluorescence lifetime τ_3 is an appropriate parameter. Assuming the latter case, and writing the beam intensities (I_j) explicitly in terms of the beam power and cross-sectional area (a) at the voxel, $I_j = P_j/a$, the power (P_v) output from a single voxel due to a pair of activation pulses may be expressed as[*]

$$P_v = \eta DV \frac{E_{fl}}{\Delta E_1 \Delta E_2} \sigma_1 \sigma_2 \frac{P_1 P_2}{a^2} \frac{T_p^2}{\tau_3} \qquad [\text{W}] \qquad (6.7)$$

The time-averaged output power per voxel is then determined by the duty cycle τ_3/τ_r, where τ_r represents the image space refresh period, and is equal to $P_v(\tau_3/\tau_r)$.

As a numerical example, we now consider parameters appropriate for a display unit employing stepwise excitation within an image space comprising an ion-doped glass[†] (Downing et al. 1996, Honda et al. 1998). Typical parameter values for such a system are given in the following partial specification:

- Absorption cross sections for each transition: $\sigma_1 \sim \sigma_2 \sim 10^{-24}$ m^2
- Excited-state spontaneous decay lifetimes: $\tau_2 \sim \tau_3 \sim 10^{-2}$ s

---

[*] The perceived voxel intensity also depends on the response of the human eye as a function of wavelength. This is described by the normalized photopic curve, which peaks at $\lambda \approx 560$ nm.

[†] Discussed further in Section 7.7.2.

- Density of the dopant ions (fluorescent centers): $D \sim 10^{26}$ m^{-3}
- Energy of fluorescent and activation photons: $E_{fl} \sim \Delta E_1 \sim \Delta E_1 \sim 10^{-19}$ J
- Voxel volume: $V \sim 10^{-9}$ m^3
- Activation beam power: $P_1 \sim P_2 \sim 3$ W
- Cross-sectional area of beams at focus (voxel location): $a \sim 10^{-6}$ m^2
- Refresh frequency: $f_r = 25$ Hz

We will also assume rectangular pulse profiles, nonsaturating radiation, and a fluorescent efficiency $\eta \sim 0.5$. The pulse duration (equal to the voxel activation time T_V) is related to the total display refresh period, the voxel activation capacity, and the degree of parallelism in the voxel addressing [recall Eqs. (2.1) and (2.2)]. For simplicity, let us assume that 10^4 voxels are required and that activation is entirely sequential. The time available to generate each voxel is thus 4 µs; which is approximately equal to T_V. This will equal T_p if we assume that the system adjustment time is comparatively short.

With these parameter values, Eq. (6.7) predicts an output power per voxel, per single activation pulse combination, of 0.02 µW. The time-averaged output power is therefore 0.05 µW per voxel [compared to the ~ 0.1 µW emitted per pixel on CRT and LCD displays (Lewis et al. 1971, Downing et al. 1996)]. An increase in the power of the activation beams could be employed to increase voxel intensity, but this may lead to saturation. Alternatively, an increase in voxel intensity may be achieved by employing a longer voxel activation time. Unfortunately, this would affect the voxel activation capacity. An increased ion concentration could also enhance image intensity.

6.6   DISTORTIONAL DEAD ZONES; OPTIMAL BEAM SOURCE POSITIONING

In Chapter 4 we discussed various types of dead zones associated with swept-volume display unit architectures. In the case of static-volume display systems in which voxels are created at the intersection of two beams, some analogous dead zones may also be present. For example, voxel elongation arises due to the varying angle of intersection of the beams; for angles tending toward 0° (Figure 6.10a) and toward 180° (Figure 6.10b) the voxels become increasingly elongated. This gives rise to the possible existence of distortional dead zones. It is apparent that the ideal angle of intersection is 90°. This type of display unit is also subject to voxel placement dead zone. In this case, the accuracy with which voxels may be positioned within an image space and the separation of adjacent (nearest-neighbor) voxels will vary throughout the image space.

In this section we analyze the distortional dead zone and derive equations indicating permissible beam source positions relative to the image space for which the voxel elongation does not exceed a specified value. This analysis is valid for all display mechanisms that comprise two intersecting beams of circular cross section. For beams with a Gaussian intensity profile, the beam boundary

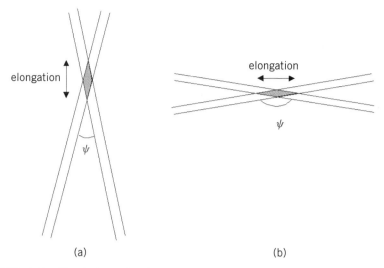

(a)   (b)

FIGURE 6.10 Voxel elongation in stepwise excitation systems arises as the angle of beam intersection approaches (*a*) 0° or (*b*) 180°.

may be considered as that within which the intensity is above a chosen threshold. The subsequent analysis is intended to highlight some of the issues that should be taken into account when designing this type of volumetric system. Other dead zones that may be associated with this type of display unit embodiment are discussed in Section 6.7.

6.6.1 Geometry

The geometry of the display unit configuration employed in this analysis is illustrated in Figure 6.11. The positions of the beam sources, and of voxel locations within the image space are described in spherical polar coordinates (r, β, ϕ). The origin of this coordinate system is assumed to be located at the center of the image space and, for simplicity, we will initially consider that the image space is spherical in form. Without loss of generality, one of the beam sources (beam 1) is considered to lie along the axis $\beta = 0$, $\phi = 0$. The position of the other beam source (beam 2) may thus be specified by its angle $\beta = \beta_{12}$ in the $\phi = 0$ plane (due to the spherical symmetry). It is also assumed that each beam source is positioned at the same distance (D) from the origin,* and that the image space has a radius R (where $R \leq D$).

From Figure 6.11 it should be apparent that the distance between beam source 1 and a location (r, β, ϕ) in the image space may be expressed as

$$d_1 = \sqrt{r^2 + D^2 - 2rD \cos \phi \cos \beta} \tag{6.8}$$

---

* In this context we are, more precisely, concerned with the location of the beam deflection apparatus.

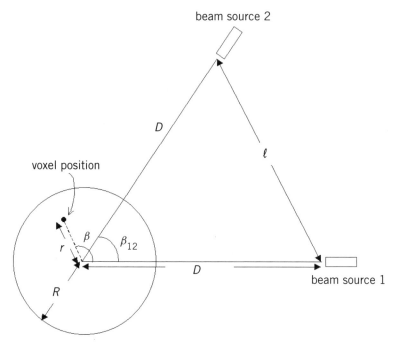

FIGURE 6.11 Geometry of the image space and beam sources employed for the dead zone analysis in Section 6.6.

and the distance between beam source 2 and the same point by

$$d_2 = \sqrt{r^2 + D^2 - 2rD\cos\phi\cos(\beta - \beta_{12})} \tag{6.9}$$

The two beam sources are assumed to be separated by a distance l which is given by

$$l = 2D\sin\left[\frac{\beta_{12}}{2}\right] \tag{6.10}$$

Therefore, the angle ψ between the beams as they intersect at a point (r, β, ϕ) is obtained from the cosine rule as

$$\psi = \arccos\left[\frac{d_1^2 + d_2^2 - l^2}{2d_1d_2}\right] \tag{6.11}$$

The angle ψ has the range $0° \leq \psi \leq 180°$.

Voxel Elongation Factor
As discussed above, voxels created at the overlap of two radiation beams become increasingly distorted (elongated) as the angle ψ between the beams deviates from

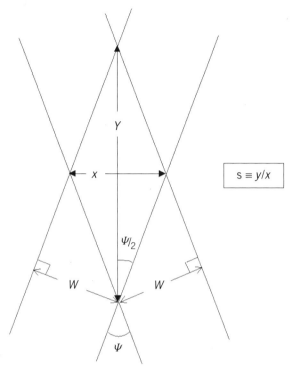

FIGURE 6.12   Cross section through the two overlapping beams in the horizontal plane. The elongation factor is defined as $s = y/x$, where y and x are perpendicular measures of the voxel size in this plane. The beam thickness t cancels out of the expression, which is obtained as $s = \cos(\frac{1}{2}\psi)/\sin(\frac{1}{2}\psi)$.

90° (Figure 6.10). To quantify this effect, we can define an elongation factor (s) to be the ratio between the length and width of the voxel in the plane of the two beams,* as illustrated in Figure 6.12. From this geometrical construction we derive

$$s = \frac{\cos(\frac{1}{2}\psi)}{\sin(\frac{1}{2}\psi)} \tag{6.12}$$

From this we may obtain

$$\psi = 2 \arccos \sqrt{\frac{s^2}{1 + s^2}} \tag{6.13}$$

---

* We assume here that the beams have well-defined boundaries, are circular in cross section with a diameter W, and that the beams completely overlap. The effect of a partial beam overlap is discussed in Section 6.7.

We can now define the greatest elongation we are willing to tolerate within the image space. If this maximum permissible elongation is denoted by s_{max}, the elongation factor for all voxels within the image space must lie in the range

$$\frac{1}{s_{max}} \leq s \leq s_{max} \tag{6.14}$$

This gives, via Eq. (6.13), a range of permissible values of the beam intersection angle ψ, namely:

$$\psi_{min} \leq \psi \leq \psi_{max} \tag{6.15}$$

where

$$\psi_{max} = 2 \arccos \sqrt{\frac{1}{1 + s_{max}^2}} \tag{6.16}$$

$$\psi_{min} = 2 \arccos \sqrt{\frac{s_{max}^2}{1 + s_{max}^2}} \tag{6.17}$$

These ideas may be illustrated by the examples below:

Example 1   If $s_{max} = 2$, the elongation factor of any voxel in the image space must lie in the range $0.5 \leq s \leq 2$, and hence the beam intersection angle must lie in the range $53.1° \leq \psi \leq 126.9°$, a range of about $74°$.

Example 2   If $s_{max} = 1.5$, the range of the elongation factor is $0.67 \leq s \leq 1.5$, and hence that of the beam intersection angle is $67.4° \leq \psi \leq 112.6°$, a range of about $46°$.

For the present time we consider only the horizontal plane ($\phi = 0°$) through the image space containing the two beam sources and therefore neglect its three dimensionality. The nature of the dead zone for $\phi \neq 0°$ is briefly discussed in Section 6.6.3. We may express the beam source distance (D) from the center of the image space as a multiple (α) of the radius (R) of the image space:

$$D = \alpha R \tag{6.18}$$

We assume that each beam source is located an equal distance from the center of the image space. Consequently, the maximum and minimum beam intersection angles (within the horizontal plane) occur at the near and far surfaces of the image space volume and at an angle midway between the two beam sources (Figure 6.13): that is, at $\beta = \frac{1}{2}\beta_{12}$, with the maximum angle occurring at $r = R$ and the minimum at $r = -R$.

We are therefore particularly interested in the value of the beam intersection angle ψ at these two extreme locations. In order that the voxel elongation be less

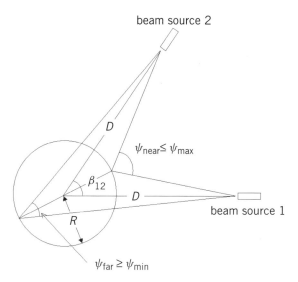

beam source 2

$\psi_{near} \leq \psi_{max}$

β_{12}

D

beam source 1

R

$\psi_{far} \geq \psi_{min}$

FIGURE 6.13   Since both beam sources are assumed to be located at the same distance from the center of the image space, the maximum beam intersection angle occurs at the near edge at $\beta = \frac{1}{2}\beta_{12}$. At this point, $\psi \leq \psi_{max}$ must be satisfied to avoid unacceptable voxel elongation. Similarly, the minimum beam intersection angle occurs at the far edge. In order that the criterion for acceptable voxel elongation be not exceeded, $\psi \geq \psi_{min}$ must also hold at this point.

than s_{max} throughout the horizontal plane of the image space, at the far surface $(\frac{1}{2}\beta_{12}, -R)$ the value of ψ must be greater than ψ_{min} [given by Eq. (6.16)] *and* at the near surface $(\frac{1}{2}\beta_{12}, -R)$ the value of ψ must be less than ψ_{max} [given by Eq. (6.17)]. Due to the reflection symmetry of this geometry about $\beta = \frac{1}{2}\beta_{12}$ (see Figure 6.13),

$$d_1 = d_2 \equiv d = R\sqrt{1 + \alpha^2 - 2\alpha \cos(\tfrac{1}{2}\beta_{12})} \tag{6.19}$$

for the near surface $(r = +R)$, and

$$d_1 = d_2 \equiv d = R\sqrt{1 + \alpha^2 + 2\alpha \cos(\tfrac{1}{2}\beta_{12})} \tag{6.20}$$

for the far surface $(r = -R)$. Since $d_1 = d_2 \equiv d$ in each case, Eq. (6.11) can be rewritten as

$$\frac{2d^2 - l^2}{2d^2} = \cos \psi \tag{6.21}$$

We now introduce the abbreviation

$$c = \cos(\tfrac{1}{2}\beta_{12}) \tag{6.22}$$

and so that Eq. (6.10) may be rewritten as

$$l^2 = 4\alpha^2 R^2 (1 - c^2)$$

For example, consider the case where $\beta_{12} = 90°$ and $D = 3R$. With these values $c = 0.71$ and $d = 2.40R$ at the near surface [from Eq. (6.19)] and $d = 3.77R$ at the far surface [from Eq. (6.20)].

By means of Eq. (6.21) it is possible to develop two expressions in terms of α and c. One of these corresponds to an inequality describing the angle between the beams at the near surface of the image space (the "near solution," $\psi \leq \psi_{max}$). Similarly, the other describes the angle between the beams at the far surface of the image space (the "far solution," $\psi \geq \psi_{min}$), where ψ_{min} and ψ_{max} are defined in Eqs. (6.16) and (6.17). If *either* of these inequalities is not satisfied, the voxel elongation exceeds the specified tolerance somewhere within the image space.

We therefore seek the range of values of α (specifying beam distance from center of image space) and β_{12} (the angle between beam sources) for which the distortional dead zone is absent. This then specifies the range of permissible beam source positions for which the chosen voxel elongation criterion is not exceeded. In the following text, for given values of β_{12} we obtain solutions for the permitted values of α.

Far Solution: Equation for $\psi \geq \psi_{min}$

The far solution specifies that the beam intersection angle be greater than or equal to a minimum permissible value, ψ_{min}. This implies that the cosine of the beam incidence angle must be *less* than or equal to $\cos(\psi_{min})$. The far solution thus corresponds to an inequality of the form of Eq. (6.21), namely:

$$\frac{2d^2 - l^2}{2d^2} \leq \cos \psi_{min} \tag{6.23}$$

where d is given by Eq. (6.20). The left-hand side of Eq. (6.23) may be reexpressed, eliminating a common factor of $2R^2$, as

$$\frac{1 - \alpha^2 + 2\alpha c + 2\alpha^2 c^2}{1 + \alpha^2 + 2\alpha c} \tag{6.24}$$

The right-hand side of Eq. (6.23) is, via Eq. (6.17),

$$\cos\left(2 \arccos \sqrt{\frac{s_{max}^2}{1 + s_{max}^2}} \right) \tag{6.25}$$

This may be simplified via the identity $\cos 2\theta = 2\cos^2 \theta - 1$. With the abbreviation

$$A \equiv \cos \psi_{min}$$

the right-hand side of Eq. (6.23) becomes

$$A = \frac{s_{max}^2 - 1}{s_{max}^2 + 1} \tag{6.26}$$

Combining Eqs. (6.24) and (6.26) and rearranging, Eq. (6.23) gives

$$f_{far}(\alpha) \equiv [2c^2 - (1 + A)]\alpha^2 + 2c(1 - A)\alpha + (1 - A) \leq 0 \tag{6.27}$$

a quadratic in both α and c. As discussed above, we solve for α, for a given value of c via Eq. (6.22). The roots of this equation are given by

$$\alpha = \frac{-c(1 - A) \pm \sqrt{(1 - A^2)(1 - c^2)}}{2c^2 - (1 + A)} \tag{6.28}$$

For chosen values of the elongation tolerance s_{max} and the interbcam angle β_{12}, Eq. (6.26) will determine the corresponding value of A. Equation (6.28) will then specify two corresponding values of α, these being the roots of $f_{far}(\alpha) = 0$.

If both these roots are negative, there exist no values of $D = \alpha R$ for which $\psi \geq \psi_{min}$ at the far edge of the image space (physically realizable solutions must be positive). Since this is the position at which the beam intersection angle is minimized, there will thus be a portion of the image space in which voxels will be unacceptably elongated.

If one root of Eq. (6.28) is positive, this provides a minimum value of α (and hence the beam source distance $D = \alpha R$) for which the minimum beam intersection angle, occurring at the far edge of the image space, is greater than the minimum permissible angle as specified by the elongation constraint. The inequality of Eq. (6.23) is therefore satisfied.

Near Solution: Equation for $\psi \leq \psi_{max}$

The near solution implies that the cosine of the beam intersection angle ψ at the near edge of the image space must be greater than or equal to the cosine of the maximum permissible angle, ψ_{max}. Eq. (6.21) therefore takes the form of the inequality

$$\frac{2d^2 - l^2}{2d^2} \geq \cos\psi_{max} \tag{6.29}$$

where d is given by Eq. (6.19). An analysis analogous to that of the far solution above yields the roots

$$\alpha = \frac{c(1 + A) \pm \sqrt{(1 - A^2)(1 - c^2)}}{2c^2 - (1 - A)} \tag{6.30}$$

If the larger of the two roots is greater than the minimum physical value of α given by Eq. (6.36) (below), then it represents a minimum value α_{min} for which the beam intersection angle at the near face of the image space is smaller than the maximum permitted by the elongation constraints.

6.6.2 Beam Source Positions That Avoid Excessive Elongation

The possible beam source positions for which all voxels in the horizontal plane are elongated by less than the chosen s_{max} are specified by the set of values of the variables $(c, \alpha) \Rightarrow (\beta_{12}, D)$, for which *both* Eqs. (6.23) and (6.29) are satisfied. Solutions exist only for a certain range of β_{12}. It may be seen from Figure 6.13 that

$$\psi_{min} \leq \beta_{12} \leq \psi_{max} \tag{6.31}$$

Values of β_{12} outside this range would necessarily result in the violation of either the near or far conditions.

   These values of ψ_{min} and ψ_{max} also correspond to *qualitative* changes in the behavior of Eqs. (6.23) and (6.29), where the turning point of the quadratic changes from a maximum to a minimum, as follows. The second derivative of $f_{far}(\alpha)$ is $2[2c^2 - (1 + A)]$ so the turning point of the quadratic curve $f_{far}(\alpha)$ changes its nature at the value of c for which this is zero (denoted by a preceding asterisk), namely:

$$^*c_{far} = \sqrt{\frac{1 + A}{2}} \tag{6.32}$$

The turning point is a maximum for $c < {}^*c_{far}$ and a minimum for $c > {}^*c_{far}$. The value of β_{12} to which Eq. (6.32) corresponds is obtained via Eqs. (6.22) and (6.26) as

$$\beta_{12} = 2 \arccos \sqrt{\frac{s_{max}^2}{1 + s_{max}^2}} = \psi_{min} \tag{6.33}$$

The second derivative of $f_{near}(\alpha)$ is $2[2c^2 - (1 - A)]$, and so the turning point of the quadratic curve $f_{near}(\alpha)$ changes its nature at the value of c for which this is zero, namely:

$$^*c_{near} = \sqrt{\frac{(1 - A)}{2}} \tag{6.34}$$

The turning point is a maximum for $c < {}^*c_{near}$ and a minimum for $c > {}^*c_{near}$. The value of β_{12} to which Eq. (6.29) corresponds is obtained as

$$\beta_{12} = 2 \arccos \sqrt{\frac{1}{1 + s_{max}^2}} = \psi_{max} \tag{6.35}$$

In practice, the value of α must be greater than 1, as the beam sources must lie outside the image space. However, there is a further constraint on the value of α, namely that it must be greater than the value α_{180} at which the beam intersection

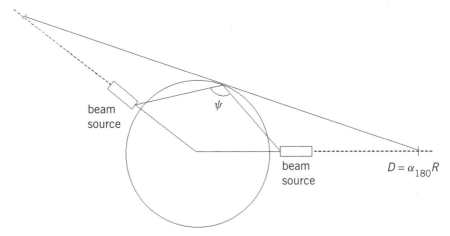

FIGURE 6.14 False solution for the near surface: If $D < \alpha_{180}R$, as illustrated here, this solution does not represent the maximum beam intersection angle attained within the image space. It may be seen that within the image space this angle will become closer, and equal, to 180°.

angle at the near face is 180°. This prevents solutions such as that illustrated in Figure 6.14, where the value of ψ within the image space exceeds the "ψ_{max}" found as the near solution. This minimum value of α is given by

$$\alpha_{180} = \frac{1}{c} = \frac{1}{\cos(\frac{1}{2}\beta_{12})} \tag{6.36}$$

Within the range of angular beam source separations specified by Eq. (6.31), the beam sources must be positioned a certain *minimum* distance from the center of the image space in order that the near and far conditions are satisfied and that the voxel elongation is acceptably small. Each of these two conditions will specify a different minimum value of α; in each case this value is the larger of the two roots given by Eq. (6.28) or Eq. (6.30). The minimum value for which both conditions are satisfied is the greatest of these values:

$$\alpha_{min} = \max\{\alpha_{near}, \alpha_{far}\} \tag{6.37}$$

Table 6.1 illustrates the nature of these solutions. Table 6.2 tabulates the roots obtained for the elongation factor $s_{max} = 2$, with the resulting α_{min} [given by Eq.(6.37)] highlighted in bold. The range of solutions is illustrated diagrammatically in Figure 6.15.

6.6.3 Practical Implications

In the discussion presented above, we dealt only with a circular 2-D cross section of image space, and in Section 6.6.1 we set the spherical polar coordinate ϕ to

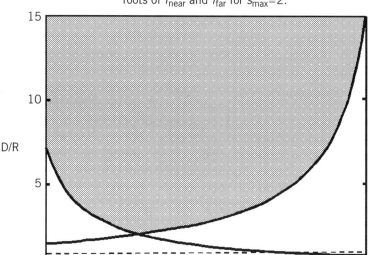

Values of D/R such that voxel elongation dead zones are avoided: roots of f_{near} and f_{far} for $s_{max}=2$.

FIGURE 6.15   Range of permissible beam positions for the example presented in Table 6.2. The highest of the two curves at each point indicates the minimum distance D of the beam sources from the center of the image space and corresponds to the numbers indicated in bold type in Table 6.2. Thus the region in this graph above both curves represents the range of acceptable beam positions. For angles β_{12} between the beam sources less than 60° and greater than 120° it is impossible to avoid voxel elongations greater than the specified tolerance $s_{max} = 2$. The dashed line indicates the image space boundary, $D = R$.

zero. In practice, of course, the image space is three-dimensional. However, if the image space is spherical, the positions of the maximum and minimum beam intersection angles still occur in the plane of the beam sources as analyzed above. This therefore provides the range of beam source locations that avoid excessive voxel elongation in an image space contained within a sphere of radius R. If the image space is cylindrical or cubic in form, the analysis above provides a solution for the spherical volume *circumscribing* this image space.

6.7   IMAGE SPACE CHARACTERISTICS FOR BEAM-ADDRESSED SYSTEMS

In this section we briefly review a number of image space characteristics that are associated with static-volume display units employing two or more deflected beams for voxel activation.

TABLE 6.1   Nature of the Roots of the Quadratic Equations $f_{far}(\alpha) = 0$ and $f_{near}(\alpha) = 0^a$

β_{12}	$c = \cos(\frac{1}{2}\beta_{12})$	$f_{far}(\alpha)$ $(\psi \geq \psi_{min})$	$f_{near}(\alpha)$ $(\psi \leq \psi_{max})$
$\beta_{12} < \psi_{min}$	$c >^* c_{far}$	Both roots < 0 Turning point a minimum	Both roots > 0 Turning point a minimum
$\psi_{min} < \beta_{12}$ $< \psi_{max}$	$^*c_{near} < c <^* c_{far}$	One root > 0 and $> \alpha_{180}$ Turning point a maximum	One root > 0 and $> \alpha_{180}$ Turning point a maximum
$\beta_{12} > \psi_{max}$	$c <^* c_{near}$	One root < 0 One root > 0 but $< \alpha_{180}$ Turning point a maximum	One root < 0 One root > 0 but $< \alpha_{180}$ Turning point a maximum

[a]Both the near and far conditions are satisfied only for β_{12} in the range $\psi_{min} < \beta_{12} < \psi_{max}$. Within this range, the greater of the two roots for each quadratic indicates the minimum value of α (and hence of the beam source distance D) for which the condition is satisfied. In order that both conditions are satisfied, and hence that the voxel elongation is less than the specified tolerance everywhere in the horizontal plane through the image space, the value of α must be larger than the largest of the two greatest roots.

TABLE 6.2   Roots of the Quadratic Equations $f_{far}(\alpha) = 0$ and $f_{near}(\alpha) = 0$ for Various Values of the Angle β_{12} Between the Beam Sources, in the Case Where $s_{max} = 2^a$

β_{12}	c	α_{180}	Roots of $f_{far}(\alpha) = 0$	Roots of $f_{near}(\alpha) = 0$
$10°$	0.996	1.00	$-0.85, -1.22$	1.05, 0.96
$20°$	0.985	1.02	$-0.75, -1.57$	1.11, 0.93
$30°$	0.966	1.03	$-0.67, -2.23$	1.20, 0.91
$40°$	0.940	1.06	$-0.62, -3.91$	1.30, 0.90
$50°$	0.906	1.10	$-0.57, -16.37$	1.44, 0.89
$60°$	0.866	1.16	$-0.54, \textbf{7.46}$	1.62, 0.80
$70°$	0.819	1.22	$-0.51, \textbf{3.05}$	1.88, 0.90
$80°$	0.766	1.31	$-0.49, 1.92$	**2.25**, 0.92
$90°$	0.707	1.41	$-0.47, 1.41$	**2.83**, 0.94
$100°$	0.643	1.56	$-0.46, 1.12$	**3.85**, 0.97
$110°$	0.574	1.74	$-0.45, 0.94$	**6.10**, 1.02
$120°$	0.500	2.00	$-0.45, 0.81$	**14.93**, 1.07
$130°$	0.423	2.37	$-0.45, 0.72$	$-32.70, 1.14$
$140°$	0.342	2.92	$-0.45, 0.65$	$-7.82, 1.23$
$150°$	0.259	3.86	$-0.46, 0.60$	$-4.46, 1.35$
$160°$	0.174	6.76	$-0.47, 0.56$	$-3.14, 1.50$
$170°$	0.087	11.47	$-0.48, 0.52$	$-2.43, 1.71$

[a]Hence $\psi_{min} = 53.1°$, and $\psi_{max} = 126.9°$. The minimum value of the normalized beam distance, α [Eq. (6.18)] for which both the near and far conditions are satisfied is highlighted in bold typeface for each tabulated value of β_{12} in the range $\psi_{min} < \beta_{12} < \psi_{max}$.

6.7.1 Other Types of Dead Zone

Static-volume systems that employ two or more deflected beams for voxel activation may also exhibit voxel placement dead-zone characteristics. In the analysis above it was implicitly assumed that the beams were perfectly aligned for all voxel positions. As with beam-addressed swept-volume systems, if each beam is to be deflected through a variable angle, it will be able to do so only to a discrete number of angular positions. Furthermore, there will be an uncertainty associated with each of these angular positions, due to the deflection apparatus itself.

Consequently, incomplete overlap of the beams is likely to occur, and this will result in changes in voxel shape, size, and brightness. In extreme cases the beams may not intersect at all. These effects will in general increase in severity toward the portion of the image space that is farthest from the beam sources. If the dependence of these effects on spatial position and system parameters can be in some way quantified, corresponding dead zones may be modeled.

6.7.2 Uniformity of the Image Space

In the case of systems based on the stepwise excitation of fluorescence, as the beam that is resonant with the first transition passes through image space it will excite the fluorescence centers all along its path, not just at the desired voxel location. Its intensity will therefore be diminished according to the Lambert–Beer law, $I_1(q) = I_0 \exp(-q/\text{constant})$, where I_1 is the voxel intensity at depth q into the image space, and I_0 is the initial beam intensity. Consequently, unless the beam intensity is varied in accordance with the depth at which each voxel is to be activated, voxel brightness may vary. The magnitude of this effect depends on the absorption cross section of the transition resonant with the beam and on the density of fluorescent centers.

6.7.3 Materials

Although display units employing the stepwise excitation of a gaseous media remain the subject of research (Kim et al. 1996), recent advances have been made in the application of doped solid media to displays of this type (Downing et al. 1996). This approach allows more freedom in the choice of the dopant ion, its concentration, and the nature of the host material than do gaseous systems.[*] A number of rare earth ions have well-defined visible and infrared energy transitions suitable for use in stepwise excitation of fluorescence-based displays. In the late 1960s, the ions Erbium (Er^{3+}) and Europium (Eu^{3+}) doped into CaF_2 crystal were investigated (Lewis et al. 1971). More recent work has centered on Praseodymium (Pr^{3+}) doped into low-phonon fluorozirconate glass. The low probability of deexcitation through coupling to phonons in the host material

---

[*] However, a gaseous medium also has certain advantages over a solid medium. Some of these are discussed in Chapter 10.

increases the excitation efficiency and hence the voxel brightness (for given activation beam intensities). This work is discussed further in Section 7.7.2.

6.7.4 Color and Intensity Scale

In the case of systems based on the stepwise excitation of fluorescence, adjustments in the beam intensities or pulse durations may affect controlled changes in the voxel intensities and may therefore be used to accommodate a gray scale. Color may be incorporated by a number of methods. Several approaches that may be suitable for Pr^{3+}-based solid displays have been proposed (Hesselink et al. 1995, Downing et al. 1996). One of these is to use additional infrared activation wavelengths to excite other fluorescent transitions within a single ion. In a suitable host material, Pr^{3+} has three fluorescent energy states that may be occupied from a common intermediate state according to the wavelength of the second activation beam (Figure 6.16). Fluorescent decay from these states is in the red, green, and blue wavelength regions, respectively.

An alternative method is to dope the host material with three different rare earth ions, each excited by, and providing fluorescence at, different wavelengths. If the required infrared wavelength of beam 1 is common to all the ions, color selectivity may be based on the wavelength of beam 2 only. In this way, red, green, and blue fluorescing transitions may be possible. However, a dopant concentration of 0.5 mol% has been determined experimentally to maximize voxel brightness in displays of this type employing low-phonon glass (Downing et al. 1996), with cross-relaxation between neighboring ions limiting the performance at higher concentrations. Therefore, for a three-color display to operate at maximum brightness, a more practical approach may be to build up the image space in alternating thin layers of the host glass, each containing single-color rare earth ions at optimal concentration (Downing et al. 1996).

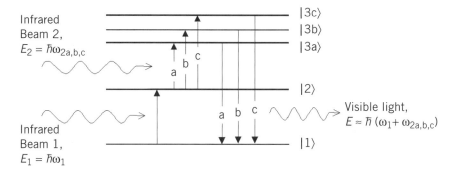

FIGURE 6.16 Possible mechanism for the provision of several colors in a Pr^{3+} based volumetric display. Three fluorescent excited states are present and are denoted $|3a\rangle$, $|3b\rangle$, and $|3c\rangle$, respectively. These are selected through a choice of wavelength of activation radiation beam 2. For example, state $|3a\rangle$ could fluoresce in the red, $|3b\rangle$ in the green, and $|3c\rangle$ in the blue regions of the spectrum.

6.8   CONSIDERATIONS ON VOXEL ADDRESSING

The voxel activation capacity of any display unit is given by Eq. (2.1). For a certain voxel time and a minimum refresh frequency (prescribed flicker considerations), the voxel activation capacity is proportional to, and limited by, the parallelism P. For example, if the refresh frequency is 25 Hz, the voxel time 10^{-6}s, and the required voxel activation capacity is $Na = 10^6$, the minimum number of voxels that must be able to be activated in parallel is 25. Two banks of 25 beams would, in principle, satisfy this requirement: one bank pumping transition $|1\rangle \rightarrow |2\rangle$, the other $|2\rangle \rightarrow |3\rangle$ (Figure 6.17).

Unfortunately, the problem is not quite so straightforward, as it is also necessary to consider the order in which voxels may be activated within the image space. In the case of display units of this type, ordering of the activation of voxels may be necessary for two reasons:

- In the case of display units that exhibit parallelism in the voxel activation process, a number of beams will be present simultaneously within the image space. Consequently, there is a possibility of beams intersecting at arbitrary locations and hence producing unwanted voxels.
- If the intermediate-state lifetime τ_2 is long compared with the time scale of voxel activation, ghost voxels may be generated. Consider the situation illustrated in Figure 6.18, where voxels 1 and 2 are generated in turn. In creating voxel 1, the material is excited to state $|2\rangle$ all along the path of beam 1 (although the voxel itself is generated only where beam 2 intersects this path). If τ_2 is sufficiently long, a significant number of fluorescent centers along the path of beam 1 remain in the intermediate state during the excitation of subsequent voxels. If voxel 2 is coplanar with voxel 1 and the beam sources, the path of beam 2 will in this case intersect the residual population in $|2\rangle$, and an undesired ghost voxel will be generated.

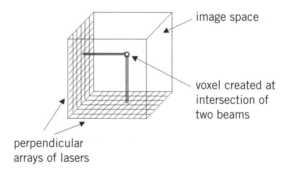

FIGURE 6.17   Two banks of beams addressing the image space in parallel greatly increase the voxel capacity of the device. However the order in which voxels are activated will be of greater importance than in systems offering lower parallelism.

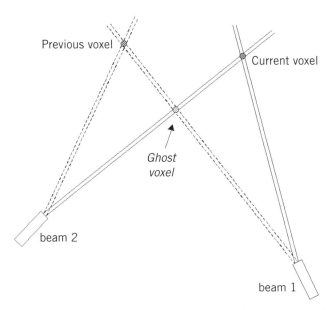

FIGURE 6.18   Ghost voxel arising in the image space of a stepwise excitation-based display and in which the intermediate-state lifetime τ_2 is long compared to the voxel time T.

The probability of voxel activation (either as a consequence of spurious beam intersections or through the lifetime of the intermediate state) increases with the voxel activation rate and therefore the parallelism of the display unit. To avoid this problem, it may be necessary to compute an optimal order for the activation of voxels throughout each image update period. This may prove to be a computationally expensive process. Analogous ordering problems (which arise for different reasons) occur in the case of the display units described in Section 7.5.5. These systems employ thermochromic and photochromic materials for voxel generation.

6.9   DISCUSSION

Although simple in concept, the actual design of a static-volume display unit is a complex process and, as with their swept-volume counterpart, involves careful consideration of a number of interrelated issues. The elimination of mechanical motion from the image space creation subsystem is very advantageous; however, in general terms this is achieved by a considerable increase in the demands placed on both the voxel generation and voxel activation subsystems.

The magnitude of the difficulties associated with the implementation of displays of this type is indicated by the very considerable time during which research has been undertaken in this area. As we will see in Section 7.2, the first recorded attempt to implement a static-volume display appears to have been made in 1912, and since then a variety of systems have been proposed and/or developed.

Some of the problems related to the development of static-volume systems are analogous to those experienced in the implementation of their swept-volume counterpart. For example, the image space associated with both beam-addressed swept-volume display systems employing rotational motion and static-volume displays employing two or more activation beams may, in principle, be adversely affected by the presence of dead-zone regions. In both cases the careful design of the image space and optimal positioning of the voxel activation mechanisms may ameliorate their impact.

The variety of approaches that may be adopted in the implementation of the three display unit subsystems and the complex issues that must be considered when attempting to predict the precise nature of the interaction between them, make the design of any volumetric display systems a difficult, time-consuming, and possibly uncertain process. However, our current level of understanding in this area is such that it is now possible to envisage the development of software-based design tools that will contribute to all aspects of this design process and particularly enable any proposed design to be largely verified and optimized with respect to a variety of characteristics. Should this approach be adopted, the performance of future volumetric display systems is likely to be more predictable and the work may therefore be conducted in a more efficient manner. In the next chapter we outline a number of approaches that have been taken in the implementation of this class of display unit.

CHAPTER 7

STATIC-VOLUME DISPLAY UNIT DEVELOPMENT

They copied all they could follow, but they couldn't copy my mind,
And I left them sweating and stealing a year and a half behind.*

7.1 INTRODUCTION

The development of systems that place no reliance on the periodic motion of
a surface for image space generation is of considerable interest. Research that
has been carried out in this area has resulted in a range of proposals and architectures that may generally be classified according to the scheme outlined in
Chapter 6 as either active matrix or passive medium display units. Research
investigations that have been undertaken concerning static-volume systems that
employ a passive medium have largely taken the form of a quest for materials that possess the appropriate passive and active states. The passive state
corresponds to one of transparency and the active state to an ability to scatter,
emit, or absorb visible radiation. Investigations into active matrix displays have
encountered their greatest difficulties in the areas of image space opacity, voxel
activation, and the physical assembly of the elements and associated connections.
This would not appear to have lessened interest in this area, and we describe a
number of previously proposed systems and one ongoing research project which
denotes a considerable advance in this type of display unit architecture.

In the following sections we examine briefly some of the research activity
that has been carried out on these two classes of display unit, within a broadly
chronological framework. This chapter is not intended to provide a complete

---

* Rudyard Kipling, *The Mary Gloster*, in *Rudyard Kipling Complete Verse* (definitive edition), Anchor
Press, 1989.

account of the considerable amount of work that has been devoted to static-volume display unit development.

7.2 PIONEERING WORK: PRE-1940

The earliest document available to the authors concerning the development of static-volume display systems originates from 1912. In this work, inventors Luzy and Dupuis (French patent 461600) describe a modified stereoscopic projection technique that permitted an image to be generated within a volume rather than on a flat screen. The underlying principle involved the use of two infrared or ultraviolet radiation sources. These were of different wavelengths and each passed through a stereoscopic plate before entering a volume (image space). The two stereoscopic plates provided a spatial intensity modulation corresponding to a different view of a 3-D scene. Within the image space the interaction between the radiation and the medium resulted in an optical change occurring in regions at which radiation from both projections was present. Within their patent the inventors write:

> Cet effet puisse être obtenu par la somme des deux radiations considérées, tandis que chacune d'elles agissant seule ne produirait aucune action.* Supposons, d'autre part, un milieu solide, liquide, gazeux ou colloidal ... ce milieu, absorbant l'énergie des radiations invisibles, pourra la restituer sous forme d'énergie lumineuse.†

A schematic illustrating the principle of operation is presented in Figure 7.1a. It should be noted that if each source of radiation is projected through an intensity modulation mask as described in the patent, then this geometrical arrangement will be prone to resolution limitations due to the occurrence of ghost voxels (see Section 6.8). These will be caused by rays from neighboring points intersecting. This may be seen in Figure 7.1b, where rays from two neighboring points are traced; they intersect at four locations within the image space, rather than at only the desired two. The size of the "region of confusion" thus created depends on the depth of the image space,‡ the focal length of the projections, and the angle between them.

The only precise proposal in the patent for a voxel generation subsystem concerned implementation of the interaction of one infrared (380 nm) and one ultraviolet (690 to 750 nm) within an image space containing selenium chloride. The concept elucidated by Luzy and Dupuis of the use of two or more invisible infrared or ultraviolet radiation sources to trigger a visible change within a volume

---

* "This effect [visible change] is obtained by the combination of the two projections, whereas each alone would have no effect."
† "Suppose also that we have a medium, being solid, liquid, gaseous or colloidal ... this medium, absorbing the energy of the invisible radiation, will be able to transform it into visible light."
‡ Note that in Figure 7.1a the two rays intersect before and beyond the image space, avoiding unwanted ghost luminescence by the absence of the image space material.

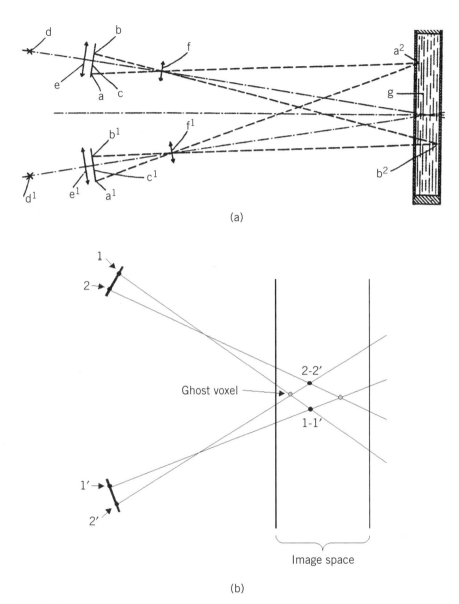

(a)

(b)

FIGURE 7.1 (*a*) Principle of operation of the display unit described by Luzy and Dupuis; (*b*) spurious interaction between radiation from neighboring points leads to a limitation of spatial resolution if the radiation from each projection impinges simultaneously on the image space. Here, two neighboring points in each projection, denoted 1 and 2, and 1′ and 2′, will give rise to ghost voxels as shown, in addition to the desired points 1-1′ and 2-2′. [(*a*) From French patent 461600].

of otherwise transparent material foreshadows much of the research into static-volume displays during the latter half of this century.

7.3   PIONEERING WORK: 1940–1950

In 1945 a static-volume display unit was the subject of a patent filed by F.S. Howell (U.S. patent 2,604,607). An illustration of this display system is reproduced from the patent in Figure 7.2. From this diagram it may be seen that the system takes the form of a novel cathode ray tube employing two electron

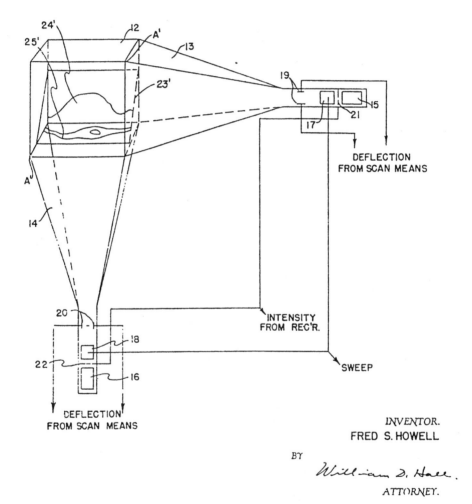

FIGURE 7.2   Display system described by F.S. Howell. Two electron beams are directed into a gas-filled image space. At the point of intersection visible voxels may be generated. (From U.S. patent 2,604,607.)

guns. The beams emanating from these guns pass into a common region (image space). The tube is reported to have been gas filled (nitrogen); however, no mention is made of the gas pressure, nor is the beam energy quantified. It is claimed that the device operated on the physical process occurring within the gas at the point of intersection of the two beams as follows:

> ... said gas having the property of emitting light in the vicinity of the intersection of said beams when the beam intensity at the said intersection exceeds a predetermined level, said gas having the further property of continuing to emit light for a predetermined time after said intensity has dropped below said predetermined level.

Since the beams do not represent two coherent sources of radiation, it is apparent that any threshold must be as a consequence of the increased beam intensity within the region of their intersection. The intensity of the emitted light is therefore related to the combined beam current and it is therefore unlikely that there exists a well-defined transition from the passive to the active state.

The author of this work placed considerable emphasis on the architecture of the control system responsible for the scanning and modulation of the electron beams and able to interface the display unit to a radar system. It is interesting to note that this work would have been carried out at approximately the same time as that undertaken by Parker and Wallis (see Section 5.2). In both cases the lack of suitable computer systems (and data storage facilities) would have necessitated a direct correspondence between the radar scanning pattern and the path traversed within the image space by the electron beams.

7.4 PIONEERING WORK: 1950–1960

In 1956 the U.S. Patent Office granted two patents that both describe active matrix static-volume display unit technologies. These systems employed an image space in the form of a 3-D matrix of conductors. The uniqueness of the individual patents most probably centered on the implementation of the control systems (prior to the availability of digital solid-state electronics) and particularly the addressing technique used to induce voxel activation. Since both these display units illustrate an implementation of the elementary example system introduced in Section 1.3, they are discussed briefly below.

7.4.1 Active Matrix System Addressed by an Electron Beam

Late in 1952, Martin Ruderfer filed a patent that described a volumetric display system intended for general use in the visualization of volume data (U.S. patent 2,749,480). Several forms of display unit are described within this patent. In one embodiment, a set of conductors that form a 3-D matrix are positioned within a gas-filled image space. Although the conductors are arranged so that they are not in direct contact with each other, they do pass in close proximity to their

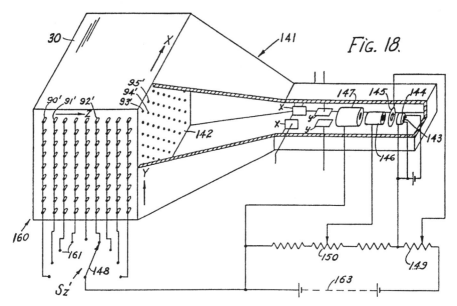

FIGURE 7.3 The use of an electron beam for voxel activation within a static-volume display unit employing a 3-D array of voxel generation centers. (From U.S. patent 2,749,480.)

neighbors. Through the application of a sufficiently high voltage between two appropriate conductors, a discharge will occur in the gas within the region at which they cross and so visible voxels may be formed. To simplify the problems that would at that time have been associated with the high-speed scanning of each possible voxel location within image space, the inventor employed an electron beam for voxel activation. This configuration is illustrated in Figure 7.3, and from this diagram it may be seen that the image space is located at the front of a CRT and that a series of horizontal conductors pass from within the CRT and through the image space. A second set of conductors divide the image space into a series of vertical planes. The activation of a voxel at a particular location within the image space is then reliant on the application of a voltage to the appropriate vertical plane and the incidence of the electron beam with the appropriate horizontal conductor. In this embodiment it is apparent that the electron beam may traverse a variety of scanning patterns. Gating of the beam ensures that it impinges on only those conductors appropriate to generation of the set of voxels comprising the image.

This approach to the implementation of an image space clearly has a number of deficiencies, which center on the occlusion of voxels located more deeply within the volume. The problem could not be easily solved at that time and would have become more acute as the intervoxel separation was reduced. Furthermore, although ingenious, this voxel activation technique unfortunately reduces the parallelism supported by the display unit to unity.

7.4.2 Active Matrix System for Echo Ranging

In 1954, a patent for the static-volume display unit illustrated in Figure 7.4 was filed (U.S. patent 2,762,031) and was assigned to the Raytheon Manufacturing Company. As will be apparent from this illustration, the image space comprises a series of approximately horizontal* conductors that pass through (and remain in electrical isolation from) a series of apertures in vertically oriented electrodes. These are described as being constructed from wire mesh. The image space is said to have been filled with a gas such as neon, and when an appropriate voltage is applied between a mesh and one of the horizontal conductors, a discharge will occur within the region in which the conductor passes through the hole in the mesh. By means of an appropriate addressing technique, images may therefore be created within the image space. As a consequence of the presence of the meshes, it is highly likely that the voxel visibility problems associated with this display unit would have been even more acute than those occurring in the case of the display unit described in Section 7.4.1.

7.5 PIONEERING WORK: 1960–1970

Articles describing the type of active matrix volumetric display systems discussed in Section 7.4 have appeared in various publications (including patents) on an occasional basis since the 1950s. However, the larger proportion of research into static-volume display systems has been directed toward architectures employing a passive medium. Displays of this type often employ a process known as the stepwise excitation of fluorescence (see Section 6.5). Following the much earlier work of Luzy and Dupuis (see Section 7.2) there appears to have been little activity in this area until the early 1960s. Since then there has been a considerable amount of interest directed to the specific objective of employing this process as a means of producing voxels within a volumetric image space. In 1961, Fajans described work directed at implementation of a volumetric display device (U.S. patent 3,123,711). At approximately the same time Zito and Schraeder (1963) also undertook an in-depth study of this methodology for display unit implementation. For clarity we break with the chronological order of events followed so far in this chapter and examine the work described by Zito and Schraeder in Section 7.5.1. Subsequently, in Section 7.5.2, we briefly examine the proposals contained within the Fajans patent.

Interest in passive media static-volume display unit architectures continued throughout the 1960s and encompassed not only the two-step excitation of fluorescence approach but also the use of photochromic and thermochromic materials for voxel generation. We believe both of these methodologies to be extremely promising. Some of the work carried out in this area is briefly outlined in Section 7.5.5.

---

* The shape of the image space was designed so as to enhance the linear perspective depth cue.

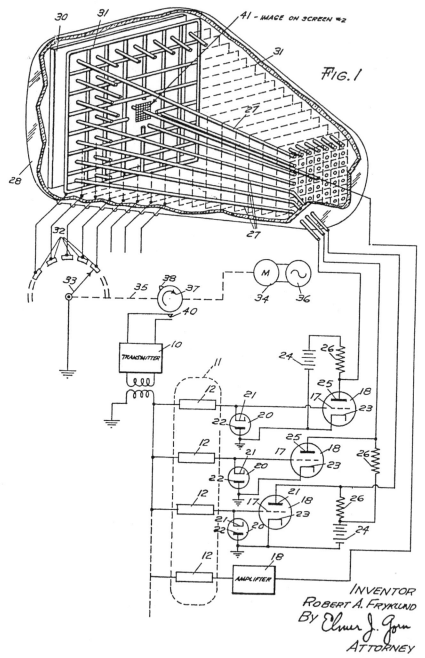

FIGURE 7.4 Gas-filled image space comprising a series of meshes (31) and horizontal conductors (27). The conductors pass through holes in the meshes. Application of a suitable voltage between a mesh and conductor will cause a visible discharge in the vicinity of the region at which the conductor passes through the mesh hole. (From U.S. patent 2,762,031).

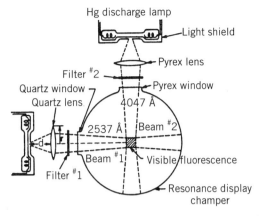

FIGURE 7.5 Apparatus used by Zito and Schraeder for the generation of voxels. This approach related to the two-step excitation of fluorescence in mercury vapor (in the presence of nitrogen gas). (Reproduced by permission; © 1963 OSA.)

7.5.1   Voxel Generation Through the Excitation of Mercury Vapor

In 1963 an article appeared in the journal *Applied Optics* (Zito and Schraeder 1963) describing work undertaken in relation to the production of voxels within an image space composed of a passive medium (a mixture of mercury vapor and nitrogen gas). The experimental apparatus described in this article is illustrated schematically in Figure 7.5. Two mercury discharge lamps were employed, the radiation from which was used to generate a two-step excitation within the region at which their beams intersected. The decay from this excited state led to the emission of light with a wavelength of 5460.74 Å (green).

The relevant energy levels and atomic transitions are shown in Figure 7.6. In essence the first of the external radiation sources stimulates the transition from the ground state to the 6^3P_1 level, and the second source stimulates the transition from the 6^3P_0 to the 7^3S_1 level. (The 6^3P_0 represents a metastable state formed by collisions between mercury atoms excited to the 6^3P_1 state and the diatomic nitrogen present.) Spontaneous decay from the 7^3S_1 results in the generation of visible radiation at a wavelength close to the peak sensitivity of the human eye.

In their extremely lucid and informative publication the authors discuss their experimental procedures in some detail and conclude that they were unable to generate voxels of sufficient intensity for use in a practical display unit. The two-step excitation of mercury vapor is also discussed at some length in another article (Zito 1963).

7.5.2   Luminous Spot Display Device

In 1961 a relatively brief but detailed patent was filed by Jack Fajans (U.S. patent 3,123,711) relating to the implementation of passive media static-volume displays. The author of this work outlines voxel generation by means of

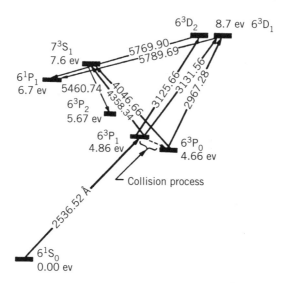

FIGURE 7.6   Energy-level diagram for mercury vapor. The 6^3P_0 represents a metastable state formed by collisions between mercury atoms excited to the 6^3P_1 state and the diatomic nitrogen present. (Reproduced by permission; © 1963 OSA.)

the two-step excitation of fluorescence and also by the photostimulation and photoquenching of phosphors. In the case of the two-step excitation process Fajans considers in some depth the use of mercury vapor and also suggests the use of various other vapors: for example, cadmium, potassium, and sodium.

The discussion provided by the author concerning voxel generation by means of phosphors is most interesting, as it provides a well-considered insight into a novel method for the implementation of a passive medium image space. The author describes the use of phosphor particles (measuring approximately 5 μm) dispersed within a transparent solid (polymethyl methacrylate) at a concentration of approximately 10 milligrams per cubic inch. A number of phosphors suitable for photostimulation are mentioned within the patent (Table 7.1). In this embodiment the generation of a visible voxel of wavelength λ_3 occurs within the region of intersection of two beams of wavelength λ_1 and λ_2. The alternative approach described by the author employs a photoquenching technique in which visible light is generated by means of a beam of wavelength λ_1. A second radiation source of wavelength λ_2 is then used to quench (disable) light output.

It is unlikely to be advantageous that an approach in which a radiation source is employed to stimulate light emission from an entire image space, and subsequently other sources are used to disable light emission from voxel locations that are not to form part of the image scene. As we have seen, the major part of a 3-D image space will be void during the depiction of any image scene. For example, in the case of a static-volume system employing an active matrix of voxel generation centers, the number of voxel locations that are activated within

TABLE 7.1   Phosphors for Use in a Passive Media Display Unit

Composition of Phosphors	λ_1	λ_2	λ_3	
Cub-Sr(S:Se):[flux]:Sm:Eu	4,600	9,300	5,700	Photostimulation
Cub-CaS:[flux]:Sm:Eu	4,800	11,700	6,600	Photostimulation
Cub-SrS:[flux]:Sm:Ce	2,900	10,200	4,800	Photostimulation
Hex-ZnS:Cu:Pb[SO$_4$]:[NaCl(2)]	3,700	7,500 or 13,200	4,880	Photostimulation
ZnS(Cu)	Below 4,200	$\approx$7,000	—	Photoquenching
Hex-9ZnS:1CdS:Cu(0.0073)	$\approx$4,200	7,200	—	Photoquenching

Source: U.S. patent 3,123,711.

an image frame will generally be very much fewer than the total number of possible voxel locations. Consequently, the total time required to generate an image by disabling activated voxels (within an image space in which all voxel locations have previously been stimulated) will be greater than the time required to generate an image by the more conventional means of activating only those voxels from which it is composed.

7.5.3   Alternative Architecture for a Passive Medium Display Unit

Most proposals concerning the implementation of display units employing the two-step excitation of fluorescence describe architectures in which voxel generation occurs within the region of intersection of two beams of circular cross section. As discussed in Section 6.6, such a configuration is likely to result in a variation in voxel shape according to voxel location. This may restrict the regions of an image space within which voxels can be activated satisfactorily. In 1964, M.R. Brown filed a patent (U.S. patent 3,474,248) that described various embodiments of a static-volume display unit employing the two-step excitation of fluorescence. As may be seen in Figure 7.7, Brown reports voxel activation occurring within the region of intersection (8) of a fan-shaped radiation beam (4) and a second beam of circular cross section (7). As is apparent, the fan-shaped beam (plane of radiation) may be scanned in the vertical direction and the other radiation source may be moved horizontally.* Voxel generation occurs within the region of intersection of the two. Consequently, variations in the angle of incidence between the radiation sources are minimized, so the shape of voxels activated within an image space will be essentially invariant with position.

The author of this patent also describes the use of a beam that may be brought to focus at different depths within an image space permitting the beam's energy

---

* The mechanical scanning system illustrated in Figure 7.7 will clearly restrict voxel activation capacity.

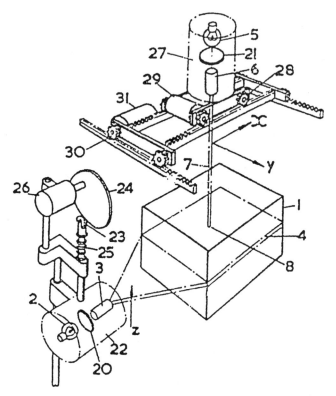

FIGURE 7.7 Apparatus described by M.R. Brown that permits voxel activation within the region of intersection of a light beam (7) and light plane (4). (From U.S. patent 3,474,248).

density to be varied. In this embodiment voxel activation is said to occur within the region of focus. This principle is described as having been used by the author to permit the simultaneous activation of two voxels. Finally, the author suggests a number of materials that he believed to be appropriate for use in the voxel generation process. These are crystals of lanthanum fluoride, calcium fluoride, strontium fluoride, and barium fluoride containing between 0.05 and 25% of trivalent thulium, erbium, or holmium ions.

7.5.4   Voxel Generation by Two-Photon Fluorescence

Although a large body of work has been carried out concerning voxel activation via a two-step excitation of fluorescence process, the related process of two-photon absorption has also been considered as an excitation mechanism. In this case, voxels are activated within the region of intersection of two beams of coherent radiation (in a suitable material). Two-photon absorption is a nonlinear optical process in which a quantum system is excited to a higher-energy state by the simultaneous absorption of two photons, with no intermediate state.* Thus a volumetric display based on this process is in many ways similar to one based on the two-step excitation of fluorescence, the primary difference being that the material within the image space has no intermediate state.

In 1967 a patent was filed by Dugay et al. (U.S. patent 3,541,542) in which this optical process was proposed as the basis for a passive medium volumetric display device. A schematic energy-level diagram illustrating the simultaneous absorption of two photons and the subsequent fluorescent decay is illustrated in Figure 7.8. For a system of this type to be excited into its fluorescent excited state, one or more photons are required, the total energy of which corresponds to the energy gap between the ground and fluorescent states. Since $E = \hbar\omega$, the frequencies of two sources of coherent radiation may be chosen such that $E_1 + E_2 \geq \Delta E$, where ΔE is the fluorescent energy gap. The inequality arises because a number of vibrational energy states are associated with each electronic state (Figure 7.8), and excitation to higher vibrational states may still produce fluorescence after nonradiative decay to the fluorescent electronic state. In regions where this energy relation is satisfied (i.e., at the intersection of the beams), the intensity of the resulting fluorescence is proportional to $I_1 I_2 \Delta t$, where I_1 and I_2 are the intensities of the two beams and Δt is the time for which the pulses of each beam overlap.

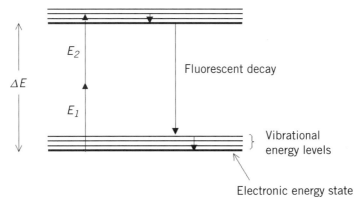

FIGURE 7.8 Schematic representation of simultaneous two-photon absorption across an electronic energy gap ΔE and subsequent fluorescent decay. Each electronic energy state has a number of associated vibrational states. Excitation by this mechanism requires that the combined energy of the photons is equal to (or, due to the vibrational states, slightly greater than) the energy difference ΔE. The intensity of the fluorescent output, following rapid nonradiative decay to the excited electronic state, is proportional to the product of the intensities of the incident radiation of each frequency.

---

* In terms of the response of an idealized three-level system to incident radiation pulses, as introduced in Section 6.5, this corresponds to the limit in which the lifetime of the intermediate state (τ_2) approaches zero (Schwarz and Blundell 1993).

The display embodiment described in the patent is based on the facts that (a) the photon energies must sum to the fluorescent energy gap, and (b) the fluorescent intensity is proportional to the product of the beam intensities. Two coherent beams at different frequencies and intensities are thus employed; one beam is of frequency $\omega_1 < \Delta E/2\hbar$, so that excitation of fluorescence is not possible by the simultaneous absorption of two photons from this beam alone, and the other has frequency $\omega_2 \approx \Delta E/2\hbar - \omega_2$. Dugay et al. proposed using as ω_2 the second harmonic of ω_1. However, although $2\omega_2 > \Delta E/\hbar$, there is still a finite probability of two-photon absorption and fluorescence, via the vibrational states mentioned above. For this reason, the intensity of beam 2 was two orders of magnitude lower than that of beam 1. Thus the intensity of background fluorescence due to beam 2 alone is below the visual threshold. When pulses from both beams overlap, the energy relation is satisfied, and fluorescence is produced proportional to the product of the beam intensities.

The authors of the patent describe a voxel activation strategy in which beam 1 and beam 2 are colinear and travel in opposite directions (Figure 7.9). Picosecond pulses along each beam are spatially separated by $2d$ and nd, respectively, where d is the separation and n is the number of voxels along the beam paths through the image space. Beam 1 contains the voxel information via the amplitude of the constituent pulses — in the simplest (binary) case, each pulse is present or absent according to whether a voxel is to be illuminated or not in the corresponding potential voxel location. Beam 1 may thus be considered as the encoding beam. Beam 2 then interrogates the information present in beam 1, with each of its pulses coinciding in turn with each of those in beam 1 at the respective voxel locations (Figure 7.9).

Pulse durations of the order of picoseconds are specified explicitly in the patent; indeed, for voxel separations of the order of 1 mm, the time interval *between* consecutive encode pulses (beam 1) would need to be about 3 ps. It is interesting to note that more recent work has applied the two-photon absorption mechanism to optical high-performance computer memory devices (Hunter et al. 1990, Parthenopoulos and Rentzepis 1990).

7.5.5 Photochromic and Thermochromic Materials for Voxel Generation

Toward the end of 1968 two patents were filed (U.S. patents 3,609,706 and 3,609,707) which detailed the implementation of static-volume display units employing a passive medium composed of photochromic or thermochromic materials. Much of the material described within these works is identical; however, the latter (by Lewis and Adelman) is more comprehensive than the former (by Adamson) and includes a discussion of the use of thermochromic materials. Within these extensive patents, a variety of display unit implementations are considered in depth. Below we outline briefly some of the ideas contained within the second patent.*

---

* A discussion of photochromic materials is provided by Martin (Petty et al. 1995).

Voxel positions

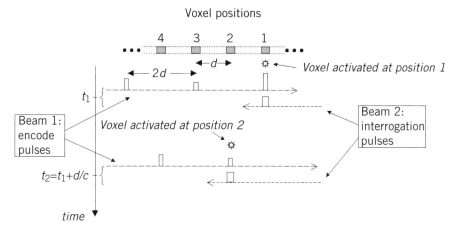

FIGURE 7.9 Voxel activation strategy along two colinear beams. Beam 1 encodes the voxel intensity by the amplitude modulation of a string of pulses separated by a spatial interval of $2d$, where d is the voxel separation along the line of intersection. Beam 2 comprises interrogation pulses at a separation of nd, where n is the number of voxels along the line. This illustration shows a line of four voxel locations; at time t_1, the first interrogate pulse from beam 2 intersects the first encode pulse from beam 1 at voxel location 1. At a time d/c later, the interrogate pulse is now coincident with the next encode pulse from beam 1, at voxel location 2.

Image Space Comprising a Photoreversible Photochromic Material

Lewis and Adelman discuss the development of an image space comprising a material having the following general characteristics:

- There are two discernible nonemissive optical states.
- One of the states above is transparent.
- One of the states above is strongly absorptive of certain wavelengths and under appropriate illumination will appear to be colored.*
- A transition from the absorptive to the transparent state may be effected by the application of radiation of wavelength λ_1.
- A transition from the transparent to the absorptive state may be effected by the application of radiation of wavelength λ_2, where $\lambda_1 \neq \lambda_2$.

A photoreversible photochromic material possessing these properties [$1'$, $3'$, $3'$-trimethyl-6-nitrospiro($2H$-1-benzopyran-2,2'-indoline)] is identified, and the preparation of an appropriate solution described. A method discussed by the authors for image generation is of considerable interest and employs two radiation sources of wavelength λ_1 and λ_2, whose purpose is defined above. Let us begin

---

* This state may be characterized by various attributes: reflection, refraction, or scattering of the illuminating source of radiation.

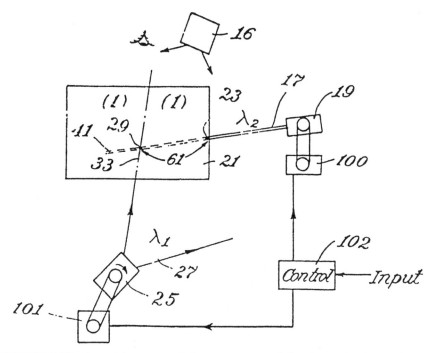

FIGURE 7.10 Essential aspects of the apparatus used for the production of images within an image space comprising a photochromic material. As the beam source (17) propagates through the medium, it causes — along its path — a change in the material from a transparent to an absorptive state. As a consequence, the beam intensity will gradually diminish. This results in the generation of a track of finite length and of diminishing intensity. A second beam source (33) is used to return to the original transparent state all portions of the image space that are not intended to contain activated voxels. In effect, this beam source may be scanned so that it acts as an eraser, removing unwanted portions of the track and permitting beam source 17 to propagate to any depth in the medium. (From U.S. patent 3,609,707).

by considering that the image space medium illustrated in Figure 7.10 is wholly in a transparent state. The beam of wavelength λ_2 (denoted 17 in the illustration) is directed into the image space and we may anticipate that along its path it will cause the medium to change into the absorptive (colored) state. However, any such change will result in an attenuation of the beam energy, and this in turn will limit the depth to which the beam is able to propagate.

To permit the beam to propagate to the depth at which a voxel is to be activated and, furthermore, so as to enable the generation of voxels (as opposed to line segments), the second radiation source of wavelength λ_1 (labeled 33 in the illustration) is brought into operation. This beam acts as an eraser returning to the transparent state those regions of the image space that are to be void. By controlling the two radiation sources in an appropriate manner, beam source 17 may be permitted to propagate to any depth within the image space and so give rise to the activation of a voxel at an appropriate location. Although this method of image generation is most ingenious, it is apparent that the order in which voxels are activated may be especially important, and in the case of complex images it may be necessary to provide a plurality of beam pairs.

Image Space Comprising a Thermochromic Material

In determining a thermochromic material suitable for use as the voxel generation mechanism, the following general characteristics may be defined:

- There are two discernible nonemissive optical states.
- One of the foregoing states is transparent.
- One of the foregoing states is strongly absorptive of certain wavelengths and under appropriate illumination will appear to be colored.
- The transition from the transparent to absorptive state is effected by raising the temperature beyond a threshold.
- This transition temperature should be both sharp and clearly defined.
- The change in optical state should exhibit a hysteresis effect. This implies that the threshold corresponding to the transition from transparent to absorptive states should occur at a higher temperature than the converse. Clearly, the difference in the temperatures of these two transitions will play a major role in determining the time duration for which a voxel activated within the image space will continue to be visible after the removal of the activation source.

Within their patent, Lewis and Adelman discuss the implementation of an image space comprising a thermochromic material such as 1,3-di-4-piperidyl propane and a method for achieving voxel activation therein. An illustration provided within the patent and used to describe the operation of the display unit is reproduced in Figure 7.11. The transition from a transparent to absorptive state is achieved by means of a dynamically focused radiation source that is depicted in this figure as being brought to focus within the region labeled 151. It is apparent that it is possible to arrange this voxel activation apparatus in such a way as to ensure that the energy density is insufficient to cause voxel activation outside the region of focus.* In principle, voxel activation may be achieved at any location within the material by varying both the beam direction and the distance at which it is brought to focus (although voxel shape will not be consistent and will vary with depth within the image space).

As mentioned previously, any material employed in this process should have a sharp transition temperature between its two states. This ensures that the time required to activate each voxel is minimized, so the largest possible voxel activation capacity may be achieved. Furthermore, the magnitude of any hysteresis

---

* Clearly, the time duration for which the radiation passes through a region and the thermal properties of the material will ultimately determine the extent of the region that is caused to pass from the transparent to the activated state.

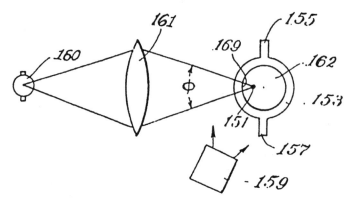

FIGURE 7.11   Apparatus employed by Lewis and Adelman for the implementation of an image space utilizing a thermochromic material (162) for voxel generation. A water jacket (153) has inlet (155) and outlet (157) and serves several purposes, including temperature stabilization of the bulk material. (From U.S. patent 3,609,707.)

in transition temperatures will affect the voxel visibility lifetime and may have consequences in the determination of the appropriate refresh frequency.

Both the Adamson and the Lewis patents provide some very interesting proposals for the implementation of static-volume display units. Although it is possible to identify a number of possible shortcomings in some of their ideas, their suggestions are likely to provide a substantial basis for further developments in this area.

7.5.6   Matrix of Elements

In 1969 a patent was filed by E.T. Maguire (U.S. patent 3,636,551) detailing a static-volume display unit that comprised a 3-D matrix of lamps contained within a transparent Lucite enclosure. In the preferred embodiment of this unit the inventor refers to the array comprising 70 by 70 by 50 voxel generation centers (a total of 245,000!). This display is equivalent to the elementary model introduced in Chapter 1. Even if we ignore the unrealistic number of lamps utilized in the implementation, the effect that their presence would have on the propagation of light through the image space would clearly be most detrimental to the image quality. (In Section 7.6.1 we see a similar approach being taken in the architecture of a display unit; however, in the latter case the display employs LEDs.) The display system was apparently intended for use in the visualization of aircraft radar information.

7.6   PIONEERING WORK: 1970–1980

In 1971 an extensive publication appeared in the *IEEE Transactions on Electron Devices* (Lewis et al. 1971) which examined issues concerning the implementation of a static-volume display system employing the two-step excitation of

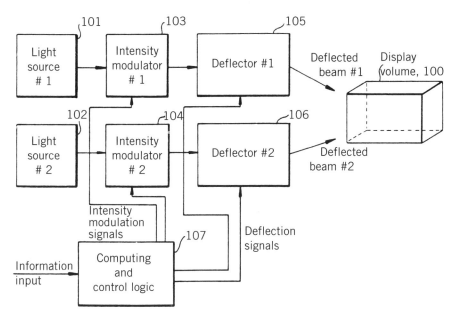

FIGURE 7.12   General form of the voxel activation mechanism apparatus for a display unit employing the two-step excitation of fluorescence. (From U.S. patent 3,829,838).

fluorescence for voxel generation. Although the authors of this work demonstrated voxel generation using transparent fluoride crystals doped with rare-earth ions, they acknowledged the advantages that would be gained through use of a gaseous rather than a solid or liquid medium (see also Barnes et al. 1974). At approximately the same time, these authors filed a patent (U.S. patent 3,829,838) in connection with their work. Although these documents describe in some detail the architecture of a suitable control system and graphics engine, the discussion presented with respect to voxel generation and activation has largely been covered in Chapter 6 and in Section 7.5. It need not, therefore, be repeated here. However, in Figure 7.12 we reproduce from the patent a diagram illustrating the general configuration of the voxel activation mechanisms discussed by the authors. In 1977, a highly informative article was published (Verber 1977) which analyzed various aspects of display units employing a two-step excitation process for the production of voxels.

7.6.1   Active Matrix Display Unit Using LEDs

A very brief paper appeared in the *IEEE Transactions on Broadcasting* in 1975 (Nithiyanandam 1975) in relation to the capture and display of volume data. The display unit is described as being constructed from a matrix of light-emitting diodes (a total of 10^6!). It appears that the image space was to have sides approximately 10 cm in length. It is unfortunate that the author of this work does not discuss major issues such as those concerning the propagation of light through the image space or deal with the electrical connectivity problems that would be associated with such a display unit implementation. In the context of an architecture of this type, it is surprising that mention is made of the ultimate application of such a display for 3-D TV. In Section 10.6 we discuss briefly the potential (or otherwise) of volumetric systems for this type of application.

7.6.2 Display Unit Employing a Particle Cloud

In 1975 a patent was filed by W.G. Rowe (U.S. patent 4,063,233) in relation to a static-volume display unit architecture, which is illustrated in Figure 7.13. This system employed an image space comprising a cloud of phosphorescent particles — voxel activation being achieved in the region of intersection of two electron beams. In this patent it is claimed that

> each individual beam's current is maintained at less than the threshold of luminescence of the particles but the combined currents may be caused to exceed that threshold by a controlled amount thereby producing light spots of variable brightness at the beam intersection point, which point may be directed into any part of the volume of the cloud.

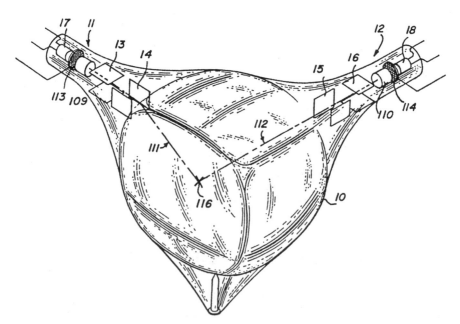

FIGURE 7.13 Apparatus used for voxel generation within a phosphor particle cloud. In this embodiment, two electron beams (111 and 112) cause voxel activation within the region of intersection. (From U.S. patent 4,063,233).

The use of multiple beams generally gives rise to registration difficulties. In view of the technique employed to maintain suspension of the phosphorescent cloud, it is likely that in the case of this technology, these difficulties would have been exacerbated. As may be seen from Figure 7.13, a conductive spike is positioned below the image space, and this is connected to a high-voltage supply (not quantified). As a consequence, a very high field gradient may be generated across the extent of the image space, and this is responsible for the continuous agitation of the particles. Although this may offer an elegant technique for maintaining the particle cloud, it is apparent that the presence of this field is likely to have a considerable impact on the linear propagation of the electron beams and would, at the very least, exacerbate beam registration difficulties.*

7.7 PIONEERING WORK: 1980–PRESENT

During the last two decades, research into static-volume display implementations has continued. In some cases, previously proposed techniques have been revisited and as a consequence of recent developments in optics, materials science, and computational power, are now perhaps more feasible. Several innovative ideas for the implementation of static volume volumetric systems are also under investigation.

7.7.1 Volumetric Systems Employing Optical Fibers

Ongoing research at the University of Texas at Dallas has led to the development of a number of prototype static-volume volumetric display systems (MacFarlane 1994, MacFarlane et al. 1994). This activity began in the summer of 1992 under the leadership of D.L. MacFarlane. The display units employ a solid image space supporting a 3-D matrix of regularly spaced voxel generation centers. The activation of each voxel is achieved by means of optical signals passed along dielectric waveguides that are embedded within the image space medium. As is apparent from the illustrations presented in Figures 7.14 and 7.15, each voxel generation center is located at the end of an optical fiber. These centers are reported as being composed of either a light-scattering or fluorescent material. Consequently, as radiation emerges from a fiber, it will stimulate the emission of light from the associated voxel generation center and so give rise to the production of a visible voxel at a predefined location within the image space.

The generation and modulation of the optical signals is achieved by means of a conventional LCD panel that is illuminated from the rear. As a consequence, it has been possible to employ conventional computer hardware in the implementation of the graphics engine. The degree of parallelism exhibited by the voxel activation mechanisms is clearly not defined by the number of fibers conveying signals into

---

* It is interesting to compare this embodiment to the display architecture described in U.S. patent 4,870,485. This describes the use of two intersecting energy beams and an image space comprising a suspension of phosphor particles.

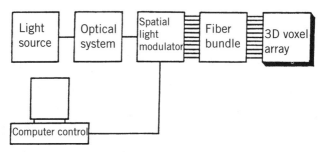

FIGURE 7.14 General configuration of the display unit developed at the University of Texas at Dallas. (Reproduced by permission; © 1994 OSA.)

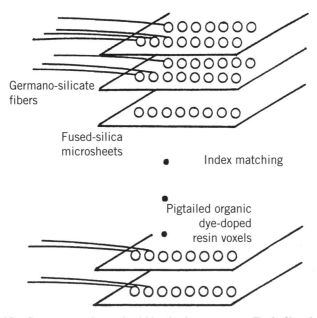

FIGURE 7.15 Components located within the image space. Each fiber is responsible for the generation of a single voxel. (Reproduced by permission; © 1994 OSA.)

the image space but rather by the parallelism of the LCD panel responsible for the generation and modulation of the optical signals. This panel may employ a row and column addressing technique and in its most elemental form will have $2N$ connections responsible for the generation of N^2 pixels. If we assume that it is possible to achieve a one-to-one correspondence between screen-generated pixels and image space voxels (i.e., there exist N^2 voxels within the image space), each electrical pathway to the LCD panel is responsible for generation of $N^2/2N = N/2$ voxels within the image space.

This approach permits the generation of a homogeneous and isotropic image space. As with other volumetric techniques, the overall image quality is greatly affected by the optical characteristics of the image space and hence its impact on the propagation of light from each activated voxel generation center. Those working on this development have undertaken considerable work in matching (as far as possible) the refractive indices of the dielectric waveguides and voxel activation centers with that of the transparent image space supporting medium.

The researchers report the production of a prototype display employing optical fibers that has a voxel activation capacity of 76,000 and a volume of 0.6 m^3. A further prototype employing ion-exchanged integrated waveguides for the transmission of the voxel activation signals is reported to have been constructed and has a voxel activation capacity of 2000 and an image space volume of 140 cm^3.

Should it be possible to develop suitable automated assembly techniques for the construction of the image space, it is most likely that very considerable improvements will become possible in both image space dimensions and voxel density. This approach will then offer a very elegant and cost-effective solution to the implementation of a static-volume volumetric display.

7.7.2   Two-Step Excitation of Fluorescence Within Solid Media

A considerable amount of recent research has been carried out in a collaboration between the University of Stanford and the U.S. Navy toward the development of static-volume displays using the stepwise excitation of fluorescence. In their most basic implementation, displays based on this process employ two infrared laser beams directed into the image space so as to trigger visible fluorescence at their point of intersection. (This process is considered in more detail in Chapter 6; earlier research in volumetric systems employing this process has been discussed in Section 7.5)

The Stanford group have concentrated on the development of a solid-state display system, similar to that pursued at Battelle Corporation in the late 1960s and early 1970s (Lewis et al. 1971, Verber 1977). However, the considerable advances that have been made in supporting technologies (that may be employed in the display unit subsystems) in the intervening two decades have enabled significant improvements in performance. Much of the work by this group appears to have focused on the development of appropriate low-phonon solid media doped with rare-earth ions that provide sharply defined electronic energy levels suitable for displays of this type (Downing et al. 1996). This has been necessary because one of the limiting factors in the performance of displays based on the stepwise upconversion process is the relative efficiency of fluorescent decay from the excited state versus other, nonvisible decay mechanisms that depopulate the state but do not contribute to the visible voxel. One such loss mechanism is that of energy release via vibrational modes (phonons) within the solid substrate. In this respect, one of the important advances to arise from this recent body of work is an increased efficiency in the conversion of the incident infrared radiation into visible fluorescence through the use of low-phonon fluoride glasses within the image space (Downing et al. 1996).

In the mid-1990s a small portable prototype display unit was demonstrated (Downing et al. 1994). This unit comprised an image space of approximately

125 cm^3 of trivalent praseodymium-doped low-phonon fluorozirconate glass, in which infrared activation beams at wavelengths of 1010 and 840 nm were used to generate monochrome voxels at their point of intersection.

Ongoing work is said to be aimed at the development of displays with increased performance in a number of important respects, including the provision of multicolor images, a larger volume of image space, and increased parallelism of voxel addressing and hence an increased voxel activation capacity (Hesselink et al. 1995, Downing et al. 1996). In the latter respect, the group proposes the use of 2-D arrays of vertical cavity surface-emitting lasers (VCSELs) to provide highly parallel voxel activation. Several methods for the production of multicolor voxels have also been suggested. These include the use of several dopant ions, each providing fluorescence of a different color, or a single ion exhibiting three suitable fluorescent states. In each of these approaches, the wavelength of (at least) one of the activation beams would select the transition and hence the voxel color. However, the use of mixed dopants can lead to problems with cross-relaxation as the optimum doping concentration is approached. This was determined empirically as 0.5 mol% (Downing et al. 1996). Perhaps a more practical technique suggested in this article was that of constructing the image space from thin slabs of the transparent host material, arranged so that the dopant ion alternates between, say, three colors. This could permit an RGB color capability.

7.7.3 Volumetric Image Space Composed of Switchable Panels

Another interesting static image space design that has been the subject of some recent research comprises a stack of 2-D panels, each of which may be rapidly switched from a transparent to a reflective state. The volumetric data set to be displayed is decomposed into a sequence of 2-D slices. In one possible implementation, a 2-D display device projects these slices into the image space in synchronization with the optical states (transmissive or reflective) of the panels. Thus the light representing the voxels within a given slice may freely propagate through intervening panels — in a transparent state — and reflect back from the panel at the corresponding depth within the image space. In this way, the full volumetric image may be depicted within the image space, provided that the switching rate is sufficiently rapid that the images appear free from flicker.

The reader will perhaps notice that systems of this type have many similarities to the translational motion swept-volume devices described in Chapters 4 and 5. In principle, a very similar graphics engine could be used for both. The major difference is that in the system outlined above the image space contains no mechanical motion, with the slice position within the image space being governed by the optical states of the panels rather than the instantaneous position of a moving surface of emission. A number of display devices based on this general concept have been proposed; two recent patents are U.S. patents 5,745,197 and 5,813,742.

7.8 DISCUSSION

Most of the ideas applied to the implementation of static-volume display units have been in existence for a considerable time. It would appear that these ideas resurface periodically and either work that has been previously undertaken is repeated or, alternatively, technological advances make a previously proposed technique more feasible. Clearly, the first of these scenarios indicates a failure on the part of researchers to undertake a proper search of the scientific literature prior to beginning their activity. However, the vast amount of material that has been published on the subject of volumetric systems (some of which is quite obscure) makes it probable that the results of any literature search will be far from complete.

The authors of this book have been unable to identify any ongoing research into the application of photochromic and thermochromic materials to volumetric systems. Given the advances that have occurred in these materials, it is most unfortunate that this promising area of study would appear to have been abandoned. The work being undertaken at the University of Texas at Dallas concerning the implementation of a system employing dielectric waveguides embedded within a solid and transparent image space medium is highly innovative. Rather than revisit previously suggested ideas, those working on this project have sought to apply entirely new techniques to the implementation of the display unit.

In this chapter we have discussed only a very small number of the static-volume systems that have been proposed to date. The reader who is interested in finding out more about the work that has been conducted on both static- and swept-volume systems is referred to the various on-line search engines that provide considerable patent information.

CHAPTER 8

GRAPHICS ENGINE: GENERAL CONSIDERATIONS

Eye, to which all order festers,
all things here are out of joint
Science moves but slowly slowly,
creeping on from point to point.*

8.1 INTRODUCTION

In this chapter we concern ourselves with the characteristics of graphics engines suitable for operation with some of the existing volumetric display unit technologies. Since the majority of these systems have voxel activation capacities of less than 100,000, we consider graphics engines in which the throughput of voxel data is no greater than this amount. By placing this limitation on performance, we can confine our discussion to systems in which the voxel descriptor throughput is, in the main, sequential and the need for parallel processing of information is avoided. Issues relating to the basic architecture of the graphics engine may then be considered without undue complication. During the course of this chapter, we occasionally refer to the more advanced graphics engines that will be required for the next generation of higher-performance volumetric display units. However, in general, discussion on the parallel throughput of voxel descriptors is deferred until Chapter 9.

The characteristics and performance of the display unit and the manner in which the system as a whole is used as an aid to visualization strongly affects the architecture of the graphics engine and the interface between it and any host

---

* Alfred, Lord Tennyson, *Locksley Hall*, in *Complete Poetical Works of Alfred Lord Tennyson*, Harper & Brothers, 1884.

computer system. To date, most volumetric displays have operated as stand-alone entities used within a research environment. Therefore, there has been little need to consider their integration with commercially available hardware systems and software packages. In the next section we highlight some of the issues associated with the application of volumetric systems to real-time control and interactive design. The authors are unaware of any volumetric technology that has been used in these roles for purposes other than evaluation. Consequently, the impact of this type of application on the architecture of the graphics engine is omitted from subsequent sections of this chapter. In the following pages we limit our discussion largely to situations in which the display unit acts simply as a peripheral device interfaced to a host computer.

In Section 8.4 we identify the major constituent subsystems within the graphics engine, and in Section 8.5 we describe an elementary graphics engine intended for use with an example swept-volume display equipped with a passive surface of emission.

In conclusion, this chapter focuses on the characteristics of graphics engines suitable for use with many of the volumetric display unit technologies proposed to date. Issues relating to more advanced systems are deferred until Chapter 9. Finally, architectures discussed in this chapter are not intended to represent optimal configurations but have been selected because of their descriptive merit and general simplicity.

8.2   APPLICATION OF VOLUMETRIC DISPLAY SYSTEMS

Image data, derived from a number of sources, may after suitable preprocessing be depicted with great clarity on volumetric systems. As outlined previously, displays of this type offer to assist in the visualization process by presenting information in such a way that it may be viewed directly and, in the case of some implementations, with very little viewing angle restriction.

In general, the current range of volumetric display unit technologies give rise to systems best suited to the display of sparse collections of simple objects in situations in which the spatial separation, the general 3-D form, or the dynamic properties of objects are of particular interest. These displays are inherently more suited to the depiction of qualitative geometrical information rather than quanti-tative data; particularly in the case of systems that permit considerable freedom in viewing orientation. For example, difficulties clearly exist in the depiction of alphanumeric information. When this information is presented within an image space, its clarity will clearly be affected by its orientation relative to each observer. A slow rotation of such information around a vertical axis so as to allow it to be read from any direction provides, in principle, a simple solution to this problem. Unfortunately, this also imposes a restriction on the speed at which information may be read and therefore assimilated. An alternative solution may be found by replicating the information on, for example, the vertical faces of a text icon (perhaps cubic in form). Unfortunately, the translucent nature of the images generated by current volumetric technologies may result in textual information presented in this manner being difficult to interpret. Both these solutions limit the amount of information that may be read with ease, so it is preferable to keep the amount of text (including quantitative information) to a minimum. As a consequence, we envisage that volumetric display systems will generally be used in conjunction with conventional displays, the latter being more suited to the depiction of alphanumeric information. In many types of application, the volumetric display therefore ideally forms a peripheral device, complementing rather than replacing the conventional computer screen. When so used, the graphics engine should interface conveniently to existing computer hardware and may perhaps use the host computer's processing facilities for the computation of voxel descriptors.

The data presented to a volumetric display may originate from measurements of real-time events such as air-traffic control.* In this application aircraft could be represented as icons and their designated flight paths by vectors. Clearly, in the case of real-time control systems, great emphasis must be placed on reliability since the failure of the system to respond to, or faithfully present, information to an operator may have catastrophic consequences. For a volumetric display to be used in such an application, it is vital that the display unit and graphics engine have a bandwidth that is sufficient to ensure no inadvertent loss of information! Uniformity of the image space is also of fundamental importance in ensuring the faithful reproduction of spatial relationships. For example, as the aircraft icons move throughout the image space, they must not disappear momentarily or become distorted as they pass close to, or through, the axis of rotation of a swept-volume display employing rotational motion. This is a major problem in the development of display units employing this image space creation technique to applications of this type.

In a further scenario, the data presented to a volumetric system may have been acquired previously and stored in a host computer, or may arise from the results of a computer software applications program. Such data may be considered as nonreal-time in their relationship to current physical events. In this case it is practical for an operator to manipulate an image and ensure that it is depicted in the most advantageous way within the image space. This may, for example, involve filtering and scaling of an image data set and perhaps include the manual positioning of the image so as to minimize the impact of adverse image space characteristics on the perceived image quality. This process is time consuming and may require the operator to have a sound technical knowledge of the display unit architecture. Unfortunately, in the case of most current volumetric technologies, manipulation is likely to be necessary for all but the very simplest of images. Clearly, this is undesirable. Consequently, the development of image spaces exhibiting isotropic and homogeneous characteristics is of considerable importance.

Volumetric systems offer to provide a convenient method for the manipulation of, for example, 3-D splines and nonuniform rational B-spline (NURBS)

---

* The consequences of the different vertical and horizontal scaling factors that may be required in such an application are discussed in Section 10.4.1.

surfaces. An operator using computer-aided design (CAD software) to generate curved surfaces is often concerned (at least in the first instance) with aesthetic appearance. The ability to prototype rapidly and interact in an intuitive and natural manner with image scene components is of great value. In employing volumetric systems for an application of this type, we are exploiting their greatest potential — as an aid for qualitative (rather than quantitative) visualization.

It is instructive at this point to consider two situations in which an operator undertaking design work may use a volumetric display system. In the first and simplest scenario the operator may wish to create a design on a conventional computer screen and observe the work as it progresses within the 3-D image space of a volumetric system. Whereas a conventional screen depicts only one (or several) views of the object under construction, the volumetric system is not so limited and therefore is likely to require the complete 3-D description of the object(s) depicted within the image space. In this case, the CAD package would need to be equipped with a device driver able to support the full data requirements of the volumetric display.

Greater complexities arise when we consider a second and perhaps more desirable way in which the volumetric display could be employed during the design process. Rather than create and manipulate image components on a conventional computer screen, the operator may wish to interact directly with them within an image space, by means of a 3-D mouse or pointing device (see Section 12.7). Any modifications made to the image scene components within the image space would immediately need to be reflected in a quantitative (and perhaps graphical) manner on the conventional display. In this case, the volumetric system would need to be highly integrated with both the software and computer hardware environment.

8.3 CONCERNING COORDINATE SYSTEMS

Volumetric image data will be assumed to be stored by a host computer in the form of an array of individual data points and/or high-level graphics primitives. Each of the points* will be defined by its location in 3-D space, and each will have an associated intensity and perhaps color descriptor (i). While the system used to describe the position of each data point in 3-D space is arbitrary, we shall assume the use of a right-handed Cartesian coordinate system (Figure 8.1a). Each image data point may therefore be represented in the form[†]

$$\{u, v, w, i\} \qquad (8.1)$$

A similar right-handed Cartesian coordinate system is assumed to be employed for the description of potential voxel locations within image space (the image

---

* Since the image data may be filtered, there is not necessarily a one-to-one correspondence between the image data "points" and the voxels activated within the image space.

[†] In fact, the source data are likely to comprise several sets of objects represented in different local coordinate systems.

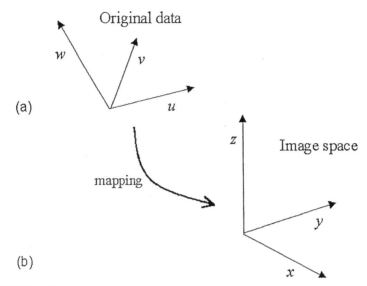

FIGURE 8.1 Right-handed coordinate systems used to describe (*a*) the original image data set and (*b*) voxel locations within the image space frame.

space frame). In practice, the technique employed for image space creation may affect the preferred position of the origin. However, unless otherwise stated, it will be assumed that the origin of the coordinate system is located at the physical center of the image space (Figure 8.1b).* Each voxel depicted within image space may therefore be represented in the form

$$\{x, y, z, i\} \tag{8.2}$$

As may be seen from Figure 8.1, we assume that the z axis is in the vertical direction. This coordinate system is a natural choice for a system employing rotational motion.

8.4 COMPONENTS WITHIN THE GRAPHICS ENGINE

For our present purposes, we define the graphics engine as the integrated combination of hardware and software subsystems responsible for the manipulation and flow of image data into the display unit. Figure 8.2 illustrates the general architecture of a simplified graphics engine that will be used as a model within this chapter. The principal functions of each of the major components are summarized below.

- The graphics processing subsystem is responsible for the transformations and mappings needed to fit image data into the volumetric image space and its

---

* Clearly, both cylindrical and spherical coordinate systems could also be used. Our choice of a rectangular coordinate system is therefore arbitrary. The type of coordinate system is best determined by considering both the physical shape of an image space and the image space creation technique.

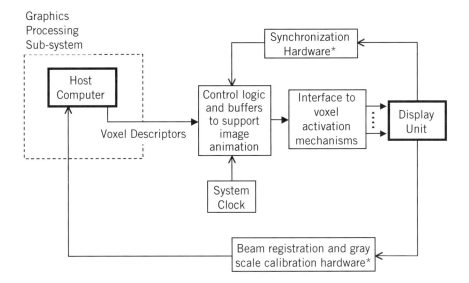

*Required by some types of display unit architectures

FIGURE 8.2 Subsystems of a basic graphics engine. Some or all of these components are present in any graphics engine intended for use with a volumetric display.

conversion into a form suitable for the voxel activation process. The image data are initially mapped into a set of Cartesian coordinates — with associated intensity and color descriptors — each corresponding to a preferred voxel location within the image space. These are then converted into coordinates suitable for use by the voxel activation subsystem.

- The elements within this set are referred to as *voxel descriptors*. As will be discussed shortly, in the case of swept-volume and some static-volume display units, the graphics processing subsystem will also be responsible for ordering the voxel descriptors.

- The storage subsystem acts as a buffer between the graphics processing system and display unit interface. Voxel descriptors may be arranged in this facility in various ways, such as in the form of a sequential list.

Many display unit technologies (particularly swept-volume systems) require an ordering of the voxel descriptors. In this case, the position of the voxel descriptor in a list may determine the time at which the descriptor is output to the display unit. In such a scenario it is arranged that this time (t_{op}) corresponds to the surface of emission passing through the appropriate location. In the case of such display architectures the voxel descriptor is assumed to take the form

$$\{x, y, i\} \tag{8.3}$$

Here the generalized coordinates x and y may reference individual voxel generation centers (for systems employing an active surface of emission) or denote the beam deflection coordinates (for systems employing a passive surface of emission).* Although voxel ordering is a general requirement in the case of swept-volume systems, it is also necessary in the case of some static-volume display unit technologies (e.g., see Section 6.8).

In the case of, for example, a static-volume display unit employing a matrix of voxel generation centers, the voxel descriptor could take several forms.

- The interface subsystem is responsible for the conversion of voxel descriptors into the form required by the display unit's voxel activation mechanism. For example, in the case of beam-addressed systems, the digital voxel descriptor values may require conversion into analog voltages.

- The synchronization mechanism is responsible for the controlled flow of voxel descriptor information from the storage facility to the display unit. In the case of, for example, a swept-volume display unit, the synchronization mechanism is responsible for determining the position of the surface of emission and ensuring that voxel descriptors are output to the voxel activation mechanisms at the most appropriate time.

- For display units employing multiple beams in the voxel activation process, the graphics engine may also be equipped with hardware responsible for ensuring the proper alignment (registration) between the beam deflection apparatus and image space coordinate system. Furthermore, it may be necessary to include a calibration system for the gray scale of each beam source.

Since the performance of the graphics engine must accommodate the peak voxel descriptor transfer rate demanded by the display unit, it is vital to ensure (at the design stage) that each of the subsystems can sustain the required throughput. As the demands of the display unit increase, it will ultimately become necessary to introduce parallelism within some or all of the graphics engine subsystems. In this chapter we assume, for simplicity, that the display unit performance can be accommodated by the sequential throughput of information. However, should the display unit support a plurality of voxel activation mechanisms, it is clearly necessary to permit parallel data flow within the display unit interface subsystem.

8.5 A GRAPHICS ENGINE FOR USE WITH A SWEPT-VOLUME DISPLAY

In this section we provide insight into the functionality and operation of the graphics engine by considering, in some detail the implementation of a basic system for use in connection with a swept-volume display unit. For clarity, the example chosen is intended to operate in an entirely sequential manner in its processing

---

* The temporal information is assumed to be determined by the position at which each descriptor is located within the buffer(s).

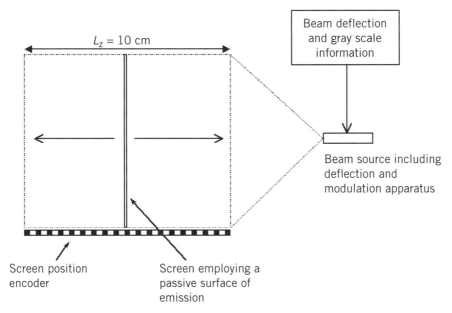

FIGURE 8.3   Example TM prototype employing the translational motion of a passive surface of emission.

and throughput of image data. To avoid the inclusion of parallelism even within the interface subsystem, the display unit selected for use in this example is equipped with only one voxel activation mechanism. We therefore consider a display unit able to depict monochrome images and operating with a single beam source that is able to address the entire image space. To simplify this display unit as far as possible, we assume that the image space is created by means of translational motion. It will be recalled from Section 3.3 that the use of beams within a swept-volume display unit indicates that the voxel generation subsystem employs a passive surface of emission. The display unit is illustrated in Figure 8.3. So as to avoid any possible confusion between the elementary model display unit introduced in Section 1.3 and the example system employed within this chapter, we refer to the latter as the translational motion (TM) prototype display unit.

8.5.1   Example Display Unit

We have not as yet identified the physical nature of either the voxel activation mechanism or the surface of emission for use with our example display unit. The manner in which these subsystems are implemented can have a significant impact on the voxel activation and voxel location capacities and also determine the maximum attainable voxel density of the display unit. As a consequence, they affect the peak voxel data transfer required of the graphics engine. Furthermore, the choice of voxel activation mechanism has consequences for the electrical characteristics of the interface between the graphics engine and the display unit.

In this discussion we assume that the display unit employs an electron beam for voxel activation and conforms to the partial specification provided in Table 8.1. Voxel generation will be achieved by means of a phosphorescent material deposited onto the supporting structure. Deflection of the electron beam to appropriate locations on the surface of emission is most easily attained by means of electrostatic fields. Such a deflection mechanism forms an integral part of some types of electron gun (see, e.g., Plate xiii).

The frequency of the motion of the surface of emission must be sufficiently high as to ensure that images are free from discernible flicker. We therefore assume that the surface of emission must complete each cycle of its motion at a frequency of at least 25 Hz. Although voxels may be drawn on the surface of emission during each half-cycle of its movement (i.e., as it moves in either direction), it will be recalled from Section 3.5.1 that image flicker cannot be reduced uniformly by refreshing voxels during each half-cycle period. However, the activation of voxels during both half-cycle periods ensures that the greatest possible use is made of the available time, and this maximizes the number of voxels that can be depicted within image space during each image refresh period.

TABLE 8.1 A Partial Specification for the TM Prototype Display Unit

Physical Dimensions

Image space dimensions:	$L_X = 20$ cm $\quad L_Y = 20$ cm $\quad L_Z = 10$ cm
Translational motion period:	$\tau_s = 1/25$ s
Distance between center of surface translation range and electron gun:	$D = 25$ cm
Number of slices:	$n = 200$

Voxel Activation: The Electron Gun

X deflection:	20 V/degree
Y deflection:	20 V/degree
Blanking voltage (required for beam extinction):	100 V
System adjustment time:	$T_S = 0.5$ µs (we assume this is limited by the amplifiers driving the electrostatic deflection plates, which form a capacitive load)
Beam acceleration voltage:	4 kV

Voxel Generation: The Phosphor

Voxel activation time:	$T_V = 0.5$ µs (this is defined as the time required to generate a voxel of satisfactory intensity at the maximum electron beam current)

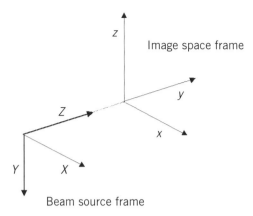

FIGURE 8.4 Beam source (electron gun) and image space coordinate systems. Although we assume in this section that these coordinate systems are simply translated with respect to each other, this will generally not be the case. Physical inaccuracies will most probably result in the two sets of axes being misaligned (see Section 9.8).

It is convenient to assume that the beam axis is normal to the surface of emission, intersects the origin of the image space coordinate system, and that the horizontal and vertical axes of both coordinate systems are parallel. However, in practice, even if great care is taken in the assembly of the display unit, there will be some uncertainty and error in placement of the electron gun coordinate system with respect to that of the image space.[*] Since these uncertainties give rise to error in voxel positioning, it is apparent that their impact will be greater in systems able to produce voxels at higher densities. This matter is of great consequence for all display units employing beams for voxel activation. However, to avoid unnecessary complication, we will assume error-free alignment.

The two pairs of electrostatic deflection plates with which the electron gun is equipped are responsible for beam deflection in the horizontal (X) and vertical (Y) directions. We associate with the voxel activation source the coordinate system illustrated in Figure 8.4. This figure also shows the coordinate system attributed to the image space, which was adopted in Section 8.3.

8.5.2 Processing of the Image Data by the Graphics Engine

Data contained within an image file form the input to the graphics engine. The output takes the form of a series of voxel descriptors that not only match the geometrical characteristics of the display unit but are also ordered in such a manner as to accommodate the temporal nature of swept-volume (and some static-volume) systems. The manner in which this conversion process is achieved

---

[*] Furthermore, the path followed by the electron beam will be affected by the presence of any stray magnetic field. This may result in significant beam positioning inaccuracy and will be of greater consequence for electron guns using lower acceleration voltages.

will vary according to the form of the source data (point-based or comprising high-level graphics primitives), animation (and interaction) requirements, and the display unit architecture. Let us briefly consider data taking the form of a set of graphics primitives that are to be displayed on our example TM display. Operations to be performed on the image data by the graphics processing subsystem prior to its depiction within image space are illustrated in Figure 8.5.

An initial task is to provide a mapping between the coordinate system employed within the image file and that assigned to the image space. This task may effectively be achieved through the use of homogeneous transformations as discussed in Section 9.8. By means of these transformations it is possible to carry out efficiently standard graphics operations such as image rotation, translation, and scaling and thereby enable an operator to manipulate* an image scene

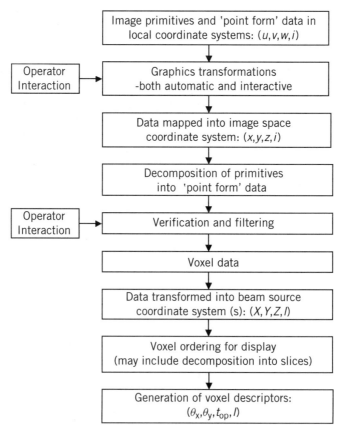

FIGURE 8.5 Processing of the image data by the graphics engine for the TM prototype display outlined in Section 8.5.1.

---

* In this case any interaction is assumed to be indirect, perhaps by means of a keyboard, mouse, or joystick. Direct interaction is discussed in Section 12.7.

depicted within the image space. The source data, denoted by the coordinate system (u, v, w), may in fact comprise several sets of objects represented in different local coordinate systems. In this model we begin by mapping the data into the image space frame (x, y, z) and simultaneously performing any necessary graphics manipulations, such as rotations, translations, or scaling. * Having united all data within the image space frame, graphics primitives must be decomposed into point form. These point-form data may not necessarily correspond directly to voxel locations within an image space.[†] For example, the density of points may exceed the bandwidth of the display unit. Some type of verification process should therefore be included so as to ensure that data will not be inadvertently omitted as a consequence of performance limitations. If necessary, a filtering mechanism may be provided which is able to reduce the number and density of points according to defined criteria. This is likely to necessitate the intervention of the operator.[‡] Whatever the precise form of this process, a set of coordinate values will ultimately be generated, each corresponding directly to the location of a voxel within the image space.

The subsequent operations of the graphics processing subsystem are largely dependent on the architecture of the display unit. In the case of the TM prototype, we must next determine the appropriate signals that must be applied to the beam deflection and modulation apparatus so as to produce each voxel at the required location within the image space. This involves a transformation from the image space frame into the beam source frame (X, Y, Z, i), where it is convenient to arrange for the X and Y axes of a Cartesian system to be parallel to the horizontal and vertical beam deflections and the Z axis along the path of the undeflected beam. The 3-D spatial representation is then reduced to one comprising the two beam deflection angles and the relative time at which the beam is to be directed to the required location $(\theta_X, \theta_Y, t_{op}, I)$. The final voxel descriptor thus comprises beam deflection, beam intensity, and timing information. In the case of a beam-addressed swept-volume display unit, the voxel output time t_{op} coincides with the surface of emission passing through the voxel position in the image space. As we will see in Section 8.5.4, this time is generally used to identify the slice within which the voxel is to be placed.

Other display unit architectures have different requirements. Consider, for example, the case of a swept-volume beam-addressed display unit that employs more than one voxel activation mechanism (see, e.g., Section 5.6.3). Each beam source is represented by its own unique coordinate system, and therefore, before it is possible to transform a voxel's coordinates in the image space frame into the beam source frame, we must clearly identify the beam source that is to be made responsible for its activation. In the case of a swept-volume system employing an active surface of emission, the transformation into a beam coordinate system (referred to above) has no relevance. For a static-volume display unit in which

---

* It is preferable to carry out graphics operations on the image primitives prior to decomposition, as this reduces the necessary computation.

† To avoid confusion we have used the term *point-form data*, not *voxel-based data*.

‡ Clearly, the arbitrary culling of data should not take place without warning.

voxel activation occurs within the region of intersection of two beams (see, e.g., Section 7.5.2) each voxel descriptor will contain two pairs of beam deflection angles, and if ordering is required, will also store a temporal descriptor.

8.5.3 Generation of Voxel Descriptors for the TM Prototype

In Section 8.5.2 we outlined a scheme permitting an image data set denoted by the coordinate system (u, v, w, i) to be mapped into image space coordinates $\{x, y, z, i\}$, and thence into the beam source coordinate frame represented by $\{X, Y, Z, I\}$. In the case of the TM prototype display, the next stage in the production of each voxel descriptor entails the calculation of a pair of beam deflection angles and the voxel output time t_{op} corresponding to the surface of emission passing through the required voxel location (in the Z direction). The gray scale* (I) (assumed to be determined in our example display unit by the magnitude of the beam current[†]) is included within the final voxel descriptor, which takes the form $\{\theta_X, \theta_Y, t_{op}, I\}$.

Clearly, the beam deflection angles depend on the spatial coordinates (X, Y, Z) of the voxel in the beam source frame (Figure 8.6),[‡] and so

$$\theta_X = \arctan \frac{X}{\sqrt{Y^2 + Z^2}} \tag{8.4}$$

$$\theta_Y = \arctan \frac{Y}{\sqrt{X^2 + Y^2}} \tag{8.5}$$

To activate a voxel at the appropriate position within image space, it is necessary to synchronize voxel activation with the motion of the surface of emission. Unfortunately, in the case of a display unit in which voxel activation takes place sequentially, each voxel must be activated at a unique time. Difficulties arise when a plurality of voxels has the same output time (t_{op}) and when differences in the output time are less than the voxel time. In such cases only one voxel may be activated at the required location and all the others will be subjected to a positioning error. This matter is discussed in some depth in the next section.

8.5.4 Voxel Ordering Considerations

In Chapter 2 we outlined the problems associated with exhaustively scanning an entire image space. Several of the early publications relating to swept-volume

---

* Note that the intensity descriptor I will be computed to accommodate the performance of the individual voxel activation mechanism(s). Also, as a consequence of the diversity of approaches that may be taken in implementing multiple color images, we have restricted ourselves to monochrome image generation.

[†] A gray scale may also be implemented by varying the voxel activation time.

[‡] Clearly, Eqs. (8.4) and (8.5) assume that both the X and Y deflection apparatus lie at the same distance from the origin of the image space coordinate system. As may be seen from Plate xiii, this is not the case for an electron gun employing electrostatic deflection.

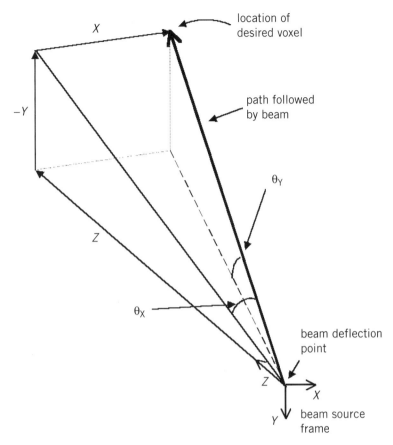

FIGURE 8.6   Geometry governing the transformations for calculation of the beam deflection angles. Coordinates are those of the beam source frame (XYZ).

architectures examine the feasibility of exhaustive image space scanning [e.g., Ketchpel (1963)] and concluded that it would result in at least one of the following:

- Poor image brightness
- Small image space volume
- Low scan resolution

As a consequence of these difficulties, swept-volume beam-addressed systems have generally employed a random scan in which the beam is directed only to voxel locations comprising the image. These are activated in a sequential fashion.* Such a random-scan approach is referred to here as a *dot graphics*

---

* In Chapter 12 we discuss in the context of current technologies the prospects for the implementation of exhaustive scanning.

technique.[*] The simplest method of implementing this type of scan involves ordering the voxel descriptors so that their output follows the direction of screen motion. This is outlined in Section 3.5.4 for a system employing the translational motion of a passive surface of emission. For voxels that are sparsely distributed in the direction of the screen's motion, this procedure works well. However, as we will see, the ordering problem becomes more complex when dense voxel distributions are encountered.

Voxel Ordering in the Case of Translational Motion

Consider a screen moving with a sinusoidal velocity profile described by Eq. (3.5). The activation of each voxel occupies a finite time [the voxel time (T)] and during this period the screen will have moved a distance (δZ), which may be expressed as

$$\delta Z = \omega_s T L_Z \cos \omega_s t \tag{8.6}$$

where $\omega_s = 2\pi/\tau_s$. Thus, if the required spatial separation of two or more voxels in the Z direction is less than δZ, a compounded positioning error is likely to occur (cf. Figure 3.10). Clearly, the inability of our example TM prototype to accommodate parallel voxel activation compounded with the velocity of the surface of emission in the Z direction and the finite voxel activation time excludes the activation of more than two voxels[†] within the same (X, Y) plane. Furthermore, should the image data set contain a large number of coordinates with similar Z values, unacceptable image distortion may occur due to compounded positional errors in the direction of screen motion. This type of problem is inherent in all beam-addressed swept-volume display units, and it becomes a responsibility of the graphics engine to minimize its impact.

The ability of a display unit to handle dense clusters of voxels may be improved by identifying the optimum order in which they are output to the voxel activation mechanism(s). Any optimum solution will clearly relate to the spatial form of the voxel data contained within a particular image frame. In its most general form, the ordering process represents a combinatorial optimization problem in three dimensions, these corresponding to the two beam deflection angles and the instantaneous screen position. Computationally, this represents an enormous exercise.

The ordering problem may be simplified by dividing the image space into a series of slices.[‡] Voxel descriptors are then mapped into the most appropriate slice. The number of voxels that may be accommodated within each slice may be limited according to the temporal duration of each slice and the individual voxel time. So as to maximize the number of voxels that may be placed within each slice, we may seek to ensure that they are output to the display unit in an

---

[*] After Harris et al. (1986). See also Fuchs et al. (1982), Sher (1993), and Schwarz and Blundell (1997).

[†] One voxel during each half-cycle of motion.

[‡] Or sectors in the case of display units employing rotational motion.

optimum manner. The 3-D ordering problem referred to above then reduces to a series of 2-D ordering tasks, at the expense of ignoring positional information in the direction of screen motion of individual voxels within each slice. This ordering process represents a form of the classic traveling salesman problem (Lawler et al. 1985) and is likely to lead to an increase in the voxel activation capacity of the display unit (Schwarz and Blundell 1997).

Identifying the optimum sequence of voxel output within each slice involves minimization of the total path traversed by the beam. This will minimize the total system adjustment time and thereby reduce the average voxel time. The increase in performance achieved by this means will depend on the spatial distribution of voxels within each 2-D slice. Consequently, this may degrade the predictability of the display unit. For example, when the voxel activation capacity of a system is quantified, should the figure include any gains that may be obtained by ordering the data within each slice? In this case it would be necessary to indicate the spatial distribution of the data used.

Combinatorial problems of this type are notoriously difficult to solve, and in practice the optimal solution can usually not be achieved (Lawler et al. 1985). However, heuristic algorithms have been devised [e.g., Golden et al. (1980); Golden and Stewart (1985)] that can provide a significant increase in the efficiency of voxel activation, as illustrated in Figure 8.7.

In principle, ordering offers to increase the voxel capacity of each slice. However, it is likely that on occasion the density of data will nevertheless exceed the maximum achievable slice population. In such a situation, a number of approaches may be adopted. In the simplest (and most unsatisfactory) solution once the maximum slice population is reached, additional voxels are culled. Clearly, this may result in vital information being lost and may have catastrophic consequences. Alternatively, once the maximum population of a slice has been reached, the controlling software may reposition additional voxels into adjacent slices. Unfortunately, this may cause a ripple effect — these adjacent slices may themselves already have reached their population limit or may become overpopulated as a consequence to the relocation of voxels. The relocation process must then be extended to encompass surrounding slices. Should the selection of voxels for relocation be made on a random basis, it is possible that an individual voxel may be moved away from its proper position within image space by an unknown and possibly large (and therefore unacceptable) distance.

Adopting the following approach may ameliorate this problem. The controlling software first maps voxels into the most appropriate slice in accordance with their temporal descriptor. As this process continues, the population of each slice is monitored so as to ensure that that its temporal extent is sufficient to permit the depiction of all the voxels contained therein — without omission. Should a particular slice exceed its maximum voxel population, it is then necessary to relocate voxels into adjacent slices. The choice of voxels for relocation would then be decided by reference to their temporal descriptor. In this way, voxels that lie closest to the boundary of an overpopulated slice are relocated, in preference to those that are more centrally positioned within the slice.

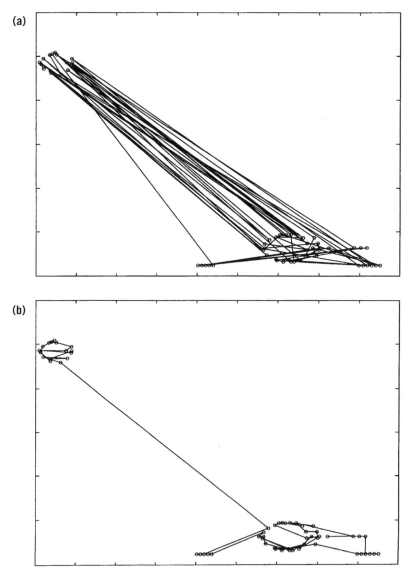

FIGURE 8.7   Voxels within an image slice, interconnected by the path followed by the beam (*a*) without and (*b*) with the inclusion of voxel ordering. The image depicted here is from an air-traffic control application; cross sections through an airplane and runway may be seen. The ordering algorithm employed considered the system adjustment time (beam deflection time) to be an integral number of small incremental deflections. For voxels separated by less than the minimal deflection distance, the algorithm had no effect; this may be seen, for example, in the beam path followed in activating the voxels comprising the airborne plane. However, the number of (unnecessary) long beam deflections has been reduced to a single extended deflection linking the two parts of the image present within this slice.

This approach fails to maximize the benefits that may be derived from any subsequent ordering process within each slice. As we have mentioned, the gains that may be derived from any ordering process are strongly influenced by the spatial form of the data. Clearly, the ordering criterion employed above does not attempt to identify the set of voxels that could be ordered most advantageously. Furthermore, after the ordering process, it would, in principle, be possible to add more voxels to fully populated slices. Should the image scene have been mapped into slices and overpopulation issues resolved prior to the optimization process, it is unlikely that the ensuing benefits could be fully exploited.

On occasion, the spatial form of the image data set may impose too great a demand on the display unit in terms of the required voxel density. Relocation of voxels between slices may fail as large groups of adjacent slices reach their maximum population levels. In such cases, the culling of voxels is likely to provide the only solution.

Ordering in the Case of Rotational Motion

In the case of swept-volume systems employing rotational motion, the ordering of voxel descriptors for output to the display unit may be achieved by dividing the image space into a series of sectors into which voxel descriptors are mapped. Should we seek to optimize the order of output of data within each of these sectors, the problem is identical to that described above for translational screen motion. However, should it be necessary to redistribute voxels into adjacent sectors, the selection of voxels is made more complicated by the nonuniform sector width.

8.5.5 Graphics Engine Hardware

In previous sections we have focused on the processing of the image data and the production of voxel descriptors suitable for transmission to the voxel activation apparatus. We now turn our attention to hardware aspects of the graphics engine and focus on issues relating to the example TM prototype display unit. As we have discussed, in the case of all swept-volume systems it is essential that the passage of each voxel descriptor to the relevant voxel activation mechanism takes place at the correct time. A synchronization mechanism is therefore needed to enable the graphics engine to maintain an accurate knowledge of the position of the surface of emission.

To avoid placing the burden of real-time data output on the host computer, it is preferable to provide a frame buffer between it and the display unit. This takes the form of RAM into which the host computer places the computed voxel descriptors. These descriptors are then clocked out of the buffer in synchronization with the position of the surface of emission. To permit image animation two distinctly separate frame buffers may be employed.* In this case the contents of one buffer may be modified by the host computer while the contents of the

---

* Other more elegant schemes are possible.

other are output to the display. Once the computer has completed updating one of the buffers, it is made available for output to the display, and the one that had previously been used by the display becomes available for update by the computer.

In Section 2.5 we introduced the voxel time parameter (T), this being the sum of the system adjustment time (T_S) and the voxel activation time (T_V). The magnitude of the system adjustment time is determined not only by the response time of the voxel activation subsystem(s) but also by the time needed to access a voxel descriptor from the frame buffer and apply this information (in a suitable form) to the voxel activation subsystem. The system adjustment time parameter is, therefore, as its name implies, composed of a number of system response times. The ordering of voxel descriptors as discussed in Section 8.5.4 may permit a reduction in the average system adjustment time by reducing the distance traversed by the activation beam as it illuminates a sequence of voxels. Otherwise, the system adjustment time is fixed for any particular combination of hardware.

The composition of the voxel activation time is determined by the nature of the voxel activation and voxel generation subsystems. In general we may consider

$$T_V = T_G + T_D \qquad (8.7)$$

where T_G represents the gating times of the voxel activation subsystem(s) and T_D the dwell time of the beam, this being the time required to produce each voxel at the required level of intensity. Considering the TM prototype, the value of T_G is determined by the time required to turn on the electron beam *plus* the time required to turn off the beam once a voxel has been activated to a suitable level of brightness. For this display unit, the dwell time represents the period during which the electron beam impinges on the phosphor-coated surface of emission.

For a given hardware and software configuration, it is apparent that changes in the voxel time may only be effected by changes in the dwell time. However, this will affect the image intensity. As will be recalled from Section 2.6, reducing the voxel time permits an increase in the voxel activation capacity of many types of display unit. As a display unit will generally operate under various ambient lighting conditions, it may be convenient to make provision for an overall scaling of the dwell time (by enabling it to be defined by the operator). This provides a mechanism by which the overall image intensity may be changed. By enabling scaling of the dwell-time parameter, it is possible to ensure that a display system operates at its maximum voxel activation capacity in a particular environment. For example, when a display unit is operated within a darkened room, it is possible to reduce the voxel time and so permit the depiction of images with a higher voxel activation capacity and therefore a higher voxel density.* So as to allow an operator to adjust the overall image intensity by this method, the voxel time should be stored as a user-defined variable. Since the voxel time defines

---

* Once more the performance of the system may be improved at the expense of predictability.

the total time required to depict each voxel, this time is used to clock voxel descriptors out of the frame buffer.

There are three other major considerations in the implementation of the graphics engine. The first relates to the method by which the position of the surface of emission is monitored, and the second to the manner in which the voxel descriptors are arranged within the frame buffers. The third concerns the form of the output hardware that interfaces to the display unit. These matters are considered below.

Synchronization of Data Flow

The synchronization hardware is responsible for ensuring that each voxel descriptor is output to the display unit at the appropriate time. This therefore forms a vital part of any swept-volume display unit. The input to this hardware is derived from one or more transducers located within the display unit. Their function is to monitor the position of the surface of emission. The transducer(s) may measure this position once (or perhaps several times) during each cycle of motion, and this information may then be used to predict the future velocity profile of the screen (over the course of the next cycle).

The simplest and most satisfactory sensing configuration generates synchronization signals as the surface of emission enters (or exits) from each image space slice (or sector in the case of rotational motion). This information is complemented with an indexing pulse corresponding to the surface of emission passing a known location during each cycle of its motion. Small variations in the overall frequency of motion of the screen have little impact on perceived image flicker. However, the graphics engine must be able to track continuously both the position and velocity of the screen in an accurate manner so as to properly position voxels and thereby avoid image distortion. There are two primary types of synchronization error:

- The inability of the hardware to accommodate fluctuations in speed during the course of each cycle of motion
- Errors in measuring the motion period of the surface of emission

Small fluctuations in velocity (which are repeated during each cycle) are likely to give rise to gradual distortions and therefore may not necessarily cause any apparent distortion of the image scene. Furthermore, increasing the screen position sampling frequency will enable the graphics engine to compensate for any such variations in speed. For systems employing rotational motion, the assumption of uniform angular velocity during each rotation is generally satisfactory, as the inertia of the screen will assist in the elimination of any small variations that may be present. If necessary, a flywheel may be added as a further precaution. However, for those systems employing translational motion the screen will generally not move with a constant velocity, so we are reliant on the predictability of the velocity profile (see Section 3.5).

A failure of the synchronization mechanism to measure the periodic time of the surface of emission properly may often be very noticeable and will take the form of either a lack of image stability and/or an image discontinuity at the start-of-motion position. This position denotes the border between the first and last image slices (or sectors) output during each cycle of motion. It therefore determines the location at which the output of an image frame begins. Should an image component span this start-of-motion position, any incorrect timing may result in a readily discernible image discontinuity. In the case of the TM prototype (in fact, any system employing translational motion), this type of image discontinuity may be ameliorated by setting the start-of-motion position at one or other extreme of the screen's range of motion. Unfortunately, this simple solution will not prevent image instability and is not applicable to display units employing rotational motion.

Arrangement of Voxel Descriptors within the Buffers

Clearly, a variety of approaches may be adopted in the arrangement of voxel descriptors within the frame buffers. The discussion below is not intended to represent an optimal design solution but one of simplicity. We assume that the frame buffer memory is divided into a series of pages into which voxel descriptors are mapped. In this case, the number of pages will equal the number of image space slices (or sectors). If we determine a minimum operating frequency for the motion of the surface of emission, define the number of slices into which we wish to divide the image space, and identify a minimum value for the voxel time, we can deduce the upper limit to the number of voxels that may be contained within each image space slice. From this information we are able to define the extent of each page within the frame buffer, this corresponding to the maximum number of voxel descriptors that may be output to each image slice.

In this simple embodiment, voxel descriptors are downloaded from the host computer and placed within the appropriate frame buffer pages. At the end of this process, pages that are not completely filled with voxel descriptors may be padded out with *null voxel descriptors*, dummy descriptors that have their intensity field set to zero and therefore do not give rise to visible voxels within image space. The presence of the null voxel descriptors is therefore to ensure that spurious voxels are not generated within image space when nonoccupied memory locations are addressed. This memory model, which is illustrated in Figure 8.8, allows changes to image objects to be effected in an efficient manner. Only pages that change between consecutive image frames need to be updated. For animated scenes, this provides the graphics engine with the ability to update the memory buffer at a higher speed than would otherwise be possible.

We now turn our attention to the events that take place during the output of an image frame to the TM display unit. We begin by assuming that the surface of emission has completed one or more cycles of its motion and that the graphics engine has acquired information on the period of the previous cycle of the screen. From this information the temporal extent of subsequent image space slices may be deduced. On receipt of an indexing signal that confirms the physical position of the surface of emission, an address pointer is loaded with the address of the appropriate RAM page, and the first voxel descriptor contained therein is passed

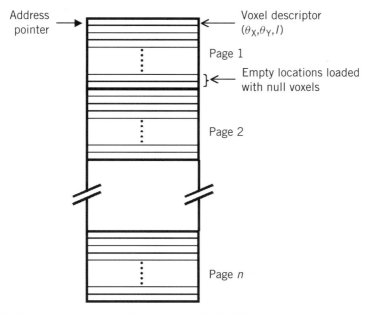

FIGURE 8.8   Storage of voxel descriptors within RAM using pages; the extent of each page corresponds to the maximum number of voxels that may be located within each slice.

to the voxel activation mechanism. At regular time intervals (T) (as determined by the voxel clock), the address pointer is incremented and subsequent voxel descriptors are downloaded to the display unit. In the background, the image space slice clock is operating, and when this clock determines that the temporal and spatial extent of an image space slice has been reached, the address pointer is loaded with the start address of the next RAM page. The output process is then repeated.

While these events are proceeding, a timing circuit responsible for measuring the periodic time of the current cycle of motion of the surface of emission is in operation. This information will be used for synchronization operations during the next cycle of motion. So as to permit image animation, the graphics engine will automatically alternate in its use of the two image buffers; exchanging their purpose at the end of each cycle of motion or image update period.

Passage of Voxel Descriptors to the Display Unit

In the case of the example TM display unit, the digital information contained within each voxel descriptor must be converted into analog form and suitably amplified prior to its application to the voxel activation mechanism (i.e., the electron gun). This process is illustrated in Figure 8.9. The $\{\theta_X, \theta_Y\}$ deflection angles together with the intensity component I of each descriptor pass through digital-to-analog converters prior to amplification. The beam deflection amplifiers are generally dc coupled so as to permit the passage of both static offset voltages and time-varying signals. The amplification stages will generally be equipped

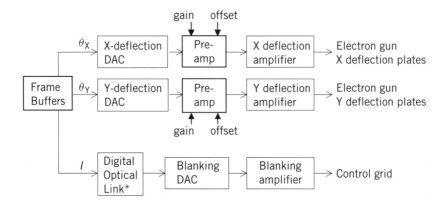

FIGURE 8.9 Interface subsystem. Voxel descriptors are converted into a suitable form before being applied to the electron gun.

with both offset and gain controls to facilitate the initial setup and calibration of the system. The output from these amplifiers is then applied to the beam deflection apparatus.

A problem associated with the use of electron guns for voxel activation arises from the required electrode potentials. A typical electron gun fitted with electrostatic deflection plates may have a beam accelerating potential of perhaps 4 kV. This is the dc potential between the cathode and final anode. The deflection electrodes normally operate at a voltage centered about the anode potential and the beam current control electrode (grid) operates at voltages below that of the cathode. As it is common practice to operate the final anode at voltages close to the ground potential (this facilitates coupling of the deflection signals to the deflection plates), it follows that the cathode electrode and grid are at a high potential (e.g., −4 kV). The coupling of the intensity modulation signal into the electron gun is therefore complicated by the need to provide electrical isolation between those signals generated by the graphics engine, which are referenced to ground potential, and the grid, which is referenced to perhaps −4 kV. As a consequence it is necessary to pass the intensity signal(s) through, for example, an optical link, to maintain electrical isolation as indicated in Figure 8.9.

The design of both the beam deflection system and intensity modulation circuits is critical, as their bandwidth must be maximized. A circuit diagram of a beam deflection amplifier able to generate a differential output voltage of approximately 700 V is illustrated in Figure 8.10.

8.6 DISCUSSION

Whatever architecture is employed in the implementation of the display unit, it is necessary for the graphics engine to transform source image data from its original

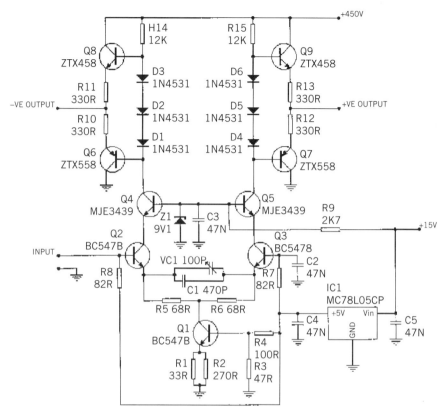

FIGURE 8.10 Amplifier with differential output suitable for use for electron beam deflection. (Courtesy of Keith Lewis).

representation (which may comprise point-form data and high-level graphics primitives described in various coordinate systems) into a common image space coordinate frame. Following the transformation and manipulation process, image data must be decomposed into point-form volumetric data. Subsequently, verification should take place to ensure that neither the volume of data nor its spatial density exceeds the capabilities of the display unit. The operator should be actively involved in any filtering process that may prove to be necessary and therefore may be confident that data will not be arbitrarily omitted from an image scene. Subsequent operations that are carried out on the data are generally determined by the architecture of the display unit.

In the case of swept-volume systems the image space is commonly divided into a number of slices (for translational motion) or sectors (for rotational motion). Should deflected beams be employed for voxel activation, ordering the data so as to minimize the total distance traversed by the beam(s) may increase the number of voxels that can be included within a slice or sector. To reduce the computational overhead, this ordering can be done in an approximate manner.

Swept-volume systems that possess minimal parallelism in the voxel generation and activation processes exhibit a low voxel activation capacity and are highly anisotropic in their ability to depict dense voxel distributions. The ordering referred to above may assist in this area, but any small advantage gained in this way will vary according to the nature of the image depicted by the system. This lack of predictability is highly undesirable. In Chapter 12 we examine the possible development of systems exhibiting a high degree of parallelism in the voxel generation and activation processes. We anticipate that this type of approach will permit a large voxel activation capacity to be achieved, and furthermore, will improve many aspects of the display unit predictability, homogeneity, and isotropy.

CHAPTER 9

GRAPHICS ENGINE:
FURTHER CONSIDERATIONS

Too exhausted with effort to eat they fall
Asleep; soft slumber seals their eyes up;
The delirium dies in a dreamy swoon.[*]

9.1 INTRODUCTION

The underlying technology employed in the implementation of a display unit affects the nature of some of the manipulations that must be carried out on the image data stream. Furthermore, it often determines the degree to which the display is capable of accepting the parallel transfer of data and thereby its potential for achieving a high voxel activation capacity. As discussed in Section 2.6 and elsewhere in this book, significant increases in the voxel activation capacity may be achieved through the introduction of parallelism in the voxel activation process. We believe it possible that the next generation of high-performance volumetric display unit will support voxel activation capacities in excess of 50 million.[†] Consequently, if we assume a refresh frequency of 30 Hz, it will be necessary for the graphics engine to provide a peak throughput of voxel descriptors, which is of the order of 1.5×10^9 s^{-1}. To support such a data throughput it will clearly be necessary to provide multiple data paths and accommodate parallel processing within the graphics engine.

---

[*] Catullus, *Poems*. Translation by James Michie, Random House, NY, 1969.
[†] This figure is intended to provide an *indication* of the magnitude of the voxel activation capacity that must be achieved to exhaustively scan an image space.

In the case of many volumetric technologies (particularly beam-addressed systems), as the degree of parallelism is increased, considerations such as the positioning and spatial separation of voxels within the image space, coupled with their rate of generation, become increasingly complex. Rather than consider an image space in its entirety, it is therefore advantageous to divide it into a number of regions that can be considered with greater clarity. With this in mind, we propose in Section 9.3 the concept of an image space comprising a number of subspaces and use this idea in Sections 9.4 and 9.5 to develop a mapping notation that may be used to describe the flow of data from a graphics engine into an image space. This mapping technique provides a general formalism applicable to all volumetric displays and is of particular value in the characterization of the performance of a display unit and for defining the interface between it and the graphics engine. The value of this approach is demonstrated by considering the impact of various display unit architectures upon the graphics engine.

In Section 9.7 we discuss a possible architecture of a graphics engine able to support the parallel throughput of voxel descriptors. The architecture of this graphics engine is developed using the concept of an image space comprised of a set of subspaces that may overlap both spatially and temporally.

In the case of display units that use a number of beam sources for voxel activation, the proper registration of the beams with respect to the image space (and consequently, to each other) is a major consideration. It is interesting to note that for such displays, registration difficulties will ultimately govern the maximum attainable voxel activation capacity. With this in mind we discuss in Section 9.8 an alignment technique that was used by the authors in the implementation of a swept-volume prototype (see Sections 5.6.3 and 11.2). Although this approach was developed for a display unit prototype operating with only two beam sources, it may, in principle, be extended to more advanced configurations.

9.2   CHARACTERISTICS OF THE DISPLAY UNIT AND THEIR IMPACT ON THE GRAPHICS ENGINE

Before proceeding, let us briefly review some of the features of a display unit that affect the architecture of the graphics engine and will be of importance to us in our subsequent discussion. We are able to identify three key areas of consideration:

- The degree to which the display unit supports parallelism in the process of voxel activation. This affects the number of data paths that must be supported by the graphics engine. We assume that the parallelism exhibited by the voxel activation mechanism(s) is supported by the voxel generation subsystem.
- The peak data rate, which may be supported by the display unit and must be accommodated by the graphics engine. This is largely determined by

the display unit's voxel activation capacity and/or maximum voxel density characteristics.

- The extent to which the voxel data must be ordered prior to its passage to the display unit. This is of particular consequence in the case of swept-volume and beam-addressed static-volume systems.

In practice, many other features of a display unit's architecture may also have a significant, and often subtle, impact on the requirements that must be accommodated by the graphics engine.

The degree of parallelism offered by a display unit technology forms a starting point when considering the architecture of the graphics engine, as this will affect the number of data paths that must exist in the interface between the graphics engine and the display unit. Within the next section, we define some necessary terminology and introduce the concept of subspace. Subsequently, we develop mappings that may be used to indicate the parallelism supported by the various types of display unit technologies.

9.3 IMAGE SUBSPACE

The expression *voxel activation pathway* will be used when referring to each data link* originating from the output stage of the graphics engine and connecting to one or more voxel activation mechanisms (e.g., electron gun, laser source, optoelectronic element, etc.). As we have seen, (previously) these voxel activation mechanisms respond to voxel descriptor information passed to them from the graphics engine by stimulating voxel generation at the appropriate locations within an image space. A single voxel activation pathway may be responsible for the transmission of voxel descriptor information to a number of voxel activation mechanisms. For example, consider the elementary system model described in Section 1.3, employing the row and column addressing technique outlined in Section 6.4. In such a configuration the number of voxel generation centers differs from the number of voxel activation pathways. Consequently, it is necessary to distinguish between the two. In subsequent discussions we also make use of the expression *voxel activation subsystem* which will be assumed to comprise a single voxel activation pathway and its associated voxel activation mechanism(s).

It is convenient to consider the physical region addressed by each voxel activation mechanism as a *subspace* of the image space (see Blundell 1998). Each activation mechanism is therefore responsible for voxel activation within a corresponding subspace and has no responsibility elsewhere within an image space. To illustrate this idea, let us contrast the example TM prototype display unit introduced in Section 8.5.1 with the elementary static-volume model described in Section 1.3. In the case of the former display unit, the entire image space is addressed by a single voxel activation mechanism—an electron gun. The image

---

* A serial or parallel connection able to transfer a voxel descriptor in its entirety.

space is therefore said to consist of a single subspace. Since there is only one voxel activation pathway through which voxel descriptors may be passed to the display unit, it is apparent that there is only one voxel activation subsystem. In contrast, the elementary system model has an image space consisting of N filament lamps, each being responsible for the generation of a single voxel. If a unique pathway connects each of these voxel generation centers to the graphics engine, there exist N voxel activation subsystems and the image space comprises of N nonspatially overlapping subspaces. Furthermore in the case of this configuration, the display unit may exhibit maximum parallelism (although the subspaces are spatially separate, they overlap completely in the temporal domain).

Alternatively, to reduce the degree of electrical connectivity between the graphics engine and the display unit, we may apply to this elementary model a row-and-column addressing technique as discussed in Section 6.4. In accordance with the definitions above the number of voxel activation subsystems is reduced correspondingly. This implies a reduction in the temporal overlap of the subspaces.

Swept-Volume Example   Let us now consider the swept-volume display unit outlined in Section 5.6.3. In the configuration where two electron guns are used to address the image space (Figure 9.1) the image space consists of two subspaces. The degree to which the subspaces overlap in both the spatial and temporal domains depends on the geometrical configuration of the electron guns and the fraction of the image update period for which it is possible for them to address the image space. Let us suppose that as depicted in Figure 9.1a, each of the guns addresses its respective subspace for only one half of the image update period, and also that the entire image space may be addressed during a single revolution of the surface of emission. In this case, there is no spatial overlap between the subspaces; although two voxel activation subsystems exist, only one may be operational at any one time. As a consequence, the information flow through each voxel activation mechanism is not continuous so neither they nor their associated pathways are used to full advantage.

With the voxel activation capacity of the display unit denoted by N_a and the image refresh period by τ_r, each of the voxel activation subsystems is responsible for the activation of up to $N_a/2$ voxels during each refresh period. For this configuration, if complete spatial and temporal separation of the subspaces is maintained, it is apparent that each of the voxel activation mechanisms must be capable of operating at a peak rate of N_a/τ_r voxels per second. In such a situation, there is no benefit in providing a unique voxel activation pathway for each electron gun, and a more elegant solution would be obtained by the incorporation of a gun selection bit within the voxel descriptor as illustrated in Figure 9.2. In the case of this method being adopted, the voxel descriptor would be passed simultaneously to both electron guns and the gun selection bit used to activate the appropriate electron gun. If it were possible for each of the voxel activation subsystems to operate at any time during the course of each revolution of the surface of emission (corresponding to neither temporal nor

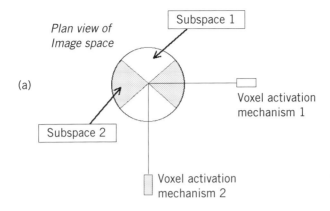

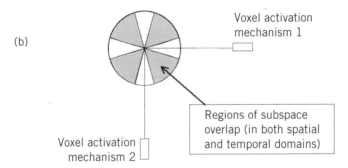

FIGURE 9.1 Beam addressing schemes for the example swept-volume display unit: (*a*) each beam addresses the surface of emission during one-half of the image space update period; (*b*) each beam addresses the surface of emission during two-thirds of the update period, so that certain regions of the image space (indicated by shading) may be addressed by both beams.

spatial separation of the subspaces), the peak voxel transfer rate to each activation mechanism would be reduced to $N_a/2\tau_r$.

For a display unit of this type, the beam scanning and gating processes are often responsible for ultimately setting an upper limit on the system performance (in terms of data throughput). In short, they generally represent the bottleneck in the graphics pipeline. In the case of this example system, the configuration of the two voxel activation mechanisms results in each having a 50% duty cycle, and they are therefore not used to their fullest potential. In fact, whereas the use of two beam sources circumvents various dead zone problems described in Chapter 4, they clearly do not provide an increase in the voxel activation capacity of the display.

Let us now consider a modification to the above example and permit the voxel activation mechanisms to each address two-thirds of the image space during each image update period (i.e., we relax our dead zone criteria) as depicted

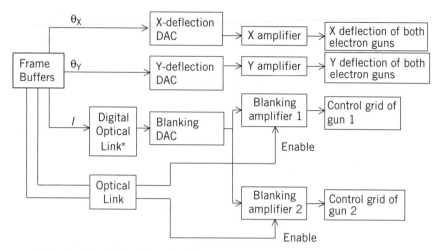

* Required for high voltage isolation

FIGURE 9.2 Use of a single voxel activation pathway able to supply voxel descriptors to two voxel generation mechanisms for a system employing two electron guns. The voxel descriptor contains a selection bit whose value is used to determine the mechanism that is to be responsible for generation of a particular voxel. Note that to reduce the capacitive and inductive loads on the output stage of the deflection amplifiers, it may be necessary to replicate these amplifiers for each electron gun.

in Figure 9.1b. The two subspaces will now intersect both spatially and tempo-rally. Clearly, this temporal intersection of the subspaces indicates that a degree of parallel voxel activation is possible. Unfortunately, this parallelism is not supported throughout the image space but rather, is confined to the region of spatial overlap between the subspaces. Within this region both voxel activation subsystems are, in principle, able to generate voxels during each image update period, and therefore image objects positioned therein may be depicted with a greater voxel density than if they were drawn elsewhere. Although this increases the overall voxel activation capacity of the display unit, it reduces the homo-geneity of the image space. The introduction of parallelism does not necessarily preclude the use of a single voxel activation pathway provided that the bandwidth of this connection is not less than the peak throughput of the combined voxel activation mechanisms.*

9.4 SUBSPACE MAPPINGS

A mapping notation provides a convenient method for representing the ideas outlined in Section 9.3, describing the flow of information into a display unit.

---

* This assumes that some type of selection data are included in each voxel descriptor so that they can be directed to the appropriate voxel activation mechanism.

Let η denote the number of voxel activation mechanisms that address a total of N_a voxels during each image update period. In accordance with our definition there must also exist η subspaces. If we assume that each subspace may contain the same number of voxels, we can represent the throughput of voxels into the image space as

$$\eta \left(1 \rightarrow \frac{N_a}{\eta}\right) \tag{9.1}$$

where the number of voxels in each subspace is represented by $N_{\text{sub}} = N_a/\eta$.

9.4.1   Examples of Subspace Mappings

This mapping provides a clear and concise representation of the number of subspaces and the number of voxels activated within each. This is illustrated in the examples provided below, where it is applied to several display unit architectures.

Example 1   In the case of the elementary model introduced in Section 1.3, the number of activation subsystems η is the equal to the voxel activation capacity (N_a). This configuration may therefore be represented by the trivial expression:

$$N_a(1 \rightarrow 1) \tag{9.2}$$

That is, there are N_a subspaces and within each there can exist a single voxel. There is no spatial overlap between the subspaces and the number of voxel activation pathways determines the extent of any temporal overlap. If the array is contained in a cube with sides of length L, the volume of each subspace may be approximated to that of a cube with sides of length $L/N_a^{1/3}$ (see also Section 9.4.2).

Example 2   Consider a display unit employing a 2-D LED matrix of voxel generation centers comprising $N_a^{2/3}$ elements and moving with a translational motion. This arrangement may be represented by

$$N_a^{2/3}(1 \rightarrow N_a^{1/3}) \tag{9.3}$$

indicating an image space composed of $N_a^{2/3}$ subspaces and within each, up to $N_a^{1/3}$ voxels may be activated. The shape and area of each voxel generation center will determine the cross section of each subspace, and the length will be equal to twice the depth of the image space.* If the image space is a cube with sides

---

* In this context, subspace length will be assumed to relate to the distance moved during the course of a single image update period.

of length L and if we assume an isotropic voxel density, the volume of each subspace may be expressed* as $2L^3/N_a^{2/3}$.

In the examples above, we have assumed that each subspace is responsible for the production of an identical maximum number of voxels. Let us now generalize our mapping notation to accommodate display units in which this is not the case. For an image space comprising N_{sub} subspaces, we may consider the generation of voxels within each individual subspace:

$$(1 \to n_1) + (1 \to n_2) + \cdots + (1 \to n_{N_{sub}}) \tag{9.4}$$

This expression may be rewritten in the form

$$\sum_{i=1}^{N_{sub}} (1 \to n_i) \tag{9.5}$$

where n_i represents the voxel activation capacity of the ith subspace.

This may be illustrated by the following example.

Example 3   Consider a display unit employing a rotating LED panel containing $N_a^{2/3}$ voxel generation centers as illustrated in Figure 5.16. Let us assume that this surface of emission is able to generate approximately N_a voxels per revolution and that these can be distributed isotropically (or approximately so) throughout the image space. The number of voxels generated by a given element clearly depends on its radial distance from the axis of rotation. For simplicity we consider only those elements located on one side of the axis of rotation. All elements at an equal radius are capable of generating an equal number of voxels. We are able, therefore, to group into columns elements that lie parallel to the axis of rotation and denote each column of subspaces by the integer j. Since at present we are only considering elements on one side of the rotational axis, it is apparent that $0 < j \le 0.5N_a^{1/3}$.

Let us suppose that the separation of adjacent voxel generation centers on the surface of emission is ξ, and the separation of voxels along each subspace is given by χ, as illustrated in Figure 9.3. Consequently, the number of voxels activated (N_j) within a subspace lying at a radius r from the axis of rotation is approximately $2\pi r/\chi$. However, as $r = j\chi$, then

$$N_j \approx \frac{2\pi j\xi}{\chi} \tag{9.6}$$

---

* As a consequence of the discrete nature of the voxel generation technique, the volume of each subspace is in reality likely to be less than this amount (see Section 9.4.2).

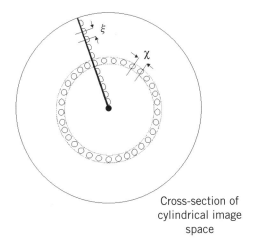

Cross-section of
cylindrical image
space

FIGURE 9.3 Surface of emission employing an LED panel moving with rotational motion. The separation of elements on the panel is given by ξ and the separation of voxels along each subspace by χ.

This configuration may therefore be represented as

$$N_a^{1/3} \sum_{j=1}^{0.5N_a^{1/3}} \left(1 \to \approx \frac{2\pi j \xi}{\chi} \right) \tag{9.7}$$

The inclusion of symmetrically placed voxel generation elements on either side of the axis of rotation results in pairs of spatially coincident subspaces and may result in a further increase in the parallelism supported by the display unit. It is therefore possible to apply a factor of 2 to the expression above so as to incorporate all elements and their associated subspaces.

Consider the situation of equal voxel spacing both radially and circumferentially (i.e., $\xi/\chi = 1$). The mapping above may then be rewritten as:

$$N_a^{1/3} \sum_{j=1}^{0.5N_a^{1/3}} (1 \to 2\pi j) \tag{9.8}$$

This indicates that the innermost column of subspaces will each be responsible for the generation of about 6 voxels, the second for about 12, the third for about 18, and so on. Consequently, in such a configuration, a surface of emission containing a matrix of 100 by 100 elements wholly located on one side of the axis of rotation would have a voxel activation capacity in excess of 3 million. Should we intend to address the image space using a rectangular Cartesian coordinate system, the chosen value of ξ/χ is critical in determining the best approximation between the required voxel activation sites (in Cartesian coordinate space) and those that are possible.

9.4.2 Effective Image Space Volume

An image space may be considered to comprise one or more subspaces. These subspaces may physically overlap or may be separated. For instance, in the case of the display unit discussed in Examples 2 and 3, the discrete nature of the voxel generation centers results in a series of subspaces that do not touch. According to our previous definitions, voxels can only be placed within subspace regions, so it follows that image space may, in some cases, be discontinuous. Those portions of image space within which subspace does not exist represent dead image space (see Section 4.2.1).

Consider an image space of physical volume V composed of η subspaces each of volume V_i and of dead image space having a total volume V_d. We can define an effective image space volume (V_e) given by

$$V_e = V_d + \sum_{i=1}^{\eta} V_i \qquad (9.9)$$

For many volumetric display unit technologies we find that $V_e \neq V$. This may be illustrated by reference to the examples above. In the case of Example 1 we see that $V_e \leq V$. In the case of Example 2, the length of each subspace is twice the depth of the image space, as the latter may be addressed twice per update period. As a consequence, $V_e \approx 2V$. In the case of Example 3, employing symmetrically placed voxel generation elements on either side of the rotation axis of the surface of emission,* we again find that $V_e \approx 2V$. It should be noted that in the case of systems in which the effective volume is found to be greater than that of the image space, we cannot deduce (from this information alone) that the display unit technology supports parallel voxel activation.

Discontinuities within the image space are not necessarily detrimental to the image quality that results. For example, if a display unit employs an array of discrete elements as voxel generation mechanisms, dead image space will surround each subspace. However, the impact of this on the perceived image quality depends on the magnitude of the intervoxel spacing relative to the voxel size and on the viewing distance. A useful analogy is that of a raster-scanned cathode ray tube display, in which the image space comprises an array of discrete pixel locations. The pixel size and the interstitial spacing are small enough to be imperceptible at normal viewing distances.

9.5 SUBSPACE IN THE TIME DOMAIN

Let us now turn our attention to the passage of voxel descriptors along the voxel activation pathways that link the graphics engine to the voxel activation

---

* The inclusion of voxel generation elements on both sides of the axis of rotation can result in pairs of spatially coincident subspaces and provides a convenient method for the generation of two-color images.

mechanisms. We will denote as σ_i the fraction of the image update period ($\tau_r = 1/f_r$) over which the ith subspace may be addressed ($0 < \sigma_i \leq 1$). We denote the temporal extent of each subspace by $T_i = \sigma_i/f_r$. The average parallelism (P) exhibited by a display unit architecture may thus be expressed as the sum of the duration of each subspace:

$$P = f_r \sum_{i=1}^{\eta} T_i \qquad (9.10)$$

where η denotes the number of voxel activation mechanisms. Alternatively, if we assume that all subspaces may be addressed for the same fraction σ_{sub} of the image update period, the parallelism expression may be simplified to

$$P = \sigma_{\text{sub}}\eta \qquad (9.11)$$

If we represent the number of voxels that may be activated within each subspace as N_{sub}, the throughput R_{sub} (voxels/second) of each activation mechanism is given by

$$R_{\text{sub}} = \frac{N_{\text{sub}} f_r}{\sigma_{\text{sub}}} \qquad (9.12)$$

Using the relation $N_{\text{sub}} = N_a/\eta$, this may be expressed as

$$R_{\text{sub}} = \frac{N_a f_r}{\sigma_{\text{sub}}\eta} = \frac{N_a f_r}{P} \qquad (9.13)$$

Combining this expression and Eq. (9.10) we obtain

$$R_{\text{sub}} = \frac{N_a}{\sum_{i=1}^{\eta} T_i} \qquad (9.14)$$

In the case of a display unit containing a single subspace, the denominator in Eq. (9.14) reduces to the display's refresh period. As the total temporal extent of the subspaces is increased so as to exceed $1/f_r$ the data rates through each voxel activation mechanism may be reduced. This applies, for example, in rotating screen systems, where the surface of emission sweeps twice through the image space in each revolution, but the image space refresh frequency is retained at the screen rotation frequency. The utility of this subspace formalism may be illustrated with reference to several examples.

Example 1 Consider a display unit employing a passive screen moving with rotational motion and addressed by two stationary beam sources: the configuration illustrated in Figure 9.1a. We assume an image refresh frequency of $f_r = 25$ Hz and a voxel capacity of $N_a = 10^4$. The number of voxel activation mechanisms is $\eta = 2$. However, since there is no temporal overlap between their respective subspaces, we have $\sigma_1 = \sigma_2 = \sigma_{\text{sub}} = \frac{1}{2}$. Equation (9.11) thus yields the parallelism $P = 1$, as expected, and the temporal extent of each subspace is $\frac{1}{50}$ s. The maximum voxel throughput rate from Eq. (9.13) is therefore $R_{\text{sub}} = 25 \times 10^4 \text{voxels} \cdot \text{s}^{-1}$. If each voxel descriptor is represented by 12 bits, the digital bandwidth required is thus $B_D = 375 \text{ kBytes} \cdot \text{s}^{-1}$.

Example 2 Consider a swept-volume system employing the translational motion of an active surface of emission containing 256 by 256 voxel generation centers. We assume that any number of these elements may be activated concurrently* and that each is responsible for the production of up to 256 voxels along the length of its 1-D subspace. For simplicity, we assume each subspace to comprise the path swept out by an element during one-half of the motion period ($\frac{1}{25}$ s) only, so that $N_a = 16.8 \times 10^6$. From this partial specification we deduce that $\eta - 6.6 \times 10^4$, $\sigma_{\text{sub}} = \frac{1}{2}$, and hence, from Eq. (9.11), $P = 3.3 \times 10^4$. The temporal extent of each subspace is $\frac{1}{50}$ s. The required voxel throughput rate from Eq. (9.13) is therefore $R_{\text{sub}} = 1.28 \times 10^4$ voxels per second. Compared with Example 1 the increased parallelism of this example system has resulted in a lower maximum throughput rate, even though the voxel activation capacity is nearly three orders of magnitude higher. If the voxel descriptors each comprise 4 bits (governing voxel intensity — the reduced voxel descriptor length relative to Example 1 reflects the fact that beam deflection information is not required), the digital bandwidth required is $B_D = 6.4 \text{ Bytes} \cdot \text{s}^{-1}$.

Example 3 Finally, consider a static-volume system such as the elementary model described in Section 1.3 or the display unit detailed in Section 7.5.6, comprising a 3-D array of 256 by 256 by 256 individually addressable voxel generation centers. For simplicity, we assume that all voxels may be activated concurrently, without restriction, at any time. In this case we have $\eta = N_a = 16.8 \times 10^6$, $\sigma_{\text{sub}} = 1$, and hence from Eq. (9.11) a maximum parallelism of $P = N_a = 16.8 \times 10^6$. The maximum voxel throughput rate is therefore $R_{\text{sub}} = 25$ voxels per second. That is, since each voxel activation mechanism is responsible for the generation of only a single voxel during each refresh period, its data throughput rate is equal to the image refresh period. If the voxel descriptor length is 4 bits, the required digital bandwidth is $B_D = 12.5 \text{ Bytes} \cdot \text{s}^{-1}$.

These examples support our intuitive notion that as the number of voxel activation pathways into a display unit are increased, the maximum data rates along each pathway may be reduced. We have given this matter considerable attention because its consequences for the architecture of the interface between the graphics engine and the voxel activation mechanisms are of utmost importance. From the perspective of cost (and practicality) we seek to minimize the number of pathways, and therefore each must be used to its maximum capacity. Periods during which a data pathway is inactive due to the inability of the voxel activation mechanism(s) to which it is connected to address the image space (as opposed to

---

* In practice, a row and column addressing technique would be employed to reduce connectivity.

inactivity arising from the spatial distribution of voxels within a data set) should be avoided. Although it is apparent that the greater part of any image space will be void, in the case of image spaces comprising numerous subspaces, one or more of these may be occupied completely by activated voxels. Clearly the extent, geometrical shape, and position of each subspace coupled with the type of application for which the system is to be employed will ultimately govern the probability of it being fully populated. To represent any image, the voxel activation subsystems must be designed so as to accommodate subspace regions in which all voxels are activated. To do otherwise would result in a reduction in the predictability of the display unit.

9.6 DATA THROUGHPUT ISSUES REVISITED

As the voxel activation capacity is increased, a display unit is likely to behave in a more predictable manner (or perhaps we should say a less unpredictable manner) and it becomes easier to characterize the performance of a display unit. In the case of exhaustive addressing, the fill factor is 100% and there is a greater chance that image components will not be affected by the location at which they are situated within an image space. Unfortunately, as we have seen, most current volumetric technologies do not support a 100% fill factor, and in the case of beam-addressed systems a dot graphics technique is commonly employed for image production. In this case, the interpretation of the voxel activation capacity, fill factor, and voxel density is by no means easy and their interrelationships can be confusing. Below we summarize (briefly) some of the issues that should considered when coming to terms with their meaning.

- The voxel activation capacity provides us with an indication of the total number of voxels that may be activated during each image refresh period. However, in the case of some architectures (e.g., many swept-volume systems), this number of voxels must be dispersed throughout the volume, and consequently, only a small fraction of the available voxel activation capacity will generally be employed for image depiction. (For a swept-volume display unit employing rotational motion, the depiction of an object such as a sphere centered about the axis of rotation may allow the number of voxels indicated by the voxel activation capacity to be employed simultaneously.) A large voxel activation capacity is essential in order to permit the depiction of detailed image components.
- To provide image clarity (especially in the presentation of spatial information), a display unit will generally be operated in such a manner as to ensure that the greater part of the image space is void. Therefore, even in the case of systems that offer a large fill factor, the percentage of available voxel locations that are activated during each image update period will remain small.

- From the perspective of the graphics engine, the rate at which voxel descriptors must be output to the display unit is of particular consequence, as this peak data transfer rate will ultimately determine the parallelism that must be incorporated within the engine. In considering this issue, it is natural to direct attention toward the maximum voxel density that is required within the image space.

- When measuring the physical density of an object, we generally assume that the object is either homogeneous in its composition or that our measurement will lead to a useful average density value. As we have seen, very few image space technologies offer to support homogeneous and isotropic voxel distributions. In determining the peak data transfer rate, a knowledge of the maximum voxel density is required, and in this case the average density (which often suffices when making physical density measurements) is of little value. Although in the case of display units that possess a uniform image space (such as the elementary display unit introduced in Section 1.3) it is meaningful to consider voxel density, the expression should be avoided for the majority of systems that have been proposed to date (which lack homogeneity and isotropy in voxel placement).

With these ideas in mind we turn our attention once more to the architecture of the graphics engine. In the next section we propose a graphics engine that exhibits parallelism in its processing and throughput of voxel descriptors and may therefore be suitable for use with some of the display units discussed in Chapter 12.

9.7 GRAPHICS ENGINE ARCHITECTURE

In general terms, the primary functions of a graphics engine intended for use with a volumetric display unit are as follows:

- The access of the volumetric data. This may be stored within the host computer, be derived from applications software, or may correspond to an incoming real-time data stream.
- The manipulation of these data into a form suitable for depiction within image space (i.e., the generation of voxel descriptors).
- The direction of voxel descriptors into appropriate subspace buffers.*
- The synchronization of the output of voxel descriptors.
- The update of subspace buffers (for image animation and frame changes).

Other tasks may include the registration (alignment) of beam sources, the control of the motion of any moving surface of emission, and the provision of facilities by which the operator is able to interact directly with image components.

---

*In our current discussions we assume that a separate buffer is associated with each subspace.

Within this section we propose a graphics engine architecture suitable for use with a highly parallel display unit.

The underlying principles that are to be adopted in the implementation of any display unit will determine its characteristics and general suitability for various applications (see Chapter 10) and will affect the requirements for the graphics engine. In this last respect we are particularly interested in the following aspects of the display unit:

- The voxel activation capacity and image refresh frequency. These quantities will affect the rate at which voxel descriptors are to be processed and passed to the display unit.
- The number of subspaces that comprise the image space (this equates to the number of voxel activation mechanisms).
- The spatial and temporal distributions of the subspaces. These are functions of the image space creation, voxel generation, and voxel activation subsystems.

From the very outset of the design process, there are very many interrelated issues that must be considered with great care. Furthermore, in dealing with these matters, it will be necessary to make many compromises during the specification and implementation of the system.

9.7.1   Data Rate Considerations

Let us consider a display unit in which there are η voxel activation mechanisms. The *maximum* parallelism exhibited by this display unit will be determined by the total temporal extent of the subspaces and the image refresh period τ_r in accordance with the following reexpression of Eq. (9.10):

$$P = \frac{\sum_{i=1}^{\eta} T_i}{\tau_r} \tag{9.15}$$

To maximize the voxel activation capacity, we should ensure that the voxel activation mechanisms operate at their highest level of throughput. By determining a convenient access and conversion speed (e.g., from digital to analog form) for the graphics engine hardware and considering the throughput of the voxel activation mechanisms, we may determine the number of these mechanisms that may be operated via a single voxel activation pathway. Clearly, in the case of two or more subspaces that have no temporal overlap, it is appropriate to connect them to a single voxel activation pathway. Such a connection scheme will place no extra bandwidth requirements on the pathway but rather will ensure that the passage of voxel descriptors along the pathway is sustained for longer periods of time. Alternatively, in the case of two or more subspaces that exhibit a degree of overlap in the time domain, it may still be appropriate to connect them to a single pathway, provided that the total maximum data throughput requirements of the mechanisms can be supported by the voxel activation pathway. Whatever connection scheme is adopted, each pathway must be able to cope with the peak rate of voxel activation and also with the possibility of one or more subspaces being completely filled with activated voxels.

9.7.2   Subspace Processors

For η voxel activation mechanisms we can envisage the use of n_p replicated subspace processors, each of which is responsible for the throughput of information into one or more subspaces, as illustrated in Figure 9.4. In the case of beam addressed systems, the rate at which voxels may be activated within an image space is generally limited by the performance of the voxel activation mechanisms. As a consequence of the limited bandwidth of these devices, the number of subspace processors is likely to be less than the number of voxel activation mechanisms.

The primary function of each subspace processor is to output voxel descriptors to one or more voxel activation mechanisms at the appropriate rate. Although each of the subspace processors illustrated within Figure 9.4 could be provided with a computational ability, this is not necessarily a requirement. In their most elementary form, the subspace processors would each consist of some form of random access memory (RAM) able to support image animation,* divided into

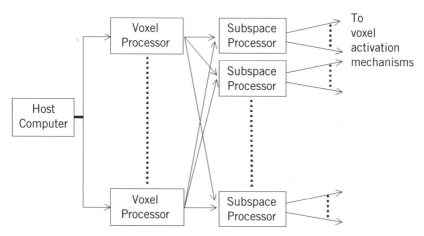

FIGURE 9.4   Possible graphics engine architecture. The subspace processors are responsible for achieving the rapid output of voxel descriptors to the appropriate voxel activation mechanisms and the voxel processors for both the generation of voxel descriptors and the direction of these descriptors to the subspace processors. The host computer may take various forms and is responsible for directing and monitoring the activities of the voxel processors.

---

* So as to avoid possible contention problems that may occur due to voxel update occurring at the same time as voxel output, a double buffering technique could be employed.

segments, each of which contains a set of voxel placement locations within a particular subspace.* Each RAM segment will be referred to as a subspace buffer. Controlling hardware within each subspace processor would be responsible for the output of voxel descriptors to the appropriate voxel activation mechanisms (at the appropriate times) and for the update of the contents of the subspace buffers by the voxel processors.

Each subspace processor must accommodate the peak data requirements of the voxel activation mechanisms to which it is connected, and their essential purpose is to provide data to the display unit at the required rate. As described above, the number of voxel activation mechanisms that may be supported by each subspace processor is determined by the peak data transfer rate that must be accommodated and the bandwidth that may be incorporated within each subspace processor. Should it be envisaged that there may be a variation in the level of utilization of subspaces (due to their physical location within an image space), this can be reflected in the interconnection between subspace processors and voxel activation mechanisms. However, this may reduce the predictability of the system and so is considered to be undesirable.

Returning to Figure 9.4, we see on the left-hand side the host computer, which may take various forms according to the nature of the incoming data. In this discussion we assume that the incoming data is in point form. Should the data be specified as a set of graphics primitives, they must be decomposed into point form volumetric data. As it is clearly more efficient to carry out graphics manipulations prior to decomposition, the graphics engine architecture as represented here is suboptimal in this regard. However, since optimization of the interface (including precise task assignment) between the host computer and the voxel processors is highly dependent on the form of the original data, it is inappropriate to complicate our present discussion with this matter. It is sufficient to note that a careful balance must be achieved between the amount of processing that must be carried out by the host computer and that which may be delegated to the voxel processors.

9.7.3 Image Update

So as to illustrate the operation of the voxel processors, let us turn our attention to the events that might occur when a new image frame is to be depicted within the image space. As may be seen from Figure 9.4, interposed between the host computer and the subspace processors are a number of identical voxel processors, each of which is able to direct voxel descriptors to the appropriate subspace processor. The data accessed or stored within the host computer is passed to the voxel processors in such a way as to ensure that it is equitably distributed. Each voxel processor is responsible for the conversion of these data points into voxel descriptors suitable for depiction within the image space (see Section 8.5). Consequently, they are responsible for undertaking graphics transformations, verification procedures, filtering, and if appropriate, the calculation of

---

* Clearly, the storage scheme adopted would be more complicated should a single voxel processor be responsible for the support of two or more subspaces that overlap in the temporal domain.

beam deflection angles and so on. All tasks that are carried out by these processors are defined and monitored by the host computer. Once the appropriate manipulations have been carried out, it is possible to determine the subspace in which each voxel descriptor is to be located and so they may be passed to the appropriate subspace processor. Finally, the frame is output to the display unit by means of the subspace processes.

For computer-generated image sequences that are to be computed in real time, it is desirable that each voxel processor be given the computational tasks associated with image creation. Appropriate program modules would therefore be downloaded from the host computer and run within each voxel processor's environment. In the case of image scenes that change between frames, it is more efficient for the graphics engine to be concerned only with changes in the data. As a consequence, the voxel processors are responsible for the addition and removal of voxel descriptors from the subspace processor memory banks. During an animation sequence, groups of voxels may move within a subspace or may be relocated into an adjacent subspace. Movement between subspaces may necessitate the deletion of voxel descriptors from one subspace processor and the generation of new voxel descriptors that are passed to another voxel processor. Naturally, the voxel processors carry out these tasks.

9.7.4   General Discussion

Clearly, many changes and improvements may be made to the basic architecture illustrated in Figure 9.4, and under certain circumstances it may be advantageous to integrate the voxel and subspace processors. Our purpose in proposing this particular architecture is to illustrate a method by which the performance of both the display unit and graphics engine may be closely matched. Furthermore, we hope that this discussion provides a useful insight into the application of subspace to the design of a volumetric system. Finally, during any development process it is important to cater for future expansion. The use of a modular graphics engine facilitates expansion and the general architecture described above may be easily adapted for use with the majority of volumetric systems.

9.8   BEAM CALIBRATION TECHNIQUE

As we have seen, many of the volumetric display units developed to date achieve voxel activation by means of one or more beams that are directed into the image space. In the case of swept-volume systems, several spatially separated beam sources may be employed so as to reduce the effects of distortional and voxel placement dead zones. In the case of static-volume systems the voxel activation mechanism may involve the intersection of two directed beams (see e.g., Section 6.5). In both swept-volume and static-volume beam-addressed display systems, increasing the number of activation beams permits voxels to be activated in parallel and this increases the voxel activation capacity. However, successful implementation of any system employing more than a single beam source requires a mechanism to ensure the accurate registration of the beam source positions and orientations with respect to the image space coordinate frame. This mechanism must not only be able to accommodate physical positioning uncertainties introduced in the production process but must also be able to account for both physical and electronic changes that may occur during the operational life of the display system. Within this section we therefore briefly discuss a calibration system which may be used to determine the spatial location and orientation of each beam source with respect to the image space coordinate frame. Once deviations from ideal beam source positions and orientations have been determined, their effect on the position of the resulting voxels in the image space may be taken into account by modification of the transformations used to generate the voxel descriptors.

In this section we deal with the specific example of a swept-volumedisplay unit equipped with a rotating planar surface of emission and employing two beam

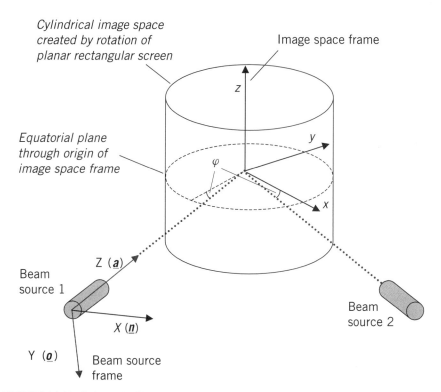

FIGURE 9.5 Schematic illustration of the architecture of the display unit discussed in the beam calibration example in Section 9.8. The position and orientation of the beam source frame relative to the image space frame is depicted for beam source 1; an analogous frame may be specified for beam source 2 (and, in general, to any number of beam sources). The Z axis of the beam source frame is equivalent to the approach vector **a**, the Y axis to the orientation vector **o**, and the X axis to the normal vector **n**.

sources, as illustrated in Figure 9.5.* As indicated in this figure, we assume that according to the display unit specification, both beam sources are located the same distance D from the center of the image space and that each addresses the image space from the same angle φ below (or above) the equatorial plane. We denote the angular screen position by the angle β and assume also that the angular position about the z-axis of each beam source is provided by the specification. Now, in practice, the positions of the beam sources will deviate slightly from those provided in the specification and illustrated in Figure 9.5. Moreover, the orientations of the beam sources will in general be such that the undeflected beam does not pass through the origin of the image space frame, as implicitly assumed in this configuration.

The scheme outlined below assigns a coordinate frame to each beam source and to the image space and seeks to determine the values of the parameters that describe a sufficiently accurate relationship between them. We consider mappings between the image space and a single beam source — the same procedure is carried out for the other. This technique may readily be generalized to the case of more than two beams and can be adapted to other display unit configurations.

9.8.1 Relationship Between Beam Source and Image Space Coordinate Frames

We define for each beam source a right-handed Cartesian coordinate frame $X = (X, Y, Z)^{\mathrm{T}}$, with the Z axis oriented along the optical axis of the beam source as illustrated in Figure 9.5. Similarly, the image space is assigned a coordinate frame, denoted $\mathbf{x} = (x, y, z)^{\mathrm{T}}$, in which the z axis is assumed to be coincident with the axis of rotation. The problem of defining the position and orientation of a beam source with respect to the image space is thus the same as determining the spatial coordinate transformation relating the two frames. Once this is known, the spatial coordinates of any voxel in the image space yield the required coordinates in the beam source frame by means of this same transformation. The beam deflections required to activate each voxel may then be computed in this frame according to Eqs. (8.4) and (8.5).

The rotations and translations required to map one coordinate system onto the other may be encapsulated efficiently by the homogeneous transformation formalism (see appendix). This enables the 3-D rotation and translations to be combined within a single 4 by 4 transformation matrix. Since our transformation relationship contains no shearing or scaling operations, the transformation relating the beam source frame $\mathbf{X}$ to the image space frame $\mathbf{x}$ may be written

$$\mathbf{x} = \mathbf{TX} \tag{9.16}$$

or

$$\mathbf{X} = \mathbf{T}^{-1}\mathbf{x} \tag{9.17}$$

---

* This general approach was developed for use with the Cathode Ray Sphere (see Section 5.6.3). A display vessel for this system incorporating three beam sources addressing the image space from below the horizontal equator is illustrated in Plate x.

where

$$\mathbf{T} = \begin{bmatrix} n_x & o_x & a_x & t_x \\ n_y & o_y & a_y & t_y \\ n_z & o_z & a_z & t_z \\ 0 & 0 & 0 & 1 \end{bmatrix} \tag{9.18}$$

is the 4 by 4 homogeneous transformation matrix linking the two frames. The vector $\mathbf{t} = (t_x, t_y, t_z)^\mathrm{T}$ describes the translation between the origins of the two Cartesian coordinate frames. The other three vectors $\mathbf{n}$, $\mathbf{o}$, and $\mathbf{a}$ together comprise the net effect of the rotational operations required to map one frame onto the other. They are known as the *normal*, *orientation*, and *approach* vectors, and in terms of the coordinate frame of the beam source, they are equivalent to the unit directional vectors of the x, y, and z axes (Figure 9.5). Once any two of these three vectors are known, the third is obtained from their vector product (see appendix). Here we determine the transformation $\mathbf{T}$ mapping the image space frame onto the beam source frame. To address the image space from the beam source frame we then use the inverse, $\mathbf{T}^{-1}$, as in Eq. (9.17).

The method outlined in the remainder of this section for determination of the elements in the transformation matrix $\mathbf{T}$ may be thought of as comprising two main parts:

- Initially, we assume that the position and orientation of the beam source relative to the image space is *approximately* known from the construction specifications of the display unit (or by physical measurements). From this information, we derive a first approximation to the transformation matrix $\mathbf{T}$.
- Once the approximate values of the elements within the transformation matrix have been obtained, more exact values are determined using a series of calibration images displayed within the image space and visual feedback from the operator.

9.8.2 First Approximation to the Transformation Matrix

From the introduction to this section, the reader will recall that we have constrained the z axis of the image space coordinate frame to be coincident with the rotation axis of the surface of emission. However, the orientation of the x and y axes remains undefined relative to the beam sources and display unit structure. This remaining (rotational) degree of freedom must be removed. As discussed in Section 8.5.5, an indexing signal is often generated by the screen's position encoder hardware and we may think of this signal as indicating the starting position of each revolution of the screen. It is convenient to define one of the remaining two axes to be coincident with the screen's surface at the time when this indexing signal occurs. We will therefore arbitrarily assume that it is the x axis that coincides with this position.

Assumptions Employed in the Derivation of the First Approximation

To derive approximate values for the transformation matrix elements, we make the following assumptions:

1. As mentioned in the introduction to this section, the scalar distance D from the beam source to the center of the image space (i.e., the distance between the origins of the two coordinate systems) and the angle φ by which the beam source lies below the xy plane are known approximately from the display vessel specification (Figure 9.5).

2. The angular position [about the screen's axis of rotation (z)] of the beam source is approximately known relative to the x axis of the image space frame (which we have taken to correspond to the screen position at which the indexing signal is generated). We denote this angular position by β_0.

3. The Z axis of the beam source frame (and hence the undeflected beam) passes through the origin of the image space coordinate frame.

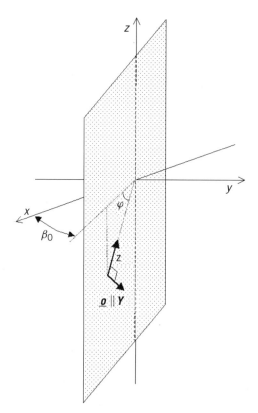

FIGURE 9.6 Geometry governing the components of the beam source orientation vector, **o**. This vector is defined as being coincident with the Y axis of the beam source coordinate system.

4. The Y axis of the beam source frame (and hence the orientation vector $\mathbf{o}$) is coplanar with the rotation axis of the image space frame. That is, voxels arising from a purely "vertical" beam deflection will remain on the z axis (Figure 9.6).

In the following, we now determine first approximations for the translational displacement between the coordinate frames of the image space and the beam source, and then for the approach ($\mathbf{a}$) and orientation ($\mathbf{o}$) vectors of the beam source frame (see Figures 9.5 and 9.6). The latter two then provide the normal vector ($\mathbf{n}$) with which all elements of the transformation matrix $\mathbf{T}$ are known.

Determination of the Translational Displacement (t)

From assumptions 1 and 2 above, the t_x and t_y components of the relative position vector t are obtained from simple geometrical considerations based on the configuration illustrated in Figures 9.5 and 9.6. These components have the form $t_x = D \cos \varphi \cos \beta_0$ and $t_y = D \cos \varphi \sin \beta_0$, where φ is the angle between the beam source and XY plane of the image space frame (see Figure 9.6). The t_z component is obtained from similar geometrical considerations as $D \sin \varphi$, so that the translation vector is thus

$$\mathbf{t} = (D \cos \varphi \cos \beta_0, D \cos \varphi \sin \beta_0, D \sin \varphi)^{\mathrm{T}} \qquad (9.19)$$

Determination of the Beam Source Approach Vector (a)

From assumption 3 above, and with reference to Figure 9.6, the unit directional vector of the Z axis is simply $\mathbf{a} = \mathbf{t}/|\mathbf{t}|$. Using Eq. (9.19), the approach vector is therefore obtained as:

$$\mathbf{a} = \frac{1}{\sqrt{t_x^2 + t_y^2 + t_z^2}} (t_x, t_y, t_z)^{\mathrm{T}}$$
$$= (\cos \varphi \cos \beta_0, \cos \varphi \sin \beta_0, \sin \varphi)^{\mathrm{T}} \qquad (9.20)$$

Determination of the Beam Source Orientation Vector (o)

The orientation vector $\mathbf{o}$ is the unit vector in the direction of the beam source's Y axis. From assumption 4 above, this is coplanar with the rotation axis of the surface of emission; the geometry is outlined in Figure 9.6. The angle φ by which the beam source is rotated away from the xy plane is known by assumption 1 above. The orientation vector within the YZ plane (indicated in Figure 9.6) is thus the intermediate vector $\mathbf{o}' = (-\sin \varphi, 0, -\cos \varphi)^{\mathrm{T}}$. The full expression for the orientation vector $\mathbf{o}$ then depends on the rotation of the YZ plane, and hence of $\mathbf{o}'$ about z.

Knowledge of the beam source position relative to the reference surface position from the display specifications provides the angle β_0 of the beam source position relative to the x axis of the image space frame (assumption 2 above).

Applying a rotation by this value about z to $\mathbf{o}'$, we obtain $\mathbf{o}$:

$$\mathbf{o} = \begin{pmatrix} \cos \beta_0 & -\sin \beta_0 & 0 \\ \sin \beta_0 & \cos \beta_0 & 0 \\ 0 & 0 & 1 \end{pmatrix} \begin{pmatrix} -\sin \varphi \\ 0 \\ -\cos \varphi \end{pmatrix} = \begin{pmatrix} -\cos \beta_0 \sin \varphi \\ -\sin \beta_0 \sin \varphi \\ -\cos \varphi \end{pmatrix} \qquad (9.21)$$

Determination of the Beam Source Normal Vector (n)
Having determined the vectors $\mathbf{a}$ and $\mathbf{o}$, the vector $\mathbf{n}$ may be then obtained from the orthogonality relation (see appendix)

$$\mathbf{n} = \mathbf{o} \times \mathbf{a} \qquad (9.22)$$

where $\times$ signifies the vector (cross) product.

Beam Deflections
Once the transformation matrix $\mathbf{T}$ of Eq. (9.16) has been established, the required beam deflections for any point in the image space may be computed. If such a point has coordinates $(x_p, y_p, z_p)^T$ in the image space frame, its coordinates in the beam source frame $(X_p, Y_p, Z_p)^T$ are given according to Eq. (9.17) by the inverse transformation (image space frame to beam source frame) $\mathbf{T}^{-1}$:

$$\begin{pmatrix} X_p \\ Y_p \\ Z_p \\ 1 \end{pmatrix} = \mathbf{T}^{-1} \begin{pmatrix} x_P \\ y_p \\ z_p \\ 1 \end{pmatrix} \qquad (9.23)$$

Now that we have determined the coordinates of the point in the beam source coordinate system, the required beam deflection angles may then be calculated by Eqs. (8.4) and (8.5).

9.8.3   Interactive Refinement of the Transformation Matrix

The determination of the transformation matrix described above is based on a number of parameters, which in practice are only known approximately from physical measurements. Moreover, several implicit assumptions were made. For example, we assumed that the planar surface of emission is not offset from the rotational axis. Should both sides of the screen be employed for voxel generation (as would be the case when using phosphor coatings for the production of multicolor images), this implies that the supporting structure must be infinitely thin. Clearly, this is invalid.

Small deviations from the assumed spatial relationship between the two coordinate frames can give rise to distortions in the resulting images and mismatch at subspace boundaries. In this example, such a boundary occurs between the regions of the image space addressed by beam 1 and those addressed by beam 2 (see Figure 9.1a). Furthermore, registration errors will become more serious as the incidence angle between the beam and screen becomes more acute.

Problems of this type are clearly visible when simple geometric structures such as cubes, squares, or spheres are depicted within the image space. For this reason, a procedure for refining the parameters employed in the transformation matrix for each beam source was developed for use with the Cathode Ray Sphere (see Section 5.6.3). This employed a simple geometric pattern such as a stack of squares, each parallel to the xy plane, depicted within the image space (see Plate vi). Any distortion or voxel mismatch was then reduced below that discernible to the observer by interactively making small alterations to the various parameters, including modifications to the beam deflection calculations to account for tilt of the target surface and the finite width of the supporting structure. With experience it became apparent that certain characteristic distortion features were indicative of an error in a particular parameter and were therefore easily remedied.

The results of one image calibration experiment were generally used as starting values the next time the calibration was performed. Many constructional deviations from specification do not vary with time; however, due to small variations and drift in parameters such as amplifier gain, it was found prudent to perform the calibration on a regular basis. In Section 11.2.2 we refer to a further development of the techniques described above, which, in principle, enables the calibration of the beam sources to be implemented in a semiautomatic manner.

9.9 DISCUSSION

The concept of subspace and the data mapping formalism introduced in this chapter enable aspects of a wide range of display unit architectures to be analyzed in a common manner. The authors believe that this concept permits a more abstract approach to be adopted during the initial stages of design process. In turn, this should assist in enabling a display unit's performance to be assessed (and weaknesses identified) prior to its implementation. As we have seen in Section 9.7, the subspace concept is also of use in the system-level design of the graphics engine and helps to ensure that the performance of both the display unit and graphics engine are properly matched.

In the case of display units employing beam sources for voxel activation, increases in parallelism (and corresponding increases in the voxel activation capacity) may be achieved by increasing the number of voxel activation mechanisms. This implies that subspaces will intersect both spatially and temporally. Voxel placement errors are particularly apparent when they occur at the boundary of two subspaces or when they occur within two (or more) spatially overlapping subspaces. In the case of such display systems, procedures must be established which enable physical misalignments in the placement of beam sources with respect to the image space to be measured and therefore accommodated. The approach described in this chapter involves considerable operator intervention and becomes increasingly time consuming as the number of beam sources is increased. The development of automatic calibration procedures is therefore of utmost importance.

CHAPTER 10

DISPLAY UNIT CHARACTERISTICS

"Our second experiment," the Professor announced ... "is the production of that seldom-seen-but-greatly-to-be-admired phenomenon, Black Light! ... This box ... is quite full of it. The way I made it was this — I took a lighted candle into a dark cupboard and shut the door. Of course the cupboard was then full of *Yellow* Light. Then I took a bottle of Black ink, and poured it over the candle: and, to my delight, every atom of the Yellow Light turned *Black*!*

10.1 INTRODUCTION

The characteristics of conventional 2-D display devices are generally described by means of various metrics. These have been devised so as not only to enable the relative performance of display systems to be compared, but also to facilitate the development of international standards dealing with matters such as image quality and safety. Unfortunately, in the case of 3-D display technologies, the diversity of techniques coupled with the variety of ways in which they affect the human visual system makes it difficult to identify a set of meaningful metrics which are universally applicable. In characterizing 3-D display systems, we are therefore faced with two problems. First, it is necessary to determine the advantages that may be derived through the adoption of 3-D visualization systems in a particular area of application. This may necessitate the introduction of trials that indicate that certain aspects of an operator's performance (such as the rate at which information may be interpreted) may be enhanced through the use of 3-D visualization systems. Second, we must identify the particular type of 3-D

---

* Lewis Carroll, *Sylvie and Bruno*. Dover Publications, 1988.

system that will produce the most beneficial results. Unfortunately, these two issues cannot be considered independently.

The establishment of performance envelopes associated with the various types of display technology and the identification of metrics that may be used to characterize these systems may assist in solving these problems. An interesting study undertaken by Bardsley and Sexton (MacDonald and Lowe 1997) involved a comparison of four types of 3-D display technology. Three of the displays were stereoscopic systems (an anaglyph display, a frame sequential display employing "active" LCD shutter glasses and a frame sequential display employing "passive" polarized glasses). The fourth display selected for use in this study was an autostereoscopic system employing a lenticular faceplate (this is a direct-viewing display system). The discussions presented by the researchers clearly highlight the difficulties in properly measuring relative display system performance. In their conclusions the authors write:

> Clearly an effective display comparison must incorporate a comprehensive range of response measures and task scenarios. Simple, isolated measures may serve to conceal critical aspects of display performance, which may be more readily apparent from a purely objective assessment.

In their work, Bardsley and Sexton identified two simple tasks that are common to many areas of 3-D visualization. The first task involved the continuous pursuit of a moving object and the second evaluated the ease of depth discrimination. A subjective evaluation of each display type by the members of the group participating in the trial was also used in the overall characterization of the display systems. Issues such as cost and system integration were not addressed in this study.*

It is likely that the relative merits and weaknesses of the various approaches to the depiction of 3-D information may best be determined by the application of these techniques to a range of visualization situations. This will produce the most reliable assessment of their "fitness for purpose." Unfortunately, at the present time the matching of a specific 3-D visualization technique to a particular application is likely to be made on quite an arbitrary basis.

The comparison and characterization of 3-D display systems within a particular technology grouping is generally less problematic than a global comparison of all 3-D display techniques and is therefore a more realistic undertaking. However, in the case of volumetric systems, the diversity of approaches to system implementation and the differing image space characteristics make this a very difficult task.

Although research into volumetric systems has been carried out over a considerable period, little (if any) progress has been made in determining generally applicable metrics suitable for display unit characterization. A few publications have started to address this issue; Williams and Donohoo (1991) considered a

---

* The similarity between the types of display discussed in this article may have alleviated the need to address these issues. Should a holographic or volumetric display have been included in this study, these issues may have been more relevant.

range of voxel characteristics, with particular emphasis on passive swept-volume systems employing laser beams for voxel activation. Subsequent work compared the performance of several swept-volume prototype systems under development in the United States (Clifton and Wefer 1993a, Wefer 1994). The experimental nature of the many prototype display systems developed to date and the consequent lack of commercial pressures may have been a major factor in alleviating the need to develop such metrics. This situation may have been exacerbated by a natural desire on the part of the general scientific community to compare volumetric display unit performance with that of conventional systems. Consequently, characteristics that are commonly applied to conventional displays have been used directly to judge the capabilities of volumetric systems. For example, the pixel capacity of a conventional computer monitor is often compared with the voxel activation capacity of a volumetric system. A lack of understanding in the interpretation of such a comparison can be most embarrassing for the volumetric researcher!

In the next section we reexamine the voxel and discuss the various attributes that may be associated with this elementary "particle." Subsequently, we examine the placement of voxels within an image space and in Section 10.4 turn our attention to the characteristics of the image space. This chapter is intended to promote discussion in connection with the characterization of volumetric display systems and illustrate the complexity of the various issues involved.

10.2   VOXEL ATTRIBUTES

Computer-generated images are generally constructed by a display device from pixels and, in certain cases, by means of voxels. Due to the widespread adoption of the pixel, voxels are often regarded as a subclass or particular instance of the pixel, taking the form of pixels that may be positioned within 3-D space. However, on closer inspection, we see that as a result of the extensive range of characteristics that may be assigned to the voxel, it is perhaps more appropriate to consider the pixel as taking the form of a voxel that has been projected on a 2-D surface.

Not only are the attributes that must be assigned to pixels and voxels somewhat different, but the two types of element are observed in a different manner. As we have seen, many volumetric display unit technologies offer great freedom in viewing orientation, and so unlike the conventional pixel, it is necessary to remember that each voxel may, in principle, be viewed from any direction. Furthermore, pixels are generated on an essentially planar surface at locations that lie close to the boundary between the surface and its surroundings, while voxels are generated at positions that may be deeply embedded within an image space. As a consequence, we must not only consider the physical nature of individual voxels but also the propagation of light through the image space and across the image space boundary, before reaching the observer.

The failure to distinguish properly between pixels and voxels often leads to a comparison being made between the voxel activation capacity of a volumetric display and the pixel capacity of a conventional system. This is generally misleading since they represent two very different quantities, and furthermore, the two types of display system have an entirely different functionality. Although the pixel capacity of a conventional computer monitor may be of the order of 1 million, the voxel activation capacities of many of the current volumetric technologies are less than 200,000. Despite this limitation, experience has shown that such volumetric systems are able to display images that are of considerable value. They are, however, best suited to the depiction of the spatial separation of objects, rather than, for example, the form of 3-D surfaces.

In the case of a volumetric system, each activated voxel forms part of the image structure. However, for conventional display systems, many of the pixels illuminated are likely to form part of the background scene and do not contribute directly to the object under consideration. Implicitly, within a volumetric image space, voxels are used in a highly efficient manner. In Section 12.2 we discuss the exhaustive scanning of an image space and subsequently examine possible methods of achieving large voxel activation capacities. Perhaps because of the large numerical difference that has existed for many years between the pixel capacity and voxel activation capacity, many working on volumetric system development have unfortunately avoided quantifying the latter.*

10.2.1 Voxel Size, Shape, and Definition

As we have seen previously, a voxel takes the form of a clearly visible change in the optical characteristics of a localized and defined region of an image space. This optical change may correspond to the emission, scattering, or absorption of light and is affected by the interaction between the voxel activation mechanism and the physical process employed for voxel generation. In the majority of display units proposed to date, emissive or scattering processes are used in the production of voxels. Both of these processes implicitly lead to the production of translucent images, and therefore the occlusion depth cue cannot easily be satisfied. Should an absorption mechanism be employed for voxel production (see Section 7.5.5), then, in principle, it is possible to depict opaque, "solid" images (see Section 10.2.3).

In the case of display systems that enforce little restriction in the viewing angle, it is important that the observed shape and intensity profile of each voxel be independent of viewing position. This requirement leads to the conclusion that voxels should be spherical in shape. Clearly, should a voxel be sufficiently small, its exact physical form is not readily discernible to the eye, and therefore its precise shape and intensity profile may be of little importance. However, consider a set of fine voxels used in the depiction of a simple image component (e.g., a line). Even if it is not possible to discern individual voxels, should each be subject to a similar nonuniformity, the effect may be cumulative and lead to readily observable anisotropic image characteristics. Potential anisotropies in voxel form are inherent in many of the systems described in this book.

---

* The "embarrassing" difference between the voxel activation capacity and pixel capacity is further exacerbated when we consider the higher refresh frequencies offered by conventional display systems.

TABLE 10.1 Main Factors of the Three Display Unit Subsystems that Impact On the Form and Size of Voxels for Four Example Volumetric Systems

Example Volumetric Technique	Impact Upon Form of Voxel		
	Image Space Creation Subsystem	Voxel Generation Subsystem	Voxel Activation Subsystem
Translating LED panel	Translational motion of surface of emission (Period and profile of motion impacts upon voxel elongation in direction of screen motion)	Illumination of LED (Shape of LED determines voxel shape in plane of screen, and also in direction of screen motion—if the dimensions of the device are significantly greater than distance moved during T_A)	Electrical stimulus to LED (Elongation in direction of screen motion due to voxel activation time, T_A)
Cathode Ray Sphere (Sections 5.4.1, 5.6.3)	Rotational motion of surface of emission (Period of motion and distance of voxel from rotation axis impacts upon elongation in the direction of screen motion)	Cathodoluminescence (Loss of definition in direction of screen motion due to finite decay time of phosphorescence; Possible loss of definition in plane of screen due to secondary emission)	Electron beam directed onto surface (Extent and intensity profile of voxel in plane of screen determined by that of beam; Elongation in direction of screen motion due to voxel activation time, T_A)
Stepwise excitation of fluorescence (Section 7.7.2)	Static volume of homogeneous material (–)	Visible fluorescent decay from static voxel (–)	Intersection of two laser beams (Beam profile and angle of intersection determine voxel shape)
Static array of optical elements (Section 7.7.1)	Static volume of voxel elements (–)	Visible fluorescent decay from static voxel (Shape of elements determines form of voxels)	UV light applied to voxel element via optical fiber (–)

In general, intrinsic voxel deformities may be due to any combination of the three component subsystems. This is illustrated in the following examples and in Table 10.1.*

Example 1 Consider the swept-volume display unit discussed in Section 5.5.2. In this system the image space is created by the rapid rotation of a helical screen, and voxels are activated by a laser beam(s) impinging on a surface of emission that is able to scatter light. Since the angle of incidence between the beam and screen varies, nonidentical voxels will be produced.[†] Furthermore, the dimension of each voxel in the direction of motion of the surface of emission depends on the voxel activation time (T_A) and the distance traveled by the surface of emission in the region of intersection during this period (Figure 10.1). Since the scattering mechanism has no intrinsic persistence after the cessation of the activation stimulus, the size and form of each voxel is independent of the voxel generation subsystem.[‡]

Example 2 Consider the swept-volume system described in Section 5.6.3. From our current perspective, the general principle of operation is much the same as that of the display unit employed in Example 1. However, in this case the voxel generation process no longer takes the form of a simple scattering of incident light, but involves conversion of the kinetic energy of a beam of electrons into visible light via the process of phosphorescence. This has an associated time constant governing the decay in emitted light intensity *after* the activation beam has been removed. Should the phosphor decay constant be sufficiently long, the surface may move an appreciable distance while this emission occurs, giving rise to voxels elongated in the direction of screen motion (see Plate vii). In this case, the voxel's shape and form are determined by all three of the display unit subsystems.

These two examples provide a clear indication that in the case of all swept-volume systems the dimension of the voxel in the direction of motion of the screen will be determined by the distance moved by the screen during application of the activation stimulus. This difficulty is exacerbated if the voxel generation process continues after removal of the activation stimulus. Should this occur, there will be a loss of voxel definition in the direction of motion, and this will be particularly apparent from certain viewing positions.

---

* As we have discussed previously, apparent voxel distortion may occur as a consequence of the passage of the light through the image space after its emission from the location at which the voxel is activated (see also Section 10.4.2).
[†] Dynamic focusing of the activation beam(s) may be used to reduce variations in these voxel characteristics.
[‡] Laser speckle may be problematic unless the surface of emission is suitably prepared.

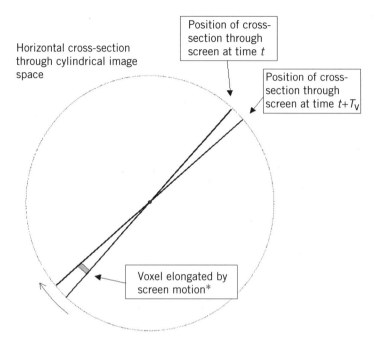

Horizontal cross-section through cylindrical image space

Position of cross-section through screen at time t

Position of cross-section through screen at time $t+T_V$

Voxel elongated by screen motion*

*N.B., Precise direction of elongation also depends upon incident beam direction

FIGURE 10.1   In swept-volume display units, motion of the screen during the voxel activation time T_V may lead to voxel elongation. The extent of any such elongation depends on the frequency of the screen motion cycle, the duration of T_V and in the case of rotational or variable-speed translational motion, on the voxel position within the image space. In the case of a beam-addressed system in which one or more laser beams scatter from the surface of emission, the direction of elongation depends on the incident direction of the activation beam.

10.2.2   Voxel Intensity and Color

For systems that offer considerable viewing freedom, it is most desirable that voxels radiate light isotropically. However, even in the case of an idealized technique permitting the production of spherical voxels that radiate isotropically, their resulting intensity profile will not be uniform across the diameter of each voxel. For example, consider the production of a translucent voxel in which light is homogeneously generated throughout its volume and, for simplicity, let us assume that this light is able to emerge from the voxel without attenuation. In Figure 10.2a, we illustrate the resulting intensity profile. This type of nonuniformity is of consequence only in the case of larger voxels whose individual profiles can be resolved. Should this be the case, image quality may perhaps be improved by arranging that adjacent voxels overlap slightly (see Figure 10.2b).*

---

* Overlap of voxels is clearly not possible for systems employing an array of voxel generation centers.

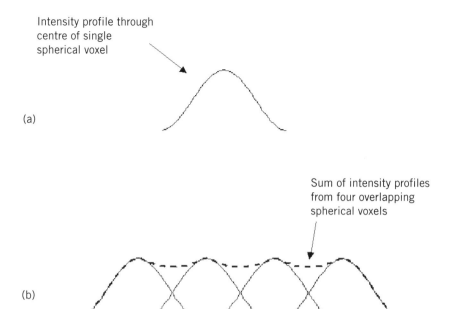

Intensity profile through centre of single spherical voxel

(a)

Sum of intensity profiles from four overlapping spherical voxels

(b)

FIGURE 10.2 (*a*) Linear intensity profile of light emitted from a spherical uniformly emitting voxel. (*b*) In composing image primitives from adjacent voxels with this intensity profile, it may be advantageous to allow for some voxel overlap in order to produce a more uniform intensity profile along the primitive.

In combination with the ambient lighting conditions, the maximum intensity of the voxels plays a part in determining the severity of flicker and therefore slightly affects the required image refresh rate.

Few of the display unit techniques proposed to date permit full color imaging. Rather, they are limited to the creation of images that are either monochrome or comprised of two or three discrete colors. Color mixing (which is essential in the production of a full color palette) presents a considerable technical challenge. Difficulties arise as a consequence of limitations either in the voxel generation subsystem (where it may prove difficult to produce the primary colors) or because of the demands placed on the combination of the three display unit subsystems. For example, in the case of systems employing a beam-addressed voxel activation technique, color mixing necessitates a very high degree of alignment between the display unit subsystems, and this must be maintained throughout the working life of the display. The performance of the beam calibration system (such as that outlined in Section 9.8, for example) will ultimately govern the ability of this type of display unit architecture to achieve color mixing.

10.2.3 Voxel Opacity

A volumetric display system able to support the production of opaque ("solid") voxels would enable the visual depth cue of occlusion (see Section 1.4.4) to be satisfied, and this is highly desirable for many applications. Furthermore, the generation of opaque images may permit the automatic creation of shadows, adding to the realism of an image scene. Any requirement for the production of opaque images affects not only the voxel generation subsystem but also the method by which an image space is created. Consider, for example, a swept-volume display employing a single beam source for voxel activation. Each voxel is activated sequentially. So as to avoid image "trails" voxels must quickly relax back into their inactive state after removal of the activation stimulus. Even if each voxel exhibited an opaque characteristic, when active this display architecture would be unable to permit the production of opaque images, *irrespective of the physical process employed to generate the voxels*. This is because only one voxel (or a very small number of voxels) is active at any one time. Therefore, the light emitted from any voxel propagates out of the image space unimpeded by others, and hence all voxels are visible.

A requirement for image opacity therefore affects not only selection of the technique employed for voxel generation but also the manner in which the image space is created. Although swept-volume systems cannot accommodate opacity, static-volume techniques employing photochromic or thermochromic materials for voxel generation offer to do so (see Section 7.5.5). In an ideal situation, it would be possible to assign to each voxel a variable degree of opacity (or conversely transparency). In this case, each voxel descriptor would contain information on opacity in addition to position, color, and intensity data.

10.3 VOXEL PLACEMENT

Voxel placement has been discussed at length in other parts of this work. However, it is instructive to review this matter briefly, as the ability to position voxels at the required locations within an image space is clearly of the utmost importance. As mentioned in Section 8.5.4, early workers in this field recognized the difficulties associated with the exhaustive scanning of an image space [e.g., Ketchpel (1963)]. As a consequence of these problems, beam-addressed systems have generally utilized a dot graphics technique for voxel activation. Unfortunately, for a display unit employing this approach, the maximum voxel activation capacity (and fill factor) is generally low. Furthermore, in the case of beam-addressed swept-volume systems, the temporal manner in which the image space is generated, coupled with a largely sequential voxel activation process, results in the maximum number of voxels that may be depicted (indicated by the voxel activation capacity) being necessarily dispersed throughout the image space. The constraints governing this dispersal depend on the screen motion. Therefore, the total number of voxels that may be employed in the depiction of many images will be significantly less than the maximum voxel activation capacity indicated.

The greater part of any image space will generally be void. Consequently, a random scan approach to voxel activation [in which the activation beam(s) move directly between the locations of consecutive voxels comprising the visible image] represents a highly efficient method of image generation.* However, as we discussed previously, this advantage is offset by a lack of predictability. This affects a number of aspects of a display unit's performance and makes it extremely difficult to prejudge the ability of a volumetric display device to depict a particular data set properly. These issues are reexamined briefly below.

For any swept-volume display, the motion of the surface of emission determines the time (t_{op}) during each image update period when each voxel may be activated (corresponding to the surface of emission passing through the appropriate region within the image space). Should one or more beams be employed for voxel activation (in conjunction with a dot graphics deflection technique), the ability to locate a particular voxel correctly will be conditional on the presence and location of others which have a similar value of t_{op}. This highlights a considerable disadvantage associated with the random activation scheme in displays of this type: namely, that *the ability to activate voxels is conditional on the activation state of others*.

A number of approaches may be employed to minimize the impact of this problem, the most elementary involving the culling of any voxel (from the image data set) which will be subject to an unacceptable positioning error. Clearly, in certain critical applications, the arbitrary omission of image data may have potentially disastrous consequences and should only take place with the knowledge of the operator. Some small relief to this problem may be gained by the introduction of an approximate ordering algorithm, as outlined briefly in Section 8.5.4. This can, in principle, increase the number of voxels that can be contained within each sector (Schwarz and Blundell 1997) but provides an additional (and considerable) burden for the graphics engine. Furthermore, the benefits that may be derived from this approach vary according to the spatial distribution of voxels within an image scene. This may decrease the predictability of the system.

Static-volume systems may also be restricted to a nonexhaustive voxel activation refresh (corresponding to a fill factor of less than 100%) and in some cases impose constraints on the order in which the voxels may be activated. For example, consider a static-volume beam-addressed system employing a stepwise excitation process for voxel generation (see Section 7.5.1). So as to avoid the creation of ghost voxels, should this type of display unit feature either more than a single pair of activation beams, or an intermediate state lifetime that is long in comparison with the voxel time T, the order in which voxels may be activated may need to be predetermined

Currently, the operator of a volumetric system is likely to have developed the system (or at the very least have been actively involved in some way with this process). An operator's considerable knowledge concerning the detailed architecture of the display will facilitate adjustments in system performance and manipulations of the image data set, which ensure that images are shown to

---

* This approach is in some respects analogous to the vector graphics approach employed on various CRT graphics devices in the late 1970s (Sherr 1992).

their best advantage. However, when we look toward the future development of high-performance volumetric display systems, the homogeneity and predictability of an image space are of the utmost importance. Only by ensuring that systems exhibit these characteristics can we hope to achieve their acceptance by the general community, who are likely to have no knowledge of the underlying techniques employed in their implementation. With this in mind, we discuss in Chapter 12 a number of approaches that may be adopted in the implementation of an exhaustive image space scanning technique.

10.4   IMAGE SPACE CHARACTERISTICS

In earlier chapters we have seen that many of the volumetric display systems proposed to date employ some form of mechanical motion for image space creation. In reviewing the history of television development, we see proposals for a range of systems that place a similar reliance on complex mechanical techniques (Abrahamson 1987). There are in fact many similarities between these two development processes, and therefore by analogy we may perhaps conclude that swept-volume systems will provide only a short-term solution in the quest for the optimal volumetric display unit architecture and that these displays will ultimately be superseded by static-volume systems.

However, attributes that are assigned to each voxel created within an image space are likely to be modified by the image space medium and its boundary. The image space itself may therefore negate efforts directed to the production of idealized voxels. Consequently, it is most important that careful attention be directed to its design, and its optical properties should be clearly understood. The techniques employed for image space creation also affect the reliability and portability of the system. Furthermore, the three subsystems in combination are likely to determine the maximum image space dimensions.

Within this section we outline briefly various image space characteristics and discuss their impact on both the perceived image quality and the manner in which they affect the suitability of a display system for various applications. The following subsections therefore provide a framework within which we can describe the general properties of an image space. Unfortunately, in practice there is a considerable degree of interdependence between each of the attributes discussed below. This increases the difficulty of characterizing a display unit and thereby accurately judging *all* aspects of its performance prior to implementation.

10.4.1   Size and Form

The optimal image space size and form are best determined by reference to the intended area of application. Unfortunately, for the majority of currently available architectures the dimensions and shape of an image space are generally determined by technical limitations associated with the image space creation, voxel generation, and voxel activation subsystems. It has not therefore been possible to conduct trials with the purpose of identifying the image space dimensions best suited to potential areas of application.

Image space dimensions must be considered in association with voxel attributes such as size, opacity, placement accuracy, and uniformity. Furthermore, in the case of architectures which, in principle, may be applied to the development of large image space volumes [e.g., a swept-volume system employing the rotational motion of an electroluminescent panel (see Section 5.4.4)], construction costs are likely to be a major consideration.

Physical Limitations of Image Space Dimensions

The low physical density of the image space associated with swept-volume and gaseous static-volume systems is most desirable. Consider a static display unit comprising a solid doped glass cube and perhaps employing a stepwise upconversion process for voxel activation. Clearly, the weight of such a display unit increases as the cube of its linear dimension, and as we saw in Section 2.4.1, for a cubic image space measuring 30 cm along each edge, its mass would be approximately 70 kg.

As the dimensions of an image space are increased, it is necessary to review and perhaps modify various voxel characteristics. For example, the size of voxels may be increased, and given the greater spatial separation between the operator and voxels activated on the far side of the image space volume, it may become necessary to increase voxel intensity. Increases in the dimensions of an image space are also likely to be accompanied by a greater voxel location capacity. To maintain the same fill factor, the requisite increase in the voxel activation capacity will provide a greater workload for the voxel activation subsystems. This may ultimately necessitate an increase in their number, and in the case of beam-addressed systems will give rise to beam registration difficulties. Any increase in the voxel capacity will also affect the peak voxel descriptor throughput that must be supported by the graphics engine.

In the case of swept-volume display systems, increasing the image space size will require higher screen driving powers and must also be accompanied by greater strength in the supporting structure. For example, consider a display unit in which the image space is created by the rotational motion of a planar surface of emission bonded to a supporting structure. As the diameter of the image space is increased, any flexing of the screen due to imperfect rigidity will exacerbate voxel-positioning errors toward the periphery of the image space. To compensate for this, a central shaft may be added, the supporting structure may be thickened, or possibly additional supporting structures may be placed around all or some of the screen edges. Although modifications of this type may ameliorate screen flexing, they will also generally result in undesirable optical interference with light emitted from within the image space. In particular, a thickening of the supporting structure or an additional support at the axis of rotation or around the edge of the screen will worsen the effect of visual dead zones (see Section 4.2). Unless most carefully designed, supports attached to the top or bottom of the

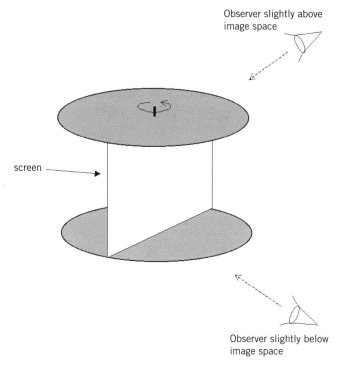

FIGURE 10.3   If additional screen stabilizing devices are employed so as to reduce screen flexing and image jitter in swept-volume systems, care must be taken to avoid compromising viewing freedom. Illustrated here is the case of a planar rotating screen, stabilized by two disks located above and below. The optical characteristics of the disks will affect the visibility of the image space.

screen may also restrict the range of viewing angles (Figure 10.3). The geometrical form of the moving surface also affects its intrinsic rigidity. A helical screen geometry is inherently more stable when rotated about its axis than is a planar screen rotated about an axis lying in its plane.

In the case of swept-volume systems, a *decrease* in the dimensions of an image space may have associated problems. For example, the extent of the visual dead zone may not decrease in proportion to any decrease in the image space volume. Consequently, the impact of the visual dead zone may be greater. Furthermore, we may expect that as the size of an image space is reduced and consequently smaller image objects are depicted therein, the observer will need to be located at close proximity to the display vessel. Nonuniformities and any lack of image definition may then be more apparent.

Influence of the Application on the Size and Form of an Image Space

The optimal size and form of an image space should be identified by consideration of the intended area of application and not by highly restrictive technical limitations. For example, consider a display unit designed for the visualization of whole-body medical data. Ideally, the image space may take the form of a tall but relatively narrow cylinder (approximating the proportions of the human body). Alternatively, should it be necessary to depict only portions of the body (for example the head in applications such as radiotherapy planning), the image space height and diameter may be approximately equal. Conversely, a display unit designed for application in military planning or air traffic control may be more effective when the height to width ratio is relatively small.

In applications such as terrain visualization, it may be appropriate to apply different scaling factors to a data set in the vertical and horizontal directions. However, in general, the scaling factors should not vary throughout the image space (see also Section 10.7). Although the application of two scaling factors may permit features of interest to be more easily discerned, when applied to the context of real-time data visualization, the ensuing nonphysical spatial relationships may hamper rapid decision making.

In Section 5.2 we discussed the work undertaken by E. Parker and P.R. Wallis and also presented extracts from a discussion held at the Radio Section of the IEE in 1948. Let us return to a comment made by C.W. Thomas:

> ... I think that it is fair to say of all these pictorial methods, which include the truly three-dimensional and perspective displays, that their volume of coverage is inherently small. ...

The relevance of this comment may easily be understood by considering the application of a volumetric display system to aircraft control near an airport. To assist in this illustration, we will consider that the display is to depict aircraft within a 10-km radius of the airport up to an altitude of 2 km. Let us assume the use of a cylindrical image space measuring 50 cm in radius and 50 cm in height. To make full use of the image space volume, two different scaling factors would be required in the vertical and horizontal directions, in this case 20,000 horizontally and 4000 vertically.

A 300-m vertical separation of two aircraft would be depicted within the image space by a distance of 7.5 cm and therefore would be readily apparent. However, a 300 m separation in the horizontal direction would correspond to only a 1.5-cm separation of the aircraft icons within the image space and may be less easily perceived unless the operator moves around the display volume. Should we wish to increase the range of operation of this system, spatial separation would become increasingly less apparent and the benefits that may be derived from the display could be negated.

Clearly, in this application we require a cylindrical image space. The diameter and height of the cylinder should be proportioned to match the visualization requirements in the horizontal and vertical directions and therefore permit the use of a single scaling factor.

10.4.2 Optical Characteristics of the Medium

The method by which an image space is generated may have a considerable and sometimes a subtle effect on the quality of an observed image. For example, an image may be subject to distortions (in terms of geometry and intensity) which vary according to the position from which it is viewed. The ability of a system to produce a high-quality image *within* the image space cannot alone be used to judge the performance of a volumetric display—we must also consider the propagation of the light from the image to the observer.

Consider the path traversed by light emitted from an activated voxel element as it passes through and exits from an image space. Should this light encounter obstacles or pass through regions that vary in transparency or refractive index, its path and/or intensity will be altered. Therefore, voxel attributes cannot be used in isolation as a means of determining a display unit's performance. These attributes must be considered in conjunction with the image space characteristics. In the remainder of this section, we discuss optical distortion effects both within the image space and at its boundary.

Optical Characteristics Within the Image Space

Consider a voxel that is activated at an arbitrary location within an image space created by means of a static-volume technique. We assume that this voxel is spherical in shape and radiates isotropically. Unless the voxel is located at the physical center of a spherical image space, the distance traveled by the emitted light as it crosses the image space will vary according to direction. Consequently, should the optical characteristics of the image space medium be nonhomogeneous, the intensity of the voxel will be affected by both the position of the observer and the voxel's location within the image space. This is in contrast to conventional display devices such as CRTs, where the light emitted from each pixel must travel a short and approximately constant distance before emerging from the display's faceplate.

It is instructive to examine various static-volume display units described in Chapter 7 and consider the impact of the image space on the propagation of light. The optical transmission characteristics of various types of image space are reviewed in Table 10.2. As may be seen from this table, problems arise due to differences in refractive index and transparency. Variations in the refractive index of materials present within the image space lead to the nonrectilinear propagation of light and dispersion effects. Nonuniform transparency will result in an image appearing to be less bright and perhaps less well defined when viewed from certain orientations. Furthermore, the presence of optically opaque material such as metallic conductors within the image space may give rise to the complete occlusion of voxels when they are viewed from certain directions.

Problems relating to propagation of light within the image space may be ameliorated by carefully matching the refractive index and optical transmission characteristics of the various components used in its construction [e.g., MacFarlane (1994)]. Unfortunately, although this will give rise to an optically homogeneous image space, its resulting refractive index will be considerably greater than that of the surrounding air. Consequently, image distortion may nevertheless occur (at the image space boundary).

Any display unit employing a gaseous image space volume will be optically homogeneous and will have a refractive index that closely matches that of the surrounding air. Light crossing the image space will therefore not be subjected to either distortion or directional attenuation. This denotes a major advantage of this type of technology over other volumetric display techniques. However, the light may still be refracted as it crosses the display vessel envelope (see below).

In the case of swept-volume display systems employing rotational motion, undesirable optical interference may occur due to interaction between the supporting structure and its surroundings. As discussed previously, should the image space dimensions be increased, there is a greater need to strengthen the supporting structure. This may increase the extent of the visual dead zone and thereby adversely affect the optical characteristics of the image space.

Optical Characteristics at the Boundary of the Image Space

An image space will generally be enclosed within a containing vessel or take the form of a single solid structure. In the case of a solid transparent medium, the image space boundary corresponds to the interface between the image space and the surrounding air. When the image space creation technique employs a rotating screen, a gas, or a liquid, the boundary corresponds to the transparent envelope encapsulating the volume.

Consider a static-volume display unit in which the image space is composed of materials that have refractive indices that differ significantly from the surrounding air. Refraction at the boundary of the image space will cause geometrical distortion in the perceived image in accordance with Snell's law. Furthermore, in the case of a planar boundary, the apparent dimensions of the image space will be reduced. This perceived diminution of the image space volume is particularly disadvantageous since to enhance the visualization of spatial information, we seek to maximize its volume. Moreover, the effect varies with viewing position. Conversely, in the case of a curved boundary, refraction may lead to the magnification of regions of an image space.*

10.4.3 The Image Space: Some Practical Considerations

Currently, practically all volumetric display systems reside within the laboratory environment and take the form of experimental prototypes. Consequently, issues regarding their long-term reliability and environmental impact have been of minor importance. However, in making the transition to commercial products, matters such as noise generation, operational life expectancy, electromagnetic interference, and general safety will be of considerable concern.

In the case of large swept-volume systems, the rapidly moving screen can produce considerable noise. In the second-generation system developed by the U.S. Navy (Soltan et al. 1994), this matter is addressed by employing a cylindrical envelope attached to, and corotating with, a helical screen. The air within the region of the helical screen is therefore trapped. Consequently, the noise that would otherwise have been generated by the pumping action of the helix is greatly reduced. Other approaches require the incorporation of an evacuated display

---

* For certain applications this may possibly be advantageous, a feature of interest may be magnified in accordance with the viewing position.

TABLE 10.2 Optical Characteristics of Selected Static-Volume Display Units*

Section(s)	Reference(s)	Technique(s)	Optical Characteristics
7.3, 7.5.1, 7.6.2	Howells 1945, Zito & Schraeder 1963, Rowe 1975	Beams intersecting in a gaseous or particle-cloud image space.	Light emitted from voxels within the image space is refracted by the transparent image space envelope. The effect on the perceived voxel placement depends on the thickness, composition and geometry of the envelope and on the position of the viewer with respect to the vessel and voxel.
7.4.1, 7.4.2	Ruderfer 1952, Raytheon Co. 1954	Electrical discharge between conducting rods/meshes.	The opacity or refractive index of the conductive elements causes scattering and occlusion of light emitted from voxels within the image space. The thickness, composition and geometry of the surrounding envelope determine refraction effects at the boundary of the image space.
7.5.2, 7.5.3, 7.5.5, 7.6	Fajans 1961, Brown 1964, Lewis & Adelman 1968, Adamson 1968, Lewis et al. 1971	Beams intersecting within a solid or gelatinous image space.	Light propagation within the image space is likely to be rectilinear. However, at the image space boundary the light is refracted, altering the perceived voxel position. This depends on the refractive index of the image space, the surface geometry of the image space boundary and the relative positions of voxel and viewer.

7.5.6, 7.6.1	Maguire 1969, Nithiyanandam 1975	3D array of lamps or LEDs.	The lamps interfere with the propagation of emitted light within the image space, whether active or inactive. Any image space envelope will further refract the light according to its thickness, composition and geometry.
7.7.1	MacFarlane 1994	3D array of voxel generation centers refractive index matched to image space.	Voxel elements and image space are refractive-index matched to enable near rectilinear propagation of emitted light within the image space. However, at the image space boundary the light is refracted, altering the perceived voxel position. This depends on the refractive index of the image space, the surface geometry of the image space boundary and the relative positions of voxel and viewer.

*Optical characteristics of selected static-volume display units which have been discussed in Chapter 7. Systems with similar image space optical characteristics have, for clarity, been grouped together.

vessel, which also serves to minimize noise (Blundell et al. 1993a). A long-term reliability issue relating to swept-volume display systems employing a passive surface of emission relates to the gradual erosion of any surface coating. In the case of systems such as the CRS (see Sections 5.6.3 and 11.2), the bonding of the phosphor particles to the supporting structure must be considered with great care, despite the low pressure within the vessel. The absence of any reliance on mechanical motion in the case of systems employing a static volume image space creation technique is likely to result in greater reliability.

The use of electron beams for voxel activation may give rise to problems associated with electromagnetic interference. Furthermore, unless active and auto-mated alignment techniques are employed, stray magnetic fields may result in unacceptable image distortion. Should the image space volume be evacuated, the risks associated with implosion will necessitate the careful design of the display vessel. This may require an increase in its thickness, and consequently, greater image distortion may ensue.

10.4.4   Viewing-Angle Limitations

Image space occlusion may arise due to the presence of apparatus and mecha-nisms responsible for voxel activation or may be an intrinsic limitation associated with the image space creation technique. For example, in the case of swept-volume systems it is natural to assume that display units employing the transla-tional motion of a screen will be more restrictive in viewing position freedom than are those that employ rotational motion. However, the physical occlusion of an image space does not in itself provide a proper indication of the freedom by which an observer may view images contained therein. Further restrictions may be caused as a consequence of the optical properties of the image space and/or the image space boundary.

10.5   COMPENSATION FOR ADVERSE IMAGE SPACE CHARACTERISTICS

In previous sections of this chapter we have discussed briefly a variety of inho-mogeneities and anisotropies associated with the production and placement of voxels and with the manner in which the image space is created. It is likely that any display unit architecture will exhibit a subset of these adverse char-acteristics. In this section we discuss issues relating to the compensation of certain deficiencies in voxel characteristics and voxel placement nonuniformi-ties. Although the impact of some adverse display unit characteristics may be reduced by suitable preprocessing of the image data set, this task may be compu-tationally demanding. However, since it provides a means of minimizing the impact of certain weaknesses that may be implicit (and therefore unavoidable) in the implementation of the display unit, the increased computational requirements may be acceptable. Computer compensation cannot be used to rectify problems associated with the propagation of light through the image space and across its boundary. Consequently, we may regard the adverse optical properties of an image space as perhaps representing the greatest weakness that may be associated with any display unit embodiment.

10.5.1 Compensation for the Variation in Voxel Attributes

In the case of beam-addressed systems, variations in voxel characteristics such as size and shape may be reduced by applying a dynamic focus to the beam(s).* A constant cross-sectional beam profile is thereby maintained, irrespective of the distance between the beam source(s) and voxel location(s). This approach will not eliminate the variations in a voxel's size and shape, which arise as a consequence of the overall interaction between the three display unit subsystems. This may be illustrated by considering two particular beam-addressed volumetric techniques: a static volume system employing the stepwise excitation of fluorescence (as described in Section 7.5.1) and a swept-volume system employing an electron beam to address a passive target surface (e.g., Section 5.6.3).

In the case of the static-volume display architecture, variations in voxel shape arising as a consequence of variations in the angle of intersection between the two beams can be eliminated only by designing the display unit in such a way that the beams intersect perpendicularly for all possible voxel locations. Although a dynamic focusing arrangement may reduce the severity of this problem, the most effective approach would involve the use of more than a single pair of activation beams.

In the case of the swept-volume example, dynamic focusing will assist in maintaining a constant beam cross section at the point at which the beam intersects with the surface of emission. However, variations in voxel shape due to either the changing angle of incidence or the screen motion are intrinsic consequences of the interaction of the display unit subsystems. They cannot be rectified by the controlling software.

10.5.2 Compensation for the Variation in Voxel Placement

As we have discussed, in combination the techniques employed in the implementation of the three display unit subsystems will often lead to a lack of uniformity in voxel placement. In certain cases, this deficiency can be corrected by manipulation of the image data set prior to its portrayal. Consider, for example, a swept-volume display unit employing rotational motion of the surface of emission in conjunction with a beam addressing technique. The maximum density of voxels that can be depicted by this image space (in the direction of screen motion) increases toward the axis of rotation. One means of ensuring that this image space supports homogeneity in voxel placement is to prefilter the image data to a uniform voxel density that can be supported *throughout* the image space. Unfortunately, this type of prefiltering of the data achieves uniformity by extending a worst-case characteristic across the entirety of the image space.

---

* In the case of certain architectures, beam focusing is used to determine the depth at which voxels are activated (see Section 7.5.5).

In the case of beam-addressed systems, another important role played by the controlling software in assuring accurate voxel placement relates to the calibration and registration of the beam(s). This process is necessary to accommodate nonprecision production techniques and errors that may arise during the operational lifetime of the system. Information derived from any calibration procedure is passed to the procedures responsible for the transformation of image data into the voxel descriptor format. One such beam calibration scheme for a swept-volume system is outlined in Section 9.8.

10.6   GENERAL-PURPOSE DISPLAY UNIT IMPLEMENTATION

The majority of display systems continue to be based on the CRT. This device, which is attributed to Ferdinand Braun (1897), found immediate application in the study of electrical waveforms (Abramson 1987).* A number of years were to pass before this display technology was integrated successfully in television systems. Since its earliest times, the CRT has undergone continuous refinement and continues to retain its position at the forefront of electronic display technologies. The widespread adoption of the CRT reflects not only its suitability for use in many areas of visualization, but also the ability of this technology to adapt to ever-increasing performance requirements. Furthermore and, most importantly, as a consequence of the development of highly ingenious mass production techniques, the cost of this type of display has gradually been reduced.[†]

By means of various techniques (e.g., the use of stereoscopic glasses), it has been possible for CRT-based displays to satisfy the binocular parallax depth cue and thereby not only enter into, but dominate, the world of 3-D visualisation. Clearly, for other types of display systems to gain widespread acceptance, it is necessary to demonstrate that they possess major advantages over existing (generally, CRT-based) systems. Given the relatively small cost associated with conventional display techniques and their flexibility in performing a range of functions, this will not be an easy task.

However, the increasing need to assimilate ever more complex multidimensional information makes it likely that more advanced types of display system will eventually gain acceptance. Volumetric display systems possess a number of advantageous characteristics, and this suggests their suitability for application in a number of areas. Although it is possible to postulate (with great enthusiasm) applications that may benefit from this type of display, it is only when rigorous trials are conducted in situ that the advantages (and disadvantages) can be assessed properly. This will necessitate the development of more advanced prototype displays that can operate reliably, interface to existing hardware, and be refined sufficiently to be operated by the end user.

Once we determine the types of applications for which volumetric systems may be suitable, it is possible to consider the prospects for the development of a single

---

* This book is highly recommended to any reader interested in the history of display technology.
[†] The future role of the CRT is discussed in an interesting article written by Senyo Sluyterman (MacDonald and Lowe 1997).

architecture able to fulfill these various roles. For example, given a display system developed primarily for the visualization of medical images, we must question its suitability (without any undue compromise in performance) for other purposes, such as the depiction of CAD images or aircraft radar data. In the remainder of this section, we consider briefly three possible applications for volumetric display systems, relating to CAD, medical imaging, and 3-D television, and highlight the diversity of requirements (See also Blundell and Schwarz 1999).

10.6.1   CAD Visualization

The design and visualization of 3-D structures together with the molding of surfaces into complex forms may be facilitated by means of volumetric display systems. In this type of application, the volumetric system would illustrate qualitative information and would be complemented by a conventional display able to depict quantitative data. The provision of an interaction device by which the operator may directly manipulate image components within the image space (see Section 12.7) would be most desirable. Clearly, any changes made to image components by this means should also be reflected immediately on the conventional display.

The diverse form of image data does not suggest that the image space should have any particular aspect ratio. Therefore, a cubic or spherical volume would probably be suitable for use in this type of application. However, the ability of the display to depict without distortion the geometrical shape of image components depicted therein is of particular importance. Severe optical distortion at the image space boundary would therefore be unacceptable, and this suggests the use of either a gaseous static-volume display or swept-volume system. For the display to perform a useful function, the voxel activation capacity need not necessarily be very high, although the ability of the system to support the depiction of images comprising several colors would be advantageous.

10.6.2   Medical Visualization

A fundamental requirement for any display system intended for use in medical imaging relates to the ability to represent the incoming data set without loss or distortion. Given the large volumes and densities typical of volumetric medical data and the lack of predictability of many of the volumetric display system technologies employing a dot graphics approach, it is likely that only systems offering an exhaustive image space scan could be used with confidence in this area of application. The particular type of data to be represented therein would indicate the form and dimensions of the image space. The display may be used as part of an instillation, and in this case portability would be unimportant. Alternatively, the system may, for example, be applied to radiotherapy treatment planning and therefore act as a computer peripheral. In this case, the display would take a physical form similar to that discussed in Section 10.6.1.

The voxel activation capacity of any system applied to medical visualization should be very high, and ideally the display should offer a 100% fill factor ensuring an ability to depict image components of considerable detail in a reliable and predictable manner.

However, it is possible that lensing effects occurring at the image space boundary could be used advantageously, particularly if a volumetric system were to be employed within an operating theater environment. The translucency of images depicted within the image space may prove to be beneficial, although this may only be assessed properly by conducting appropriate trials. Similarly, the advantages that may be derived by permitting direct interaction with the image space (see Section 12.7) may only be assessed by the application of volumetric system prototypes to specific areas of medical visualization.

10.6.3   3-D Television

A number of publications have suggested that volumetric systems will provide the next logical advance in television. However, when we consider the practicalities of such a proposal, this would appear to be most unlikely. Some of the associated difficulties are:

- Requirement for image opacity
- Requirement for full color
- Ability to support very high data rates (in both image depiction and data transmission)

Although it may ultimately prove to be possible to meet these technical requirements, there are several other factors that must be considered which present a far greater obstacle in the implementation of 3-D television:

- Capture of the volumetric data sets
- Enforced viewing orientation associated with television. This contrasts with the practically unrestricted field of view offered by a number of volumetric technologies.

A further contrast between conventional television and volumetric imaging relates to the overall viewing perspective. Volumetric systems are limited to the presentation of a "God's eye" view of any scene, whereas conventional television is not so restricted. In certain areas of entertainment, volumetric systems may be used advantageously. For example, an overview of a sporting event may provide considerable clarity by depicting the spatial separation of artificially created figures (icons) and the true trajectory of a ball in three dimensions. When used in this type of application, volumetric systems would not be providing the next generation of television display but would, rather, offer an entirely new method of entertainment. In short, conventional television and any volumetric display system developed for entertainment are likely to have a different functionality.

10.7   LINEAR PERSPECTIVE AND HIDDEN-LINE REMOVAL

In the case of conventional computer graphics, the depth cues of linear perspective and height in the visual field (see Section 1.4) are generally employed to provide a sensation of depth. For example, if two objects are known by the observer to be of approximately the same size, the depiction of one as smaller (and vertically offset) results in its being perceived as further away. This approach allows display devices such as the CRT to provide an essentially unlimited depth. However, this technique cannot be used in the case of those volumetric display technologies that impose minimal restrictions on viewing orientation. For example, consider the application of different scaling factors to two similar objects. Although this would enhance the apparent depth of a scene when observed from one particular orientation, the linear perspective cue would be incorrectly satisfied from all other viewing positions.

In the case of volumetric systems that offer considerable freedom in viewing position, conventional hidden-line removal techniques cannot be employed to compensate for a lack of image opacity. Hidden-line removal assumes a particular viewing orientation, and therefore should this approach be applied to a volumetric image, viewing freedom would be severely restricted.

10.8 DISCUSSION

The development of a set of metrics by which volumetric display systems may be characterized is an extremely difficult process. Difficulties arise not only as a consequence of the diversity of approaches that may be adopted in the implementation of these systems but also because of the variety of ways in which they may be employed. The matter is further complicated by many of the currently proposed technologies failing to exhibit predictability. As we have seen, the performance of a display unit may vary according to the spatial distribution of the voxels contained within an image scene. This is particularly serious in the case of beam-addressed swept-volume systems that employ a dot graphics scanning technique.

In certain areas of application (e.g., medicine and real-time control), it is most important that the performance of a display device be accurately defined. Any lack of predictability will result in a lack of confidence on the part of those using display systems of this type and may perhaps have catastrophic consequences. The quest for predictability is therefore of the utmost importance and should be the focus of all future research in this area. It is, in fact, quite likely that a low-performance display system whose characteristics are well defined (and which do not vary according to the form of the image data) would be considerably more useful than a higher-performance system that does not possess these attributes.

The lack of predictability exhibited by many volumetric display architectures may be clearly illustrated by reference to what is perhaps the most basic metric, the voxel activation capacity. For many display unit configurations, it is possible to quote a value for voxel activation capacity. However, the spatial form of an image scene and its position within an image space will determine the total number of voxels that may be employed in its depiction. Naturally, this uncertainty brings into question the value of this particular metric when used in connection with certain types of display unit architectures.

CHAPTER 11

IMPLEMENTATION OF VOLUMETRIC SYSTEMS

Young man, you can be grateful that my invention is not for sale, for it would undoubtedly ruin you. It can be exploited for a certain time as a scientific curiosity, but apart from that it has no commercial value whatsoever.[*]

11.1 INTRODUCTION

The broad range of techniques that may be applied to the architecture of a volumetric display unit generally makes it difficult for those entering this area of activity for the first time to select the most appropriate approach. Ideally, an optimal architecture could be determined by a careful theoretical analysis of various configurations or by rapidly prototyping systems. Unfortunately, as we have seen, the implementation of any volumetric display unit involves the close integration of a diverse range of technologies. Consequently, without considerable experience, it is difficult to predict reliably (and completely) the image quality and overall performance of any proposed display prior to its construction. Furthermore, the prototyping of a display unit is likely to involve a very considerable amount of time and effort and may require contributions from a range of highly skilled scientists and engineers.

In this brief chapter several swept-volume display unit architectures are discussed. The first of these is the Cathode Ray Sphere, which was the subject of the authors' work in New Zealand for a number of years. Rather than concentrating solely on technical issues relating to this research work, we prefer to provide a more personal account which, for a variety of reasons, emphasizes the work done in the early years of this project. Technical aspects of more recent work are

---

[*] Auguste Lumière, pioneer of cinematography, 1895.

recorded in a number of scientific publications and patents that are referenced where appropriate. In Section 11.3, Rüdiger Hartwig discusses implementations employing a helical-shaped screen. This is followed by a brief contribution from Robert Batchko, who describes the approach that he adopted in the construction of the laser-addressed multiplanar display apparatus and in Section 11.5 we draw upon material contributed by Parviz Soltan and his research team in San Diego. Finally, in Section 11.6 we discuss briefly issues relating to the commercialization of volumetric technologies. This discussion is based largely on aspects of the authors' own experiences in this area.

11.2   RESEARCH RELATING TO THE CATHODE RAY SPHERE

Swept-volume display unit architectures employing cathodoluminescence for voxel generation and utilizing one or more electron beams for voxel activation have been the subject of research activity for a considerable time. This approach was adopted on a number of occasions during the late 1950s and early 1960s and gave rise to several outstanding architectures (see, e.g., Sections 5.3.1, 5.3.3, and 5.4.1).* Unfortunately, at that time the lack of appropriate solid-state electronic components coupled with the absence of the necessary software tools and processing hardware, made it impossible to utilize the capabilities of these various display units.

In 1989, one of the authors (Blundell) initiated in New Zealand a university research project directed at the development of a volumetric display unit. The generic title *Cathode Ray Sphere* (CRS) was adopted for the various prototype systems that were subsequently developed. This series of display units embodied many of the ideas suggested by the earlier pioneers of volumetric display unit development and employed the traditional rotating phosphor-coated screen addressed by a number of electron beams. See Plates i-iv, vi-xi, and xvi-xvii. In this section we provide a brief insight into some of the earlier work that was undertaken on this project.

11.2.1   First Prototype System

An essential task at the outset of any research activity concerns a review of all previous related work. This generally provides ideas, highlights difficulties, and often forms a basis for subsequent work. Unfortunately, stringent financial limitations may play a greater part in directing research and determining the paths that may, in practice, be followed. This is particularly true in the case of experimental scientific research that does not conform to the current direction of mainstream activity and which may not lead to results in an easily predicted period of time. Consequently, the decision to use conventional CRT technologies in the implementation of the author's first prototype display was not made on

---

* Despite the passage of 40 years, the work undertaken by Ketchpel (see Section 5.4.1) is perhaps the best known, and his publications are still frequently cited in current scientific literature.

scientific grounds but was dictated by financial restrictions and a need to make the best use of the facilities that were available.

The MkI CRS (see Plate xi) represented the culmination of the first year of intensive effort (Blundell and King 1991). The metallic vacuum vessel shown in this illustration was constructed from stainless steel scrap and fitted with three glass viewing ports, together with a long vertical drive-shaft tube. An external electric motor rotated a set of magnets around the top end of the drive-shaft tube. The field from these magnets was used to create a torque on a soft-iron bar located within the vacuum vessel. A long vertical drive-shaft connected this bar to the screen (located within the bulbous part of the vacuum vessel). The considerable length of the drive-shaft tube was intended to separate the magnetic coupling clearly from the region swept out by the screen and so ensure that stray field did not disturb the rectilinear propagation of the electron beams.*

Particular attention was given to the deposition of the phosphor coating on the supporting structure (glass). Erosion of the phosphor was considered to represent a potential problem (despite the evacuation of the display vessel), and consequently, in the earlier prototypes, the phosphor was impregnated into the glass. The choice of suitable phosphors was limited by a requirement for short persistence and a need for the material to withstand the high temperatures required in the bonding process (ca. 500 °C). Fortunately, calcium tungstate offers to satisfy both these needs (von Ardenne 1939). Although this P5 (BJ) phosphor is not very efficient in its conversion of the incident electron energy into visible light (blue), the decay in light output after the removal of the excitation source is rapid (90% fall within 25 µs). Initially, this impregnation approach to the bonding of the phosphor proved to be problematic; however, after practice, highly durable screens were produced consistently.

The dot graphics technique discussed earlier (see Section 8.5.4) was used for image generation in all the various CRS prototypes. This approach necessitates the use of electron guns equipped with electrostatic deflection apparatus (see Plate xiii). Guns of this type are readily available and are most frequently employed in conventional oscilloscope tubes. Unfortunately, financial restrictions precluded the purchase of new electron guns, and it was therefore necessary to recycle guns from a large stock of old oscilloscope tubes (type 5BP1 of World War II vintage) which were, fortuitously, available. Reusing electron guns is by no means an easy task, as in the case of guns employing oxide-coated cathodes, the cathode coating is easily damaged (and the cathode emission thereby reduced). Oxide cathode coatings are (following activation) particularly sensitive to oxygen and

---

* This type of drive system (i.e., using magnetic coupling) was adopted in most of the subsequent prototypes. However, it became possible to dispense with the long drive-shaft tube. By proper design, the amount of field leakage may be minimized, and furthermore, the rotation of the external magnets is in synchronism with the rotation of the screen. Any interaction between the magnetic field and the beams therefore causes a gradual distortion of the image, this distortion being essentially constant between successive frames. It does not give rise to image "jitter" and is therefore not particularly noticeable. In the case of more advanced prototypes (permitting color mixing), greater care must be taken in this area.

water vapor, and therefore the removal of an electron gun from its glass housing followed by its insertion into the CRS vacuum vessel had to be carried out within an inert and dry atmosphere. The techniques employed for the extraction of an electron gun from an oscilloscope tube may be summarized as follows:

- The exterior of the tube was cleaned and a flange fitted to its neck (this flange allowed the electron gun later to be connected to the CRS vacuum vessel).
- The tube was placed within a sealed container. This was flushed continuously with dry argon. At all times the cathode was kept warm through the application of a voltage to the heater filament.
- A small tungsten carbide drill was operated within the container and a hole bored through the neck of the tube.
- Once the argon had entered the tube, a heated wire was used to cut the neck.
- During the process of reassembly into the CRS, the cathode region of the gun was flushed continuously with dry argon.

Clearly this was a time-consuming process (6 to 8 hours), and considerable practice was required to perfect the technique and so gain satisfactory results. On several occasions the last stage of the process (a gradual reduction of the heater voltage as the pressure within the CRS vacuum vessel was reduced) was inadvertently overlooked. Consequently, as the thermal conductivity of the gas surrounding the heater filament was reduced, the filament's temperature increased. This resulted in the heater fusing and the process outlined above had to be repeated. Over a six-month period, 50 or more electron guns were recycled in this way for the evaluation of various configurations of the MkI CRS. Subsequent prototypes demanded greater electron gun performance, and fortunately it became possible to adopt a more usual approach to their acquisition.

The MkI CRS did not represent a significant technical breakthrough [it was, in fact, very similar to the system that had been developed many years beforehand (Ketchpel 1963), see Plate xii] but rather, reflected the application of considerable enforced ingenuity in its implementation. The vacuum vessel was constructed from scrap, comprising numerous welded joints and seals which when pumped continuously by an aging (almost antique) vacuum system was able to reach a pressure of 1.5×10^{-6} torr (the pressure gauge was available only in retrospect). The electron guns were extracted from oscilloscope tubes manufactured some 40–50 years previously, and the phosphor was obtained in the form of the mineral sheelite (calcium tungstate), donated freely by a gold prospector on the west coast of New Zealand. The MkI CRS displayed 3-D Lissajous figures and very simple computer-generated objects such as squares. While developing this display it became apparent that the use of electron beams for voxel activation was fortuitous in terms of cost and performance (particularly the high deflection velocity). Furthermore, later work necessitated the accurate calibration of the beam deflection coordinate system with respect to the image space reference frame (see Chapter 8). This matter is discussed briefly below, and as we will see, the use of electron beams facilitates this process. The lack of initial funding which dictated the use of low-cost technologies (and hence electron guns) was perhaps, in retrospect, beneficial.

11.2.2 Subsequent Developments

The next CRS prototype (denoted the MkII system) employed an all-glass display vessel and appears in a number of the plates incorporated into this book (see also Figure 11.1). The increased viewing freedom offered by this system represented a considerable step forward. Furthermore, the use of new electron guns resulted in a reduction in the voxel time and improvements in both image intensity and image definition. Two electron guns positioned 120° apart (in subsequent prototype systems, this was reduced to the optimum 90° angle) enabled the distortional and voxel placement dead zones to be circumvented. The acquisition of a new vacuum system (and associated pressure gauge) improved reliability, and the use of several phosphors (bonded in the conventional manner) enabled the production of color images. In the case of the monochrome display prototypes, the low conversion efficiency of the impure P5 phosphor made it necessary to use P31 phosphor (green), which is more efficient in its light output but has a slightly longer decay time.

Gradual refinements were made to the CRS technology through the development of a number of prototypes. The metallic center shaft, which on earlier prototypes passed through the image space, was eliminated. Consequently, the clarity of images located on the far side of the image space with respect to the observer was improved. In 1994, one of the CRS prototype systems was shipped to Singapore and demonstrated at the IEEE Tencon '94 conference (Blundell and Schwarz 1994). In parallel with the development of the various CRS prototype display units, a number of graphics engines were implemented and this work culminated in the last CRS prototype exhibiting a voxel activation capacity in excess of 100,000 (with a refresh rate of 25 Hz).

One of the most important areas of work undertaken within the CRS project concerned beam calibration. Although a manual calibration system may be appropriate for experimental systems, ultimately an automated approach must be adopted. Furthermore, as the voxel activation capacity and voxel density are increased and a color mixing capability is introduced, it is of the utmost importance that beam-addressed systems (such as the CRS) incorporate automatic recalibration mechanisms. To this end, the screen was provided with a transparent conductive coating by means of which the electron beam current could be detected. Beam deflection algorithms were developed which enabled the beam(s) to follow various trajectories, and the time at which the beam crossed the edge of the rotating screen could be measured (and related to the angle of the screen). Information derived from a number of measurements was then used to adjust the values of the elements in the transformation matrix. Despite considerable success in this area, further experimentation is needed to refine the technique properly.

The display vessel developed for the MkIII CRS is illustrated in Plate x. This display employed three electron guns positioned below the equator of the glass

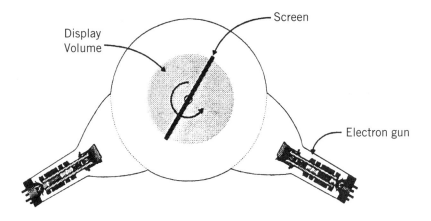

PLAN VIEW

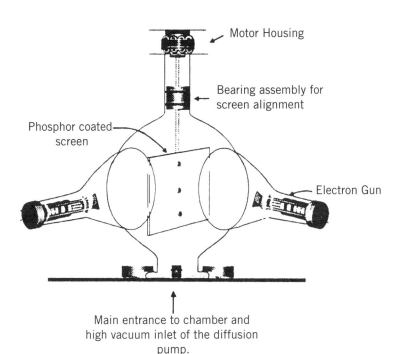

SIDE ELEVATION

FIGURE 11.1 MkII Cathode Ray Sphere. Two electron guns address a planar screen. The screen drive is located above the display vessel and the system is pumped continuously by the combination of rotary and diffusion pumps located below.

sphere. In this position the guns were less obtrusive and their impact on viewing freedom was minimized. A range of application software was developed for the various CRS prototypes, and image data taken from a number of sources have been depicted within the CRS display systems. These data include medical CT data, computer-aided designs, crystal structure visualization, mathematical objects (such as the well-known Lorenz attractor), and satellite orbit data.

11.3 CONSIDERATIONS REGARDING HELIX LASER TECHNOLOGY*

The basic principle of operation of the HL3D technology may easily be understood and is outlined in Section 5.5.2. In this section some basic considerations are discussed and the evolution of this technology is described.

11.3.1 Characteristics

Some of the advantages associated with the combination of helix and laser as a means of achieving volumetric 3-D representation are briefly described below. The geometry of the helical surface allows a relatively simple conversion of the Cartesian coordinates of each image element (denoted x,y,z) into voxel locations within an image space. The x,y values may, in principle, remain unchanged, but the z coordinate (which is assumed to be parallel to the axis of rotation) is converted into a temporal value (denoted t). Due to the geometry of the surface, the relationship between z and t is linear. The temporal descriptor indicates the time at which the laser beam must be activated so as to generate each voxel at the appropriate location within the image space.

The material from which the helix is fabricated may be selected so that light is emitted uniformly in both the upward and downward directions (see Section 4.4.2). Furthermore, the 3-D image space can in principle be "totally translucent." In this case, any number of observers are able to see each individual voxel from any position around the volume. Generally, the transparency of the image space and hence the freedom in viewing position are determined by the nature of the material used for the construction of the helix and its form (see Section 11.3.3).

The proportions of the 3-D cylinder swept out by the helix can be adapted according to the nature of the application. For example, a broad flat cylinder may be found to be most suited for use in air traffic control applications. Similarly, the overall volume of the image space can be adapted according to requirements.

11.3.2 Development

The essential content of the patent filed in 1976 (see Section 5.5.2) concerned the use of laser light for the generation of volumetric 3-D images on the surface of a rotating helix. Looking back at the developments over the past decades, the specific

---

* The material in this section was contributed by Rüdiger Hartwig.

version of helix described in this patent document can still be regarded as optimal. A number of volumetric displays continue to be based on this idea and are now equipped with highly sophisticated laser deflection technology. The advance from the original patent to the present embodiments of this technology was not easily achieved and the main phases of evolution are illustrated in Figure 11.2.

1982 saw the first public demonstration of a HL3D system. At that time it was still necessary to prove the feasibility of this approach to volumetric imaging. Consequently, simple deflection systems employing rotating prisms and vibrating mirrors were sufficient for the deflection of the laser beam and succeeded in generating simple 3-D images [e.g., Lissajous figures that moved through the image space (Brinkmann 1983)].

In about 1986, scientists at Texas Instruments started to use the HL3D technology, and R.D. Williams introduced their 3-D display OmniView in 1989. Subsequently, in approximately 1990, scientists of the U.S. Navy started to develop HL3D systems for submarine navigation and air traffic control. In 1991 they introduced their first generation 3-D display (see Sections 5.6.6 and 11.5).

11.3.3 Helical Surfaces

During the years various forms of helix have been developed and used. These are outlined briefly below. The original form of helix described in the 1976 patent is illustrated in Figure 11.3. This is single bladed and is supported at its periphery by a corotating cylinder. If the helix is fabricated from semitransparent material, voxels produced on its surface can be seen simultaneously from all positions around the volume. A subsequent simplified version was used in 1982 for the first publicly demonstrated HL3D system (Cross 1982, Brinkmann 1983). In this case the helix is supported by a central rotation shaft. The presence of a center shaft may occlude portions of an image located on the far side of the image space relative to the observer. The screen is illustrated in Figure 11.4.

In 1989, Texas Instruments introduced a double-bladed helix (Williams and Garcia 1988). In this case, the helix may be regarded as ascending over 180° and then descending through the remaining 180° (Figure 11.5). When rotating this helix does not pump air in the vertical direction. Transparency may be poor, and as a consequence viewing freedom is severely restricted. In 1991 the U.S. Navy introduced a symmetrical double-bladed helix (Soltan et al. 1994) (Figure 11.6). Unfortunately, as a consequence of the presence of the two blades, transparency (and therefore viewing freedom) around the periphery of the image space may again be limited. Whatever the helical form, the vertical speed is not affected by position on the surface or by the number of blades employed. Vertical speed depends simply on the height of the helix and the frequency of rotation.

Images generated on double-bladed helices may be refreshed twice during each cycle of rotation. Consequently, it is possible to reduce the frequency of rotation in comparison to systems employing only a single blade. However, viewing angle restrictions may, to some extent, offset this advantage and these restrictions have led workers to revert to the original single-bladed configuration (see DE 26 22 802 C2).

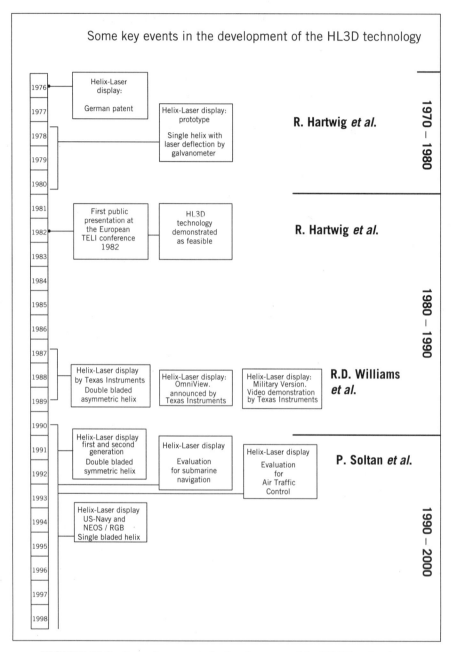

FIGURE 11.2   Some key events in development of the HL3D technology.

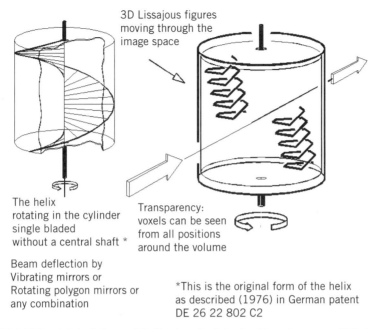

3D Lissajous figures
moving through the
image space

The helix
rotating in the cylinder
single bladed
without a central shaft *

Transparency:
voxels can be seen
from all positions
around the volume

Beam deflection by
Vibrating mirrors or
Rotating polygon mirrors or
any combination

*This is the original form of the helix
as described (1976) in German patent
DE 26 22 802 C2

FIGURE 11.3   Original form of helix described in the Hartwig patent. This is single bladed and is supported around its periphery by a corotating cylinder.

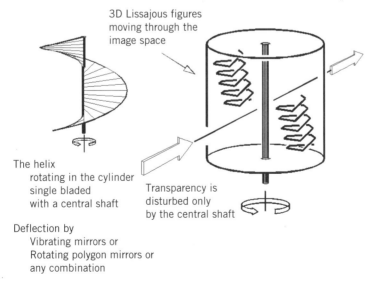

3D Lissajous figures
moving through the
image space

The helix
rotating in the cylinder
single bladed
with a central shaft

Transparency is
disturbed only
by the central shaft

Deflection by
Vibrating mirrors or
Rotating polygon mirrors or
any combination

FIGURE 11.4   Simplified screen used by Hartwig in 1982, which employs a central supporting axis.

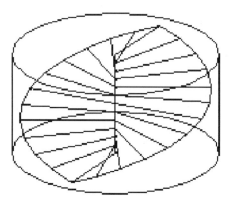

Form of asymmetric double-bladed helix
(transparency is reduced)

FIGURE 11.5 Double-bladed helix introduced by Texas Instruments around 1989.

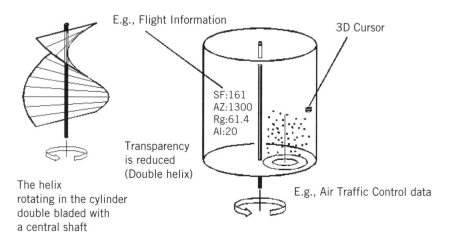

FIGURE 11.6 Double-bladed helix introduced by the U.S. Navy in approximately 1991.

11.3.4 Laser Deflection and Scanning

The performance of a display system is often judged by the number of voxels that may be activated during each refresh period. In the case of the HL3D displays, rapid deflection of the laser beam is therefore critical and consequently sophisticated optoelectronic scanners are now generally used for this purpose. The number of deflection systems can be increased to raise the number of voxels that may be activated during each frame. Each half (180°) or even each quadrant (90°) of the image space may then be addressed by a separate deflection system.*

---

* By clearly separating the subspaces beam registration difficulties may be minimized.

The original concept for the generation of images within a HL3D display was simply to extend 2-D scanning techniques to encompass the third dimension. This would have involved achieving a set of planar raster scans. This approach was not adopted for two reasons. These related to the unacceptably high scanning speeds involved (see Section 2.7), and since the larger portion of any image space is empty, exhaustive raster scanning of the entire image space is inefficient. Since raster scanning is too inefficient for HL3D technologies random scanning provides an alternative solution.

11.4   LASER-ADDRESSED MULTIPLANAR DISPLAY APPARATUS*

Robert Batchko invented the "rotating flat screen fully addressable volume display system" (U.S. patent 5,148,310). This is a swept-volume display unit that employs the rotational motion of a beam-addressed planar surface of emission. In this implementation, dead zones are avoided by employing an optical projection system that corotates with the screen (see also Section 5.3.2). The approach is described in some detail in the above patent and is outlined briefly below.

Two problems are commonly inherent in angular multiplanar (i.e., swept volume) volumetric displays employing a scanned image projected onto a rotating screen. The first concerns dead zones or nonaddressable spots (Batchko 1994) associated with a region within the image space that cannot be addressed by the image source (e.g., electron beams, lamps, LEDs, or lasers). In the case of some rotating helical screens, a dead zone occurs in a cylindrical column which is centered about the axis of rotation. This was, for example, apparent in Texas Instruments' OmniView (in which the split-half-helix screen was fixed to an opaque cylindrical rod connected to the motor drive system). Another more dramatic example of this type of dead zone occurred in an earlier system developed by Texas Instruments EP 0491284 A1 (see Section 5.6.5) in which a flat screen was mounted at an oblique angle to the rotation axis. In this case, the volume swept out by the screen omitted two conic regions centered about the rotation axis. These cones could therefore not be addressed. The second problem referred to above concerns self-occlusion (Batchko 1994) and results from the instantaneous position of the curved screen blocking an observer's view of a portion of the illuminated image space.

The system developed to address these problems of dead zones and self-occlusion incorporated a planar screen positioned so as to be coplanar with the rotation axis. Mirrors, positioned on, and radially displaced from the rotation axis permit addressability of the full screen surface throughout its rotation (Figure 11.7). The image source is projected along the rotation axis. In this approach the entire surface of the screen and hence the complete cylindrical display volume, are fully addressable. Further, self-occlusion is avoided by the flat screen surface in combination with the binocular disparity of the eyes. This display architecture was given the acronym LAMDA (laser-addressed multiplanar

---

* The material in this section was contributed by Robert Batchko.

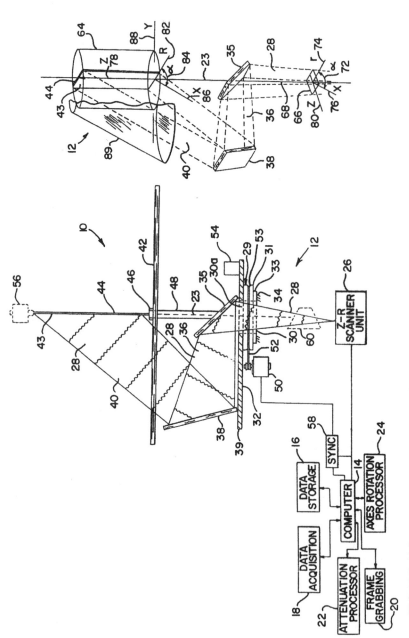

FIGURE 11.7 In R.G. Batchko's rotating flat screen fully addressable volume display system, a flat screen (44) rotates about a rotation axis lying in the plane of the screen. A scanned image is projected up along the rotation axis. Axial and radial mirrors (35 and 38) direct the scanned image onto the screen. The screen and both mirrors rotate in unison, allowing the scanned image to strike the screen at a constant angle throughout its full rotation cycle. In this fashion, the entire screen, and hence the entire image space swept out by the screen, can be addressed. (From U.S. patent 5,148,310.)

display apparatus) (Batchko 1992). In 1994 a more advanced LAMDA system was developed (Batchko 1994).

11.5 U.S. NAVY HL3D SYSTEMS*

The Space and Naval Warfare (SPAWAR) Systems Center, San Diego (SSC) Simulation and Human Technology Division has developed two generations of a 3-D volumetric display system for depicting data, information, and scenes in a three-dimensional volume. These are based on a computer-controlled laser optics system that projects three laser beams simultaneously onto a 36-in.-diameter, 18-in.-high double helix spinning at 10 Hz.

From 1990 to 1992, the first generation prototype of the 3-D display system was produced. During the period 1992 to 1995, an improved second-generation prototype was developed and successfully demonstrated. Current interest in developing 3-D volumetric displays arises as a consequence of the close matching of the characteristics of volumetric images with the human visual system. Examples of applications include the following:

1. *Air Traffic Control.* Air traffic monitoring in 3-D and in real time has been demonstrated at SSC San Diego's Laser Display Laboratory with data obtained from the Navy's Identification Friend or Foe (IFF) antenna. This naval 3-D display system offers to depict 3-D data in color for group viewing of air traffic.
2. *Submarine Navigation.* This illustrates a further application requiring the integration of several different types of two-dimensional data to create a mental picture of 3-D surroundings. With the requirement that submarines operate in littoral waters, awareness of the ocean floor has increased. In the case of the Navy's new 3-D display, sonar data of the ocean floor topography has been incorporated and used to demonstrate the possibility of navigating a submarine in and around a varied marine environment, as well as to track the movement of other objects in the water (e.g., submarines, mines, torpedoes), and to display all this information on one 3-D display.

11.5.1 Technical Objectives

First-Generation System

This system features a 13-in. diameter double-helix. The system utilized a single laser (green) and a single acousto-optic (AO) scanner. It was later upgraded to use lasers of several colors and additional scanners. This early device, while limited in capability (4000 voxels per color at a 20-Hz refresh frequency), proved that the technology was feasible and had potential applications to a number of military platforms and commercial users.

---

* This section is based upon material provided by Parviz Soltan, Mark Lasher and colleagues.

The Second-Generation System

The second-generation system was significantly upgraded with multiple color, RGB (red, green, blue) capability. The system also boasted a relatively noise-free 36-in. diameter by 18-in. high double helix providing a 20-Hz refresh rate with 256 levels of intensity. This second-generation design increased the voxel activation capacity to 40,000 voxels per color.

11.5.2   The Optical System

A high performance point-to-point laser scanning system is required to display a large number of voxels in the image space. Acousto-optic random-access laser scanners are used to address four quadrants of the rotating helix. A random access scanner provides an efficient means of addressing the image space. With proper synchronization, only those voxels that comprise the image are illuminated, while dark areas remain unaddressed. Both galvanometer mirrors and acousto-optic deflectors can be used for this purpose. However, the access time for galvanometers (1 ms) is too long to be useful for volumetric systems. Acousto-optic scanners have effective access times that are 1000 times less.

Figure 11.8 illustrates a single-channel AO deflection system. This consists of an AO intensity modulator followed by two AO deflectors for independent X and Y deflection. The system used had an access time of 5 μsec, corresponding to the 5 μsec required for the acoustic wave to fill the aperture of each deflector and achieve the desired deflection of the laser beam to each new location. This implies a maximum time per voxel of $1/5 \times 10^{-6}$ s^{-1}, or 200,000 voxels per second. With a 20 Hz refresh rate, this results in a voxel activation capacity of 10,000. However the usable voxel activation capacity is somewhat less than this, as the laser light must dwell on each location in order to generate voxels of a sufficient intensity.

One approach to increasing the voxel activation capacity is to couple several scanners together in parallel. A diagram illustrating the optical layout of such a system and employed with a 36-in. diameter helix is presented in Figure 11.9. The green scanner uses two Nd:YAG lasers (100 mW CASIX Model DP110),

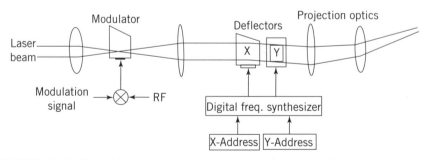

FIGURE 11.8   The optical arrangement for a single channel AO deflection and modulation system.

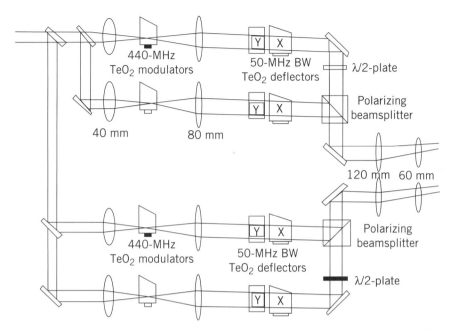

FIGURE 11.9   Four channel random access scanner arrangement (see text for details).

one for channels A and B, the other for channels C and D. Each beam is then focused into a 440-MHz TeO$_2$AO modulator. Subsequently, the beams are recollimated to a 3-mm diameter and pass to the X and Y TeO$_2$AO deflectors. The red scanner uses four helium-neon lasers (20 mW Melles Griot Model 05-LHP-925), one for each channel. The superior efficiency of the red scanner allows the use of low-power lasers. The blue scanner is similar to the green except that it currently uses an argon ion laser. However solid-state blue lasers are now commercially available. These lasers will facilitate the design of transportable volumetric display systems.

Standard dichroic mirrors were used to combine the red, green, and blue channels. The projection optics consist of a series of three lenses. The first (300 mm) and third (120 mm) lenses provide the approximate overall angular magnification (2.5×) and focusing of the scan. The second lens (200 mm) provides fine adjustment for expanding or contracting the image size. These optics help reduce the required distance between the scanners and helix, while maintaining an acceptable spot size. The helix is mounted as a separate structure that is physically apart from the laser scanner. The scanned beams are relayed by means of a series of mirrors so as to be transmitted vertically downward onto the helix (see Plate xiv). The surface of the helix is coated with a reflective white barium sulphate paint. The electronics subsystem consists of the RF drive and switching electronics and a custom interface card which routes the image data from the host computer and sequences it to the appropriate channels in the scanning apparatus.

11.5.3 Transportable Volumetric Display

The project goal was to produce a system that occupied a smaller physical volume and consumed less power while maintaining a useful voxel activation capacity and offering color capability. A transportable system was therefore developed incorporating two colors (red and green). The voxel activation capacity was 80,000 (40,000 per color). The scanner assembly consisted of two red-green scanner subassemblies, each capable of 10,000 voxels per color at a 20 Hz refresh frequency. As with the 36 in. system, each quadrant of the helix was addressed by two scanners, one for red and one for green.

To provide as compact a system as possible, the helix surface material was changed to translucent 20 mil polystyrene which enabled the laser projection apparatus to be located beneath the surface. The helix diameter was reduced to 16 in., allowing a much shorter distance between the laser scanners and the image space. One of the disadvantages of the 36 in. reflective helix is the appearance of occlusion zones when one of the helix blades blocks a portion of the displayed image. The use of a translucent helix ameliorates this problem. The display uses two 100-mW diode-pumped, doubled Nd:YAGs for green voxel generation and eight single-mode red diode lasers. Table 11.1 shows the system specifications for this transportable volumetric display.

In 1995, a Cooperative Research and Development Agreement (CRADA) was signed with NEOS Technologies, Inc. and RGB Technology, Inc. to develop a commercial version of the volumetric display that incorporated these changes. The goal was to produce a system that was transportable and reasonably affordable for both military and government applications. The first such system was

TABLE 11.1   Transportable HL3D Display: System Specifications

Item	Characteristic
Display surface	Axially symmetric
	Translucent single helix
Rotational axis	Vertical
Rotation frequency	20 Hz
Display volume	25,000 cm^3
Volume dimensions	
$\quad$ D = cylinder diameter	D = 40 cm
$\quad$ H = height	H = 20 cm
Illumination method	Passive translucent screen with scanned laser
	$\quad$ beams projected from bottom
Scanning method	Acousto-optic
Voxel diameter	0.7 mm
Voxel refresh per rotation	1
Frame rate	20 Hz
Primary colors	Red, Green
Voxel activation capacity	40,000

deployed on the USS Stennis for the Joint Warrior Interoperability Demonstration (JWID 97) where it was interfaced to a LINK 16 tactical data feed.

11.6 TECHNOLOGY TRANSFER

There can be few other areas of applied research in which experimentation has continued for such a long duration without some form of technology transfer and commercialization taking place on a significant scale. Furthermore, given the long period during which volumetric systems have been under investigation, it may be supposed that this area of research would have gained a level of maturity in which terminology, performance measurements, and general principles would have become widely standardized. An examination of the very large amount of material published on this subject shows this certainly not to be the case.

There are numerous ways in which it is possible to implement a volumetric display device, and in general, even those with only a rudimentary knowledge of science may easily comprehend the underlying principles of operation. It is therefore not difficult for those possessing even a very basic technical or scientific background to propose a novel method of implementation. However, although such a method may be easily conceptualized, it is extremely difficult (even for those who have considerable experience in this area of research) to determine, in advance, the ultimate performance characteristics of any proposed volumetric technology. The real challenge is therefore not the invention of a novel architecture, but rather in foreseeing the weaknesses of any proposed system and accurately judging the ultimate image fidelity. In Chapters 5 and 7 we have outlined a variety of volumetric display unit implementations. Each represents a realistic embodiment in principle, but in some cases it is quite difficult to assess the homogeneity and isotropy of the image space with respect to voxel placement and particularly to foresee variations in image quality with viewing direction. In the case of many techniques, these issues are obscured by subtle interactions between the display unit subsystems.

As we have seen, volumetric displays often employ subsystems that comprise mechanical components, electronic systems coupled to a variety of transducers, software control, and image processing modules, and so on, all of which must operate harmoniously and in real time. The integration of such diverse technologies is seldom achieved easily and generally represents a time-consuming undertaking. Furthermore, the modification of such highly integrated systems can be most difficult. The problems that are likely to be experienced when prototyping a volumetric system should therefore not be underestimated.

It is apparent that a number of volumetric systems have been constructed in prototype form without proper consideration being given beforehand to such critical issues as:

- Individual voxel characteristics
- Voxel placement characteristics

- Voxel activation capacity
- Optical characteristics of the image space

In many cases an immense amount of work has been expended on system development, and while the resulting displays have been able to depict volumetric images, the ultimate image quality has been below expectation. Consequently, prototype systems have eventually been abandoned after attempts to improve system performance and image characteristics have failed (or resulted in considerable increases in the overall complexity of the system). This explains to some extent the large number of systems that have been proposed but not developed beyond the prototype stage.

The development of new technological products may occur in a number of ways. For example, a breakthrough in science or engineering may give rise to one or more entirely new (and perhaps previously unconsidered) products. Alternatively, research programs may be intentionally directed toward the development of products that address specific needs which may have arisen as a consequence of our ever-changing requirements. Neither of these scenarios describes the long-term research activity that has been directed toward the development of volumetric systems. There is, in fact, considerable similarity between the manner in which research into television was conducted during the first few decades of this century and the more recent search for a satisfactory volumetric display system. The history of television is recorded in a number of most interesting texts [e.g., Abrahamson (1987)]. In the introduction to his book, Abrahamson writes:

> The pioneers involved were treading on ground that had never before been covered. They were creating a new form of communication that man had intuitively desired for thousands of years. . . . No one person invented television; most of the inventors were ahead of their time and technology; some were idle dreamers, others were practical men who could turn their ideas into machinery. Ideas did not always occur in a logical order. . . .

The development of television receivers was made more difficult by the diversity of technologies employed in their implementation (a number of systems made great use of electromechanical mechanisms), many techniques were investigated (only to be abandoned due to unforeseen performance limitations), and most important, scientists, engineers, and inventors embarked on an obsessive quest whose purpose they themselves defined.

Clearly, there are certain dangers associated with the development of a technological product whose ultimate usefulness cannot be assessed properly in advance. For example, although it is possible for scientists working on a development to postulate with great enthusiasm potential uses and advantages that may be derived from their creation, the reality may be somewhat different. To gain acceptance, new products must prove their worth. Furthermore, they must overcome a natural desire by end users to continue with the tried and tested approach. Under such circumstances, there may be a tendency to overclaim the capabilities of a new product, and after a time those involved in the undertaking may even begin to believe their own publicity!

It is often difficult to assess in advance the advantages that may be derived through the application of a display technology to a particular area of activity. A new visualization methodology may assist in various ways; for example:

- A display technique may depict information with greater clarity, thereby enabling certain features to be more easily discerned.
- A display technique may enable spatial data to be interpreted more quickly. In the case of real-time data this may allow faster operator response. Alternatively, the increased rate at which data may be assimilated may permit an operator to process greater volumes of data in a shorter time. For example, in radiotherapy treatment planning this may be advantageous, particularly if it results in a more accurate interpretation of the data.
- A display technique that permits direct operator interaction with 3-D data sets (via a pointer or 3-D interaction device) may enable a designer to form complex shapes in a natural and intuitive manner. In the area of computer-aided design this may be highly beneficial.

The potential of a display technique to be beneficial in one or more of these areas may only be assessed by having a detailed understanding not only of the display technology itself (including its limitations) but also of the nature of the application. It is also important to remember that the introduction of the visualization technique is likely to modify the manner in which tasks are to be performed. The practicality of any new working practices must therefore also be considered. For example, consider the application of a volumetric system to air traffic control. As we discussed in Section 10.4.1, the volume of coverage of a volumetric system is inherently small. Should we seek to enhance the volume of coverage by increasing the dimensions of the image space, it will ultimately be necessary for the controller to move around the volume constantly so as to discern accurately the spatial separation of objects.* This is likely to be highly inconvenient and therefore does not represent a feasible solution (Blundell and Schwarz 1999).

It is apparent that technical progress is still required to develop a volumetric display architecture whose overall performance, predictability, and image quality will stand up to critical and objective examination. The large-scale commercial exploitation of any particular volumetric technology will also represent a major undertaking in which performance and safety issues will be most important. The greatest difficulties are likely to concern user acceptance and the need for the integration of volumetric systems with existing hardware and software products. Current trends within the computer industry are unlikely to simplify these matters.

---

* Alternatively, the information depicted within the image space could be rotated slowly around a central axis. However, varying the observer's viewpoint with respect to the data (either by movement of the observer or by rotation of the data) is likely to lead to disorientation, and naturally this could be extremely hazardous.

11.7 DISCUSSION

Research into volumetric systems has given rise to a significant number of excellent scientific publications and has received considerable attention from the general media. Unfortunately, this area of activity has never been regarded as a mainstream area of scientific and technical research. During the course of this century many other areas of applied research have been regarded in a similar light, and although this makes it more difficult to undertake the necessary research, it does not reflect adversely on the success that may eventually ensue.

In the text quoted above (Abramson 1987), Abramson compares the rapid evolution of cinema with the slow development of television and writes:

> Television took a long time gestating. It grew with the technology of the times; its disciplines came from a broad spectrum of the arts and sciences that were more often than not unrelated. ... The early history of television is made up of discoveries that at the time seemed absolutely unrelated to each other. To find a bridge for them was impossible. Time was needed to fit the pieces together into a sort of jigsaw puzzle.

Perhaps it is only by studying at length the work that has been undertaken to date on volumetric systems that one is able to form a mental picture of the scope of this approach to visualization. By gaining this overall perspective, it is possible to have a better insight into the most appropriate future course of investigation. This contrasts with the situation to date in which researchers (including these authors) have concentrated largely on their own ideas for a volumetric display system implementation. Ironically, this approach may relate to the individualistic nature of those who choose to work outside the track of mainstream science.

CHAPTER 12

TOWARD A HIGH-DEFINITION VOLUMETRIC DISPLAY

They seemed to be hardly railway children at all in those days.*

12.1 INTRODUCTION

In previous chapters we have attempted to define terminology and formalize concepts relating to volumetric display system research and have examined aspects of both historical and current work carried out in this area. We now turn our attention to the future and look toward the next generation of high-definition volumetric display systems. Ultimately, these systems should be capable of depicting full-color high-quality images within a display volume of an appropriate size. These objectives will not easily be realized and represent a considerable challenge from both a scientific and a technical perspective.

In Section 12.2 we reexamine issues relating to the exhaustive scanning of an image space. Although we do not believe that this technique is essential for all high-definition displays, exhaustive scanning is desirable since it increases the voxel activation capacity (and hence the fill factor) and so increases the predictability of an image space. In this chapter we propose several display unit architectures which have been selected to illustrate the ideas presented previously in this book. In each case we suggest associated strengths and weaknesses. The display unit described in Section 12.4 is of particular interest since it uses a combination of rotational and translational motion to create the image space.

For certain types of application, an operator may wish to interact directly with the volumetric image depicted within an image space by means of some type of

---

* E. Nesbit, *The Railway Children*. Puffin Books, 1994.

3-D cursor or specialized interaction device. For example, it may be desirable to manipulate image objects contained within a volumetric image space and have any modifications reflected immediately in some manner (e.g., quantitatively) on a conventional computer display. When used in this way, it is evident that the volumetric system must be highly integrated within the computing environment and cannot be considered simply as a peripheral device to which information is passed. Although the authors are not aware of any significant recent research on direct interaction with volumetric images, it is hoped that interest in this area will increase. The topic is discussed briefly in Section 12.7. The various display units, scanning techniques, and interaction devices proposed in this chapter are intended to illustrate some of the concepts introduced earlier in the book. They do not necessarily reflect optimal configurations.

12.2   EXHAUSTIVE SCANNING OF AN IMAGE SPACE

As the voxel activation capacity approaches the voxel location capacity, it is possible to have greater confidence (as far as voxel placement is concerned) in the predictability of an image space. Consequently, an image component should, in principle, be less affected by its location. Unfortunately, many of the display units discussed in Chapters 5 and 7 provide a fill factor (see Section 2.9) that is quite small. This is generally caused by a lack of performance in the voxel activation technique, coupled with a failure to accommodate parallelism in this subsystem. In the case of beam-addressed systems the performance limitations arise as a consequence of three factors: beam deflection, beam blanking,* and beam energy.

In the case of the next generation of volumetric systems, we suggest that the image resolution required will necessitate a very high voxel location capacity. To take proper advantage of the resolution, there must be a corresponding increase in the voxel activation capacity. Given that the majority of current systems (which have a relatively small voxel location capacity) are unable to accommodate a large fill factor, the problem in the case of future systems will become much more difficult to solve. This matter may be addressed by providing a high degree of parallelism in the voxel activation process (corresponding to the division of an image space into a number of subspaces that exhibit the greatest possible overlap in the temporal domain).

An exhaustive scanning technique offers to be advantageous since it guarantees that every potential voxel location can be activated during each image update period. This is beneficial when considered in terms of the predictability of an image space. However, the reader will realize that in the main, an image space will be (and should be) void. Therefore, should an exhaustive scan be employed, the voxel activation mechanisms will be disabled for the majority of each image update period. This clearly denotes an inefficient utilization of this subsystem.

---

* In the case of beam-addressed systems, the beam modulation signal must operate at the highest bandwidth.

Further, the use of exhaustive scanning returns us to a situation in which the voxel activation capacity is linked directly to the dimensions of the image space (see Section 2.9). As a consequence, the maximum extent of this volume may be dictated by the performance of the voxel activation subsystems.

An alternative approach applicable to beam-addressed systems concerns the exhaustive scanning of a number of localized regions of an image space. Within these regions high-definition images may be depicted. For example, we may permit the updating of perhaps ten regions of an image space and allow these regions to be located freely (they may be disjoint or may, in principle, be amalgamated so as to form larger volumes). This approach represents an intermediate solution with the following results:

- The voxel activation capacity is independent of the image space dimensions.
- Each region may be scanned at high resolution and detailed image components may therefore be depicted.
- The freedom in positioning of the scanned regions permits larger scanned volumes to be constructed. Consequently, the size of an image object is not necessarily constrained by the extent of an individual scanned region.

We refer to this methodology subsequently as a localized scanning technique, and before discussing this approach further in Section 12.6, we concentrate on issues relating to the implementation of the exhaustive scanning technique.

12.3   BEAM-ADDRESSED SYSTEMS

In systems employing multiple beam sources it is vital to ensure that any beam deflection mechanisms are aligned properly with respect to each other. The entire image space may then be represented by a single uniform coordinate system.* Although gradual and minor geometric distortions of an image may pass unnoticed, discontinuities or regions in which portions of an image become blurred are usually readily apparent. Such adverse effects often occur at the border of two adjacent subspaces or in regions where they overlap spatially. In the latter case, it is particularly important to ensure that the required placement accuracy is maintained throughout the regions of overlap. Spatial overlap between subspaces therefore places very considerable demands on any alignment mechanism. For a given image space volume, the accuracy of voxel placement should rise as the voxel activation capacity (and therefore the voxel density) is increased. The difficulties of positioning voxels correctly within spatially overlapping subspaces are then exacerbated.

In Section 9.8 we discussed a calibration scheme for use in connection with a system employing several scanned beams. In principle, the refinement of this

---

*This ignores nonuniformities that may arise as a consequence of, for example, the image space creation technique.

technique could permit proper registration of a larger number of beam sources. In the following sections we outline possible techniques that could be employed to achieve the exhaustive scanning of an image space. Although these systems employ image spaces comprising a number of subspaces, we exclude their spatial overlap. The configurations discussed below may not be directly realizable; they are simply intended to highlight concepts that have been discussed in this work.

12.3.1 Beam-Addressed System Employing a Helical Screen

The deflection apparatus represents an inherent weakness of any display unit that employs one or more beams for voxel activation. It is therefore appropriate to consider the possibility of a beam-addressed system able to operate without the need for any conventional form of beam deflection mechanism. Suggestions relating to the implementation of such a display unit architecture are discussed in this section.

Figure 12.1 illustrates a rotating helical screen and below this structure is a disk containing a 2-D array of light-emitting elements. These devices are arranged to form a series of concentric circles, and we assume that when activated each is

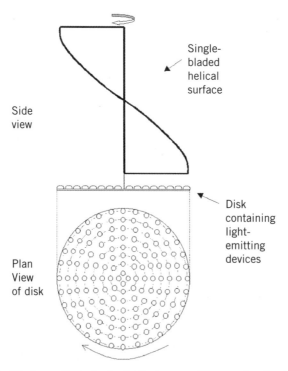

FIGURE 12.1 Display unit employing helical screen. The voxel activation mechanisms are situated (below this screen). These consist of light-emitting devices — the beams emanating from which pass vertically upward. To reduce the number of these devices, the disk rotates at high speed (see the text for details).

able to produce a collimated beam of light that propagates vertically upward. Each device is therefore able to give rise to a visible voxel within the region where the light beam impinges on the screen (this is assumed to be equipped with a passive surface of emission able to scatter the incident light). If we assume, for example, that the single-bladed helix rotates at a frequency of 30 Hz and that each of the elements may be switched at a frequency of 30 kHz, then during the course of a single rotation of the helix each element may output a vertical column consisting of up to 1000 voxels. Each element is therefore able to give rise to voxels located along the length of a vertically oriented subspace. Each subspace is spatially separated from the others, but in principle, they may all be coincident in the temporal domain (indicating a high degree of parallelism in the voxel activation process). We have therefore identified an architecture in which an image space created by the rotation of a helical screen can be exhaustively scanned.

Clearly, an essential problem with the configuration described above concerns the number of optoelectronic elements required to achieve a useful resolution throughout an image space of reasonable dimensions. For example, if we assume that the image space has a diameter of 80 cm, that there exist 400 concentric rings of optoelectronic elements (corresponding to a separation of 1 mm) and that the spacing of elements within each ring is approximately 1 mm, we can deduce that the disk must contain more than 400,000 elements!

In view of the natural rigidity of the helical shape a system employing this image space creation technique offers to accommodate a large image space volume. Unfortunately, in the configuration described above, the number of optoelectronic elements required within the array will make it unfeasible to implement an image space with a large diameter. Furthermore, since the helical structure is particularly suited to the creation of an image space whose height is considerably less than its diameter, any restriction placed on the image space diameter will also affect height. The achievable image space volume is therefore limited by this embodiment, and an important advantage offered by the helical screen is negated.

In the example provided above we suggested a switching time of 30 kHz for the optoelectronic devices. Readily available components can support a higher bandwidth and we may therefore consider the feasibility of extending the extent of the subspace supported by each component. This would enable each element to produce a larger number of voxels during each image update period while maintaining the same voxel activation capacity. Should this prove possible, it would, in principle, permit a reduction in the number of optoelectronic elements within the array.

Let us therefore consider the possibility of introducing a rotational motion to the array of optoelectronic elements illustrated in Figure 12.1. In this case, the array would rotate about the same axis as the helical screen. The frequency of rotation is arranged to be considerably greater than that of the screen, and the extent of each subspace rises as a consequence of the difference in the rotational speeds. The number of elements needed in each concentric ring may therefore be reduced.

In a further less easily conceptualized embodiment, it may be possible to reduce the number of concentric rings of elements (without making any reduction in the voxel density) by arranging for the axis of rotation of the array of elements to be offset from that of the helix. These proposals rely on the difference in rotational frequency of the array of elements and the screen. Naturally, the screen and the array of elements would rotate in opposite directions.

The increase in the size of the subspace supported by each element, coupled with a reduction in the voxel activation time,* will affect image brightness adversely. In selecting suitable devices it will be vital to ensure that they not only exhibit suitable switching properties, but are able to produce the required output power. Although this configuration offers a display unit enabling a 100% fill factor and permits, in principle, a minimization of the number of optoelectronic elements needed to attain this goal, it does not address the problem of the visual dead zone,† or the dead image space (at the center of the volume). Furthermore, the reliance placed on the very high speed of rotation of the array of elements may prove to be unacceptable.

In Chapter 5 we outlined the architecture of a number of display units that employ either a rotating active surface of emission or voxel activation mechanism(s) which corotate with the screen. In both cases it is necessary to transmit information at a rapid rate from the static external hardware to the moving components. In configurations such as those outlined in Sections 5.3.2 and 5.3.3, this was achieved by means of electrical commutators, and in the case of the system described in Section 5.5.3, by means of a serial optical data link. For the display architecture described above, voxel descriptor data must be passed to the rotating disk. In Section 12.4 we propose a display unit employing a rotating active surface of emission. In both cases the information link must be able to support a high data transmission bandwidth. Although, in principle, it may be possible to identify a single optical transmitter and receiver able to support the data transmission rate required, it is instructive to discuss the implementation of a parallel optical link able to sustain an essentially unrestricted data throughput.‡

Transmission links suitable for the parallel transfer of data are illustrated in Figures 12.2 and 12.3. The first of these figures shows two concentric cylinders centered about the axis of rotation of a planar screen (comprising an active surface of emission). In the following discussion we assume that the outer cylinder is stationary and that the inner cylinder corotates with the screen. (In the case of the architecture described above, the inner cylinder would corotate with the disk on which the voxel activation mechanisms are mounted.) The outer stationary

---

* The large difference in angular velocity between the screen and disk will cause voxel elongation in the direction of the disk's rotation. This may be minimized through adoption of a reduced activation time.

† However, since the display employs a passive surface of emission and helical supporting structure it is possible for the screen to be very thin. Consequently the impact of the visual dead zone will be minimized.

‡ This would be complemented by a commutator able to pass the dc voltages needed to power the moving circuitry.

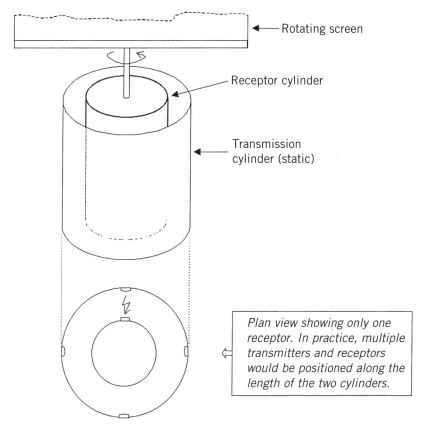

Rotating screen

Receptor cylinder

Transmission cylinder (static)

Plan view showing only one receptor. In practice, multiple transmitters and receptors would be positioned along the length of the two cylinders.

FIGURE 12.2   Parallel data transmission link. The inner cylinder containing the reception devices corotates with the screen and the outer cylinder (containing the transmission devices) is fixed. From the perspective of the receptors, transmission devices appear to pass overhead. These optoelectronic devices are arranged so that for each receptor, as one transmitter falls towards the horizon, another rises.

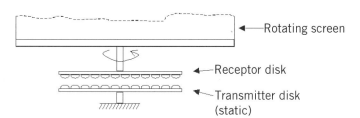

Rotating screen

Receptor disk

Transmitter disk (static)

FIGURE 12.3   Parallel data transmission link in which the optoelectronic devices are mounted on two disks. The operation of the link is similar to the embodiment depicted in Figure 12.2.

cylinder is equipped with an array of light-emitting devices such as LEDs, and the inner rotating cylinder with an array of receptors. These devices are configured so that the transmitters emit in an inward radial direction, and the receptors point outward (away from the axis of rotation).

Let us now simplify our discussion by considering the operation of a portion of the cylinder comprising a 1-D array of devices located around its circumference. From the perspective of a receptor, an emitter will appear to rise on one side, move overhead and sink beneath the opposite horizon. This process will repeat as each of the transmitters passes overhead. By equipping the cylinders with a suitable number of transmitters and receptors we can ensure that at any time there is always one (and no more than one) transmitter able to communicate with each receiver.* Operation of the complete array of elements located on the two cylinders follows from the description above. By monitoring the relative positions of the two cylinders, the controlling hardware can determine which transmitter element is able to communicate with each receiver at any moment. Various schemes may be used for the encoding and reconstruction of data passed across the link. Furthermore, by introducing bidirectionality in communication it is possible to introduce error correction mechanisms.

To ensure that only one transmitter is able to communicate with each receptor at any one time (and to accommodate any mechanical misalignments) it may be appropriate to arrange for brief periods of dead time (i.e., each transmitter is allowed to sink completely below the horizon before the next rises). Although this would reduce the possibility of error that would be caused by two transmitters attempting to communicate with the same receptor, it will result in a reduction in the rate of information transfer. It is possible to envisage a number of solutions to this problem that need not be discussed here.

In a second configuration illustrated in Figure 12.3, the two cylinders are replaced by two disks on which the optoelectronic devices are mounted. One of these disks rotates with the screen and the other is stationary. The link operates in a manner similar to the previous configuration, and the arrangement of devices ensures that for the majority of the time, each transmitter is able to communicate with a receiver.

12.3.2 Beam-Addressed System Employing a Planar Screen

It is instructive to continue our discussion concerning the exhaustive scanning of an image space created by the rotational motion of a passive surface of emission, by considering a voxel activation mechanism suitable for use with a planar screen. Let us therefore examine the configuration illustrated in Figure 12.4. From this diagram it may be seen that a set of beams are employed to scan simultaneously through the image space in a direction parallel to the axis of rotation. If the beam scanning frequency is considerably greater than the frequency of rotation of the

---

* We should naturally take into account the angles through which the two types of device are able to reliably operate.

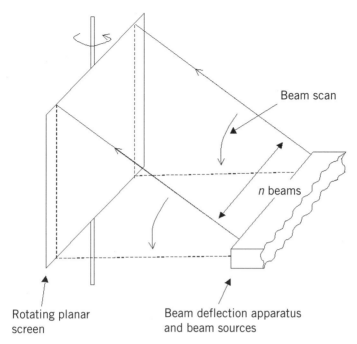

FIGURE 12.4   In this configuration a set of n beam sources scan simultaneously through the image space (for clarity, only two beams are illustrated). As may be seen, the direction of the scan is parallel to the axis of rotation of the planar screen. The scanning frequency is considerably greater than the frequency of rotation of the screen.

screen, each beam will produce a multiple line scan across an approximately planar region of the image space. The extent of each of these planes will vary according to their position in the image space.

Clearly, the component of the screen's velocity along the direction of propagation of the beams varies continuously throughout each rotational cycle. As a consequence, should the beams sweep at a constant rate, the scanned lines generated within the image space will not be positioned evenly. Furthermore, both positional and distortional dead zones will exist within the image space. Falling back on the standard solution of introducing a second beam-scanning mechanism located at an appropriate position may assist in alleviating these dead zone problems.

In the case of a system employing devices such as laser diodes, beam deflection may be achieved by means of a rapidly rotating multisided mirror. Although this scanning technique assists in minimizing the misalignment between a set of simultaneously scanned beams, the introduction of a second scanning system is likely to complicate the alignment process. The nonuniformities mentioned previously indicate that there will be local variations in the maximum achievable voxel density. Should each scanning mechanism contain n beam sources, the parallelism supported by the display unit should be n and not twice this value, as indicated by the presence of the second scanning system. Any attempt to increase the parallelism beyond n (corresponding to a spatial overlap of the subspaces) will exacerbate the nonuniformity of the image space.

Electron beams could, in principle, replace the laser beams mentioned above. In this case, a bank of closely spaced electron guns, each equipped with a single pair of electrostatic deflection plates, could perhaps be employed. Furthermore, it may be possible to generate a set of deflection waveforms able to provide a more regular spacing of the scanned lines produced within the image space.

The geometry of the voxel activation mechanisms with respect to the image space results in a nonequitable sharing of the workload by the various beam sources. For example, the innermost and outermost beams would be responsible for the generation of voxels across a very narrow cross section of the volume. For most of the time, these beams would not be active. However, the other light sources would be responsible for the generation of voxels across more extensive regions (Figure 12.5). This denotes a significant inefficiency of the configuration.

The outermost beam is responsible for scanning across the region of image space that is moving with the greatest linear velocity (in the direction of the beam's propagation). In principle, this scan must therefore be achieved at the greatest speed over the shortest time. The voxel intensity considerations referred to in Section 12.3.1 are equally applicable to this embodiment. This embodiment serves to highlight various complications, and if implemented, the application of the many correction factors needed to achieve a tolerably uniform scan of the

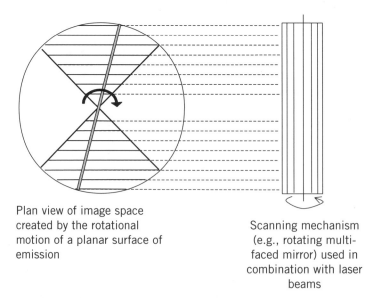

Plan view of image space created by the rotational motion of a planar surface of emission

Scanning mechanism (e.g., rotating multi-faced mirror) used in combination with laser beams

FIGURE 12.5 Plan view of the image space. The variation of the component of the screen's linear velocity in the direction of propagation of the beam sources results in the generation of a series of nonuniform scans within the image space.

entire image space would lead to a complex architecture. Furthermore, should the scanning apparatus be located on the equator of the image space, they will be particularly obtrusive. We may therefore consider moving this apparatus to a lower position. However, the generation of a uniform scan will then be even more difficult, and consequently it may even prove necessary to add a third scanning mechanism.

Having highlighted various problems associated with this implementation, let us turn our attention to an alternative approach. In the next section we propose a display unit that employs voxel generation centers dispersed throughout an image space volume.

12.4 HYBRID DISPLAY UNIT

In this section we examine the implementation of a swept-volume display unit that employs a set of voxel generation centers. Voxel activation will be assumed to take place by the application of electrical stimuli. This system will employ the rotational motion of the entire image space and, in the region close to the axis of rotation, there will also be a small-amplitude translational motion. As will become apparent, our intention is to take advantage of the benefits associated with both an active surface of emission and with rotational motion. At the same time we attempt to minimize the impact of various deficiencies associated with these techniques (see N.Z. Patent 314700).

12.4.1 Architecture

We begin by considering a display unit employing the rotational motion of a 2-D array of optoelectronic devices (see, e.g., Sections 5.4.4 and 5.5.3). This type of architecture exhibits, in principle, the following attributes:

- There is considerable parallelism in the voxel activation and generation mechanisms.
- There is a 100% fill factor (i.e., the voxel activation capacity may be equal to the voxel location capacity).
- Each voxel generation center is responsible for the production of multiple voxels.
- The subspaces may overlap completely in the temporal domain but are spatially separated.

These characteristics are highly desirable in the implementation of any high-definition display unit. Unfortunately, this configuration also has a number of associated weaknesses:

- The presence of a visual dead zone
- The presence of dead image space around the axis of rotation

- Nonequitable voxel production by the various voxel generation centers (as a consequence of the relationship between the subspace length and the radial distance of each center from the axis of rotation)

In the following discussion it is our intention to eliminate (or reduce) the impact of these problems while retaining the advantages of this configuration. However, since the implementation of any volumetric display unit involves a series of compromises, the architecture discussed below also has a number of associated weaknesses. Depending on the nature of the intended application, these may be tolerable.

As we have seen previously, the visual dead zone is caused by three major factors:

- The difference between the refractive index of the supporting structure and that of the surrounding image space volume
- The optical occlusion that may be caused by the presence of a center shaft
- The optical occlusion that may be caused by the optoelectronic devices (and their associated connections) which form the surface of emission

The first of these may be eliminated by the provision of an image space of uniform refractive index. Let us therefore consider that the image space comprises a rotating cylinder of transparent material within which the voxel generation centers are embedded, as illustrated in Figure 12.6*a*. The cylindrical volume therefore forms the supporting structure. This configuration also removes the need for a center shaft and thus resolves the second of the problems listed above.

It is, however, most unlikely that the refractive index and opacity of the voxel generation centers will match that of the material from which the cylinder is constructed. To reduce the impact of the third difficulty listed above we may envisage the dispersal of the voxel generation centers throughout the image space medium, as illustrated in Figure 12.6*b*. This offers the further advantage that the connections necessary to provide the voxel activation stimuli to each of the centers are also dispersed, so their impact on the transmission of light will be minimized.

As an alternative to the use of an electrical stimulus for voxel activation, we may consider the use of optical fibers for transmitting signals to the voxel activation centers (an example of a display unit employing this method for voxel activation is provided in Section 7.7.1). A close matching of the refractive indices and the transmission characteristics of the fibers, the voxel generation centers and the image space medium will serve to minimize the impact of the visual dead zone. Should there be any significant optical mismatch of the various components, their dispersal throughout the image space will be beneficial.

The use of rotational motion permits each voxel generation center to produce multiple voxels during each image update period. The number of centers required to permit the exhaustive addressing of the entire image space is considerably reduced (in comparison with the requirements for an equivalent static-volume

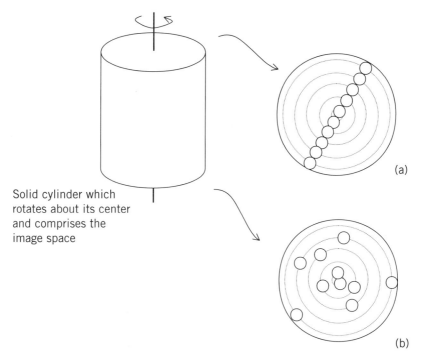

FIGURE 12.6 Image space consisting of a cylinder of transparent material. This cylinder rotates and the voxel generation centers are embedded within it. (*a*) The voxel generation centers located within a plane; (*b*) the dispersal of the voxel generation centers throughout the medium.

display). In practical terms, we may envisage the image space to be constructed from a number of thin cylindrical laminations. Following the insertion of the voxel generation centers, these laminations would be bonded together and could then be shaped so as to produce a cylindrical or spherical image space.

Unfortunately, the impact of the visual dead zone has not been reduced without penalty. Since the refractive index of the image space is now significantly greater than that of the surroundings, distortion will occur at the image space boundary. As mentioned previously (see Section 10.4.2) it is possible that in a limited number of applications any such lensing effect may be advantageous. However, in applications in which the 3-D form of an object is of interest, this type of distortion may not be tolerable.

By forming the image space from a solid transparent structure, the need for a center shaft (associated with the use of a rotating screen) has been removed, and consequently, the dead image space located around the axis of rotation is also eliminated. Unfortunately, the situation toward the center of the image space remains problematic. Difficulties arise as a consequence of the increasing density of voxel generation centers at locations close to the axis of rotation. Problems that may be associated with differences between the optical characteristics of these centers (or their associated connections) and the image space medium cannot be reduced by dispersal. For example, at the very center of the image space there will exist a column of elements located along the axis of rotation. At radial locations close to this position, the density of elements and connections will remain large. The central region of the image space may therefore continue to cause image distortion and perhaps occlusion.

In our quest for a high-performance display unit we are therefore faced with two possible solutions. The most desirable of these involves an attempt to match carefully the optical characteristics of all the components within the image space. Should this be possible, the problem is solved. Unfortunately, we may encounter difficulties in achieving this goal, and as a precaution, we consider an alternative solution. This is discussed below and is particularly instructive, as it draws on the concepts of subspace discussed in Chapter 9.

12.4.2   Introduction of Translational Motion

The display unit described above is referred to in the title of this section as a hybrid system. This name has been used to indicate that the image space will be created using a combination of rotational and translational motion. Although the practical realization of this approach may be problematic, it is instructive to consider the consequences of simultaneous application of these two forms of movement.

The display unit described in Section 12.4.1 has two associated problems:

- As we approach the axis of rotation, there is an increasing density of voxel generation centers (and connections). As a consequence of this congestion, optical distortions caused by the difference in the refractive index between the centers and the image space medium cannot be reduced by dispersal of the elements.
- The peak workload performed by each voxel generation center varies in accordance with the distance of each center from the axis of rotation. In an ideal situation all elements would be responsible for generation of the same maximum number of voxels during each image update period. In this case, however, it is apparent that the length of each subspace is determined by its distance from the rotational axis, and therefore elements close to this axis are responsible for the production of a smaller (maximum) number of voxels than those which are farther away.

We will therefore consider two related objectives:

- Achieving a reduction in the density of voxel generation centers at distances close to the axis of rotation
- Achieving an increase in the workload of voxel generation centers located at distances close to the axis of rotation

Let us consider a single column* of n evenly spaced voxel generation centers located at a radial distance r from the axis of rotation. We assume that the column spans a length L, and therefore the intervoxel separation will be L/n. The rotational motion ($\omega = 2\pi f$) of these elements permits the generation of n subspaces each of length $2\pi r$. For convenience we assume that the desired inter-voxel spacing in the direction of motion is also L/n. Consequently, the maximum number of voxels created by each element during a rotational cycle will be approximately $2\pi nr/L$. The time T between the activation of two voxels that are adjacent (in the direction of motion) may be expressed as $T \approx L/2\pi nrf$.

Suppose that we now reduce the number of voxel generation centers by 50% (retaining their equal spacing) and apply a translational motion having a sinusoidal velocity profile and amplitude L/n. Should the period of this motion equal T, the vertical separation of adjacent voxels will continue to be L/n despite the reduction in the number of voxel generation centers. By increasing the amplitude of the translational motion, it is possible to further reduce the number of these centers without any reduction in the adjacent voxel separation. In practice, the translational motion would be applied to a cylindrical core region of the image space, as illustrated in Figure 12.7. This technique increases the length of each subspace within the core region and therefore permits elements to be responsible for the production of a greater number of voxels. Consequently, the density of voxel generation centers may be reduced.

Unfortunately, the frequency of the translational motion required in any useful display would be very high. However, by increasing the rotational frequency of the voxel generation centers (while retaining the same image update period), the frequency of translational motion may be reduced. This may be illustrated by a simple example in which the rotational frequency is arranged to be twice the image refresh frequency, as illustrated in Figure 12.7. In this case, the frequency of translational motion should equal the image update frequency. It is instructive to consider the impact of making further reductions in the number of voxel generation centers within the core region. By varying the amplitude and frequency of the translational motion and increasing the rate of rotation of the image space core, we are able to maintain the voxel density, despite the reduced number of voxel generation centers.

In summary, from Figure 12.7 it may be seen that the image space consists of two structures, an outer cylinder and inner core. The outer cylinder rotates in the normal manner and is subject to no translational motion. The number of voxel generation centers in the inner core is reduced in the vertical direction (although a uniform spacing is maintained), and this is subject to both rotational and translational motion (the frequency of rotation is higher than that of the outer cylinder). Since our objective is to reduce the density of voxel generation centers within the region close to the axis of rotation, it is not necessary to subject the entire image space to a translational motion (in view of the considerable mass of the image space, this would be most undesirable). We therefore restrict the

---

* For simplicity in this explanation we disregard their dispersal.

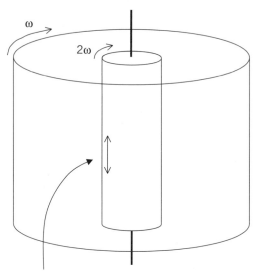

Small amplitude of translational
motion of the inner core

FIGURE 12.7 Possible hybrid display unit architecture. The inner cylinder rotates at a high frequency and is subject to a translational motion. The outer region is subjected to only a rotational motion. In this example, the set of elements contained within the inner core rotate at a frequency equal to twice the image update frequency (the outer region rotates in the normal manner at the image update frequency). In this case, the required translational motion should take place at the image update frequency.

translational motion to the most critical portion of the image space, a cylindrical column centered on the axis of rotation.

Many other possibilities exist for the application of the techniques discussed above, and this represents an interesting area for future consideration. This approach also highlights the benefits that may be derived by considering an image space to comprise a set of subspaces.

12.5 STATIC-VOLUME DISPLAY UNIT

Conceptually, static-volume display systems employing an array of elements contained within a transparent structure represent the simplest type of volumetric display unit. For this reason, this type of architecture was chosen for the elementary model introduced in Section 1.3. However, when we consider the actual implementation of such a display, we are faced with a number of problems that become more severe as the size of the image space and the voxel resolution are increased. In summary, these difficulties relate to:

- The large number of voxel generation centers and interconnections required
- The physical assembly of the image space

- The maintenance of the required optical image space characteristics despite the large number of voxel generation centers and interconnections contained therein

The one-to-one correspondence between voxel generation centers and voxel locations within an image space indicates that each element is not used to its greatest advantage. This is an inherent weakness in this type of display. In the case of any volumetric display, the voxel generation subsystem must fulfill a set of stringent requirements relating to its response to the activation stimulus, its optical characteristics in both the active and inactive states, and its ability to operate in harmony with the image space creation technique. The latter requirement applies particularly to swept-volume systems. In Chapter 7 we discussed a variety of static-volume architectures, and from this account it was apparent that the selection of the appropriate materials needed for voxel generation represents the greatest difficulty in the implementation of systems that employ a passive medium.

Implicitly, swept-volume systems are unable to produce opaque images; this characteristic is limited to certain static-volume technologies. It will be interesting to observe progress in this area during the next few years, particularly in the case of embodiments employing photochromic and thermochromic processes. Both of these techniques permit the production of opaque voxels, and in a number (if not the majority) of areas of application this is likely to be highly advantageous. It is to be hoped that in the near future we will see a resurgence of interest in static-volume displays that are able to provide "solid" full-color images.

12.6 LOCALIZED SCANNING TECHNIQUE

In Section 12.2 we briefly proposed the use of a localized scanning technique that permits one or more regions of an image space to be scanned at a high resolution. In principle, by enabling these regions to be positioned freely within an image space, it is possible to amalgamate them. This permits the construction of larger scanned volumes so that the volume of a single scanned region does not necessarily limit the maximum size of individual image objects (see also Bahr et al. (1996). Let us consider an embodiment in which each voxel activation mechanism is responsible for generation of the scan within a single localized region. Since the extent of each of these regions may be much less than the dimensions of the image space, the demands placed on any beam deflection apparatus will be reduced. The signals applied to the beam deflection apparatus will produce a rapid and repetitive scan (needed to sweep out the region), and superimposed on this will be a constant offset that determines the location of the scanned volume within the image space.*

---

* If, for example, the rotational motion of a screen is employed for the creation of the image space, as the location of a scanned region is changed it will be necessary (to maintain the geometry of the scan) to make appropriate changes to the scan waveform(s).

Should the localized scanned regions be positioned adjacently, proper registration between them is likely to be problematic and will require careful consideration.* This approach is best suited to situations where it is possible to maintain separation between the scanned regions (a lack of registration will then be of less consequence). For example, localized scanning may be particularly beneficial when a display system is to be used to show the spatial separation of objects (such as aircraft or projectiles) that move within the image space. These objects could be depicted in considerable detail and perhaps have associated alphanumeric information (see Section 8.2). The motion of each object through the image space could be facilitated by the motion of its associated scanned region.

This approach could be refined further by the addition of a conventional dot graphics scanning technique able to support the depiction of other information throughout the image space. For example, we may envisage a system that enables perhaps 10 localized (and freely locatable) scanned regions to be generated within the image space. These could be complemented by a dot graphics scan responsible for the production of low-resolution information that may be dispersed throughout the volume. Should the system be intended for the depiction of aircraft, each localized scan may, perhaps, give rise to a single airplane, and the dot graphics scan may be responsible for the generation of a terrain map, grid lines, and/or other appropriate information. In this form, the display would be less suited to, for example, medical visualization, where the operator is likely to require a system able to support the depiction of large single entities drawn at high resolution. This would necessitate placing the scanned regions in close proximity and would give rise to the alignment problems referred to above.

12.7 DIRECT INTERACTION WITH AN IMAGE SPACE

Interaction with conventional display systems is normally achieved by means of a keyboard and mouse. The latter provides a highly intuitive mechanism well suited for use with menu-driven user interfaces, and it also facilitates the manipulation of graphical images. With the widespread acceptance of this type of interaction hardware, it is natural to consider extension of its use to other types of display system. For example, in the case of volumetric systems, a 3-D mouse could in principle be used to direct a cursor to any location within an image space. Once the cursor has been positioned at the appropriate location, mouse (or keyboard) buttons may be used to effect appropriate selections.

The use of a conventional mouse assumes a particular orientation of the operator with respect to the display system. For those types of volumetric display that offer wide freedom in viewing position, there is (and should be) no preferred operator position. In the case of this type of display, a mouse motion to the right (which results in a corresponding cursor movement) will result in the reverse

---

* In constructing a larger scanned region from two or more smaller regions, it is advisable to avoid spatial overlap. In this case the shape of each of the smaller regions may require careful consideration.

cursor motion when the operator moves to the far side of the display vessel. Although this may be resolved by incorporating sensors able to detect the orientation of the mouse (and hence the operator) with respect to the image space, it presents an additional complication.

Limited trials carried out by the authors in conjunction with the Cathode Ray Sphere (see Section 5.6.3) have indicated that the rapid and accurate positioning of a cursor within an image space by means of 3-D mouse requires some considerable practice. Although this matter requires further and more rigorous experimental trials, it is possible that more intuitive methods of interaction may exist. It is apparent that the usage of conventional systems may be somewhat different to that of volumetric displays. In the case of the former, the display is frequently employed by a single user. However, as we have seen, volumetric displays offer considerable freedom in viewing orientation, and so this type of system may be well suited to situations in which there are a number of users (positioned, for example, around the image space). Naturally, in this type of situation it would be appropriate to provide each of the viewers with a device able to highlight (or manipulate) items of interest. This type of interaction may include a provision for the movement of one or more cursors within image space (e.g., each observer may be provided with a unique and clearly identifiable cursor). However, only when volumetric systems are in common use will it be possible to assess the benefits that may be derived from multiple operator interaction, and so properly determine the best form of interaction device.

The integration of mice, which are able to transmit their orientation with respect to the image space, poses little technical challenge. In this section we prefer to promote discussion on a more difficult problem: the use of hand-held pointers that may be used by a set of observers to highlight and/or modify objects depicted in an image space.

Let us turn to the possible use of some type of hand-held pointer and discuss briefly various possibilities for implementation of such a device.

12.7.1 Image Space Pointer

In this section we consider briefly the manipulation of a volumetric image by means of a single pointing device. For such an interaction device to operate correctly, two essential conditions must be fulfilled:

- The operator must be able to observe the location within the image space at which the pointer is directed.
- The controlling hardware must be able to detect the location within the image space at which the pointer is directed.

To reduce the possibility of errors caused by hardware (or software) malfunctions or by inaccuracies in the calibration of the pointing device, it is preferable to devise an approach in which the first of these conditions is achieved by means of information arising from the second. A pointing device may be considered to operate in an active manner (by the emission of information into the image space) or in a passive manner (by the reception of information emitted from the image space). We will assume the use of a passive pointer. This pointer will be assumed to be equipped with an optoelectronic device (and associated optical system), permitting it to respond to the light emitted from an image space.*

We now turn to various issues that will determine both the nature of the required hardware and the method by which the pointer will be used to interact with the image space.

Manipulation of Individual Voxels

Consider a pointer intended to permit the manipulation of individual voxels within an image scene. As we have seen, many volumetric display unit architectures fail to support parallelism in the voxel activation process. Voxel activation therefore occurs sequentially, and each voxel is activated at a unique time. In this case, voxels may be identified by the time (as measured from the commencement of the output of an image frame) at which they exist in an active state within the image space.

The information provided by the pointer may be used to change the intensity and/or the color of the detected voxel and so give the operator an indication of the location selected. In Figure 12.8 we illustrate a readily apparent deficiency of this scheme. This diagram depicts two voxels that exist along the line of sight of the pointer. Depending on the design of the hardware, the pointer may select

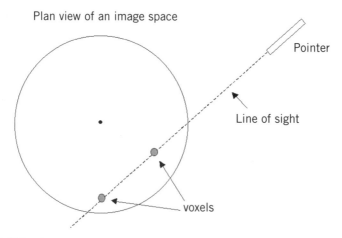

FIGURE 12.8   Two or more voxels exist along the pointer's line of sight. Although in the case of a display unit exhibiting no parallelism in the voxel activation process, these voxels will be activated at different times, there may still be confusion in the detection process.

---

* To assist in the detection of light emitted from the image space it will be desirable for the pointer hardware to also monitor the strength of the ambient lighting.

either of these voxels or may even oscillate in its selection between the two. This situation may be resolved by providing the operator with a facility enabling the selection of either of the voxels (e.g., by means of a selection button attached to the pointer). This approach could be generalized to permit the operator to scroll through any number of voxels existing along the pointer's line of sight.

The identification of voxels according to the time at which they exist in an activated state during each image update period is less straightforward in the case of display units that support parallelism in the voxel activation process. Since more than one voxel may be activated at any one time, it is not possible to distinguish between them by using only temporal information.

It is unlikely that a pointer held at some distance from an image space could easily and conveniently be directed toward a particular voxel and held steady in that position. The pointer represents a low-accuracy interaction device, and as the size of individual voxels is reduced (and their density increased), this problem will become more acute.

Finally, the sensing of light emitted by individual voxels is not a trivial task, as we should remember that the voxels will vary in their light output (we naturally assume that each voxel has an associated gray-scale value). As a consequence of ambient lighting and the possible close proximity of voxels of higher intensity, voxels of lower intensity may be difficult to detect. We are therefore faced with three considerable difficulties:

- The problems associated with the detection of individual voxels in a display unit supporting parallelism in the voxel activation process
- The lack of accuracy associated with a hand-held device
- The difficulties associated with the detection of individual voxels of varying intensity

In view of these difficulties, it is preferable to consider an alternative approach: the implementation of a pointer intended to direct the motion of a cursor within an image space.

Interaction via a Cursor

Let us consider the provision of one or more cursors within an image space. Each cursor will (and should) be constructed from a cluster of voxels. By employing high-intensity voxels positioned in close proximity, it becomes a relatively simple task to distinguish a cursor from other image components. Alternatively, rather than base the detection of the cursor solely on its overall intensity, it may be possible to improve this process by including information concerning the time at which its constituent voxels are activated.

Clearly, the size of the cursor could be modified according to the requirements of the operator, and a button located on the pointer could control its motion along the third axis. The use of a cursor therefore overcomes the problems referred to previously in connection with the detection and manipulation of individual voxels. However, we must still consider the implementation of a technique that enables the motion of the pointer to be determined and which produces a corresponding motion of the cursor.

Tracking the Motion of the Pointer

The smooth and rapid tracking of the motion of a passive pointer employed in conjunction with a volumetric display unit is problematic. The difficulties occur primarily as a result of the limited rate at which images may be redrawn. If a refresh rate of 30 Hz is assumed, we may infer that it is possible to make 30 attempts per second to identify the direction of a pointer's motion. Ideally, any attempt to locate the pointer's position through the activation of voxels within the image space should be transparent to the operator.

Following a movement of the pointer, any attempt to relocate its position by redrawing the cursor in all directions until it is redetected by the pointer is likely to be unsatisfactory. The process will cause visible flickering within the image space, and should the pointer motion be rapid, the detection process will occupy too much time. In principle, this problem may be resolved by the provision of a CCD-based imaging system within the pointer where the controlling hardware would attempt to maintain the location of the imaged cursor at the center of this array.

It is apparent from the brief discussion above that proper implementation of a pointer-based interaction device is nontrivial, and the type of application may determine the nature of the optimal interaction technique. In certain applications a pointer-based tool may be found to be beneficial. For example, it would appear that the provision of a CAD facility in which a designer could shape and otherwise manipulate surfaces within 3-D space using a pointer-based device would be advantageous. Considerable work needs to be undertaken to determine not only the best type of interaction device but also to assess the advantages that may be derived from direct interaction within an image space.

12.8 TOWARD THE FUTURE

In view of the challenges posed by the implementation of volumetric systems and the interesting nature of this type of work, the continuation of research in this area seems assured. Unfortunately, considerable uncertainty surrounds the application and acceptance of this type of technology.

For a number of years the development of computer systems has been driven by market dominance, and technological merit has sometimes taken second place. From a technical perspective, very few aspects of computer system development can be predicted with confidence 10 or 20 years in the future. It is equally difficult to predict the types of visualization devices that will gain widespread acceptance, although we are able to postulate an increasing need for systems that can depict increasingly complex types of information. In the case of many computer-related technologies it is possible to quantify in an exact manner the benefits that may be derived from some technical advancement. The benefits offered by a particular visualization technique are often less readily apparent and may perhaps only be assessed retrospectively.

The lack of widespread interest in volumetric display system technologies may indicate that such displays are not needed, or alternatively, that it is foreseen that they cannot be made to operate at the required level of performance. There have, after all, been numerous areas of development activity, which have been abandoned after the expenditure of considerable amounts of money and effort either because of intrinsic technological problems or because more effective alternative solutions were identified.

As we have seen during the course of this book, many problems associated with the implementation of volumetric systems can be solved and the prospects for the development of the next generation of high-performance displays are most promising. Furthermore, volumetric systems offer to provide a unique solution in visualization of various types of information.

The reader is left to judge whether volumetric display systems will accompany curiosities such as the Poll Giant into the archives of past endeavors.

APPENDIX

HOMOGENEOUS TRANSFORMATIONS

A.1 3-D AFFINE TRANSFORMATIONS

Consider an object in 3-D space, described by a number of coordinate points $\{x\}$, where

$$x = (x_1, x_2, x_3)^{\mathrm{T}} \tag{A.1}$$

A number of operations may be applied to these points to alter the position or form of the object. For example, the object may be translated from one position to another, or be rotated about some combination of axes. Equally, the object may be scaled to a different size, be scaled by different amounts in different directions, or be deformed by a shear operation (Figure A.1). Any combination of the operations above is known as an *affine transformation*. This group of transformations preserves the parallelism of lines but not the magnitude of angles or lengths (Foley et al. 1990), as may be verified by a glance at Figure A.1.

In general, an affine transformation is applied mathematically to each point by way of a matrix multiplication and a vector addition:

$$\mathbf{x}' = \mathbf{A} \cdot \mathbf{x} + \mathbf{t} \tag{A.2}$$

where the matrix $\mathbf{A}$ encapsulates the shear, rotation, or scaling operations, the addition of the vector $\mathbf{t}$ performs the translation operation, and $\mathbf{x}'$ is the vector containing the coordinates of the object point resulting from these transformations. The forms of the affine transformation matrix $\mathbf{A}$ for single rotations of θ_{x1} about the x_1 axis, θ_{x2} about the x_2 axis, and θ_{x3} about the x_3 axis are[*]

---

[*] We assume a right-handed coordinate system.

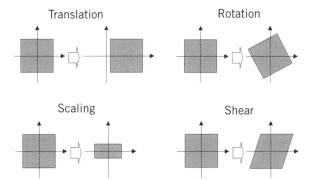

FIGURE A.1   The effects in 2-D of translation, rotation, scaling, and shear operations.

$$A_{\text{rotn},x1}(\theta_{x1}) = \begin{pmatrix} 1 & 0 & 0 \\ 0 & \cos\theta_{x1} & -\sin\theta_{x1} \\ 0 & \sin\theta_{x1} & \cos\theta_{x1} \end{pmatrix} \tag{A.3}$$

$$A_{\text{rotn},x2}(\theta_{x2}) = \begin{pmatrix} \cos\theta_{x2} & 0 & -\sin\theta_{x2} \\ 0 & 1 & 0 \\ \sin\theta_{x2} & 0 & \cos\theta_{x2} \end{pmatrix} \tag{A.4}$$

$$A_{\text{rotn},x3}(\theta_{x3}) = \begin{pmatrix} \cos\theta_{x3} & -\sin\theta_{x3} & 0 \\ \sin\theta_{x3} & \cos\theta_{x3} & 0 \\ 0 & 0 & 1 \end{pmatrix} \tag{A.5}$$

respectively. Scaling operations are represented by the scale multipliers along the matrix diagonal:

$$A_{\text{scale}}(S_{x1}, S_{x2}, S_{x3}) = \begin{pmatrix} S_{x1} & 0 & 0 \\ 0 & S_{x2} & 0 \\ 0 & 0 & S_{x3} \end{pmatrix}$$

and shear operations by off-diagonal elements.

The consecutive action of several transformations described by different instances of $\mathbf{A}$ (let's call them $\mathbf{A}_1$, $\mathbf{A}_2$, and $\mathbf{A}_3$) may be concatenated to form a single matrix describing the total transformation:

$$\mathbf{A}_{\text{tot}} = \mathbf{A}_3\mathbf{A}_2\mathbf{A}_1.$$

However, it is crucial to remember that in general, the overall transformation $\mathbf{A}_{\text{tot}}$ depends on the order in which the transformations are performed. For rotational transformations, the reader may readily verify this by applying two rotations in opposite order to a handy object, for example an (empty) coffee cup. The physical fact that rotations in 3-D space do not commute is reflected in the mathematical fact that the corresponding matrix multiplications do not commute.

General affine transformations, which also include translations [i.e., of the form given in Eq. (A.2)], may also be concatenated, but due to the presence of translations, the resulting expression will, in general, be the sum of several matrix-vector transformations, as illustrated in the following example.

Example Consider the application of the following three primitive affine transformations to point $\mathbf{x}$ in 3-D space:

1. Rotation about the x_1 axis [described by matrix $\mathbf{A}_1$, being of the form of Eq. (A.3)]
2. Translation parallel to the x_2 axis ($\mathbf{t}$)
3. Rotation about the x_3 axis ($\mathbf{A}_3$)

Applying the first of these transforms $\mathbf{x}$ into $\mathbf{x}' = \mathbf{A}_1\mathbf{x}$. The second then results in $\mathbf{x}'' = \mathbf{x}' + \mathbf{t}_y = \mathbf{A}_1\mathbf{x} + \mathbf{t}$, and finally, the third gives rise to the final result $\mathbf{x}''' = \mathbf{A}_3\mathbf{x}'' = \mathbf{A}_3(\mathbf{A}_1\mathbf{x} + \mathbf{t}) = \mathbf{A}_3\mathbf{A}_1\mathbf{x} + \mathbf{A}_3\mathbf{t}$. The composite transformation clearly involves more than one term. If, for example, operations 1 and 2 were applied in the reverse order, the final transformed point would be $\mathbf{x}''' = \mathbf{A}_3\mathbf{A}_1\mathbf{x} + \mathbf{A}_3\mathbf{A}_1\mathbf{t}$.

A.2 HOMOGENEOUS TRANSFORMATION FORMALISM

A.2.1 Overview

Application of the transformations discussed in Section A.1, using the formalism above, involves in general a matrix multiplication and a vector addition operation. However, the same set of transformations may be performed by a single matrix multiplication by employing the *homogeneous transformation* formalism. This is achieved by representing the points in 3-D space by 4-element vectors ("4-vectors"), and operating upon them with a 4×4 transformation matrix. Given an initial 3-vector $\mathbf{x}$, the transformed 3-vector is obtained by the following procedure:

1. Add a fourth element of unity to $\mathbf{x}$ to obtain a 4-vector (denoted $\mathbf{X}$): $\mathbf{x} = (x_1, x_2, x_3)^\mathrm{T} \rightarrow \mathbf{X} = (x_1, x_2, x_3, 1)^\mathrm{T} = (X_1, X_2, X_3, X_4)^\mathrm{T}$.
2. Multiply $\mathbf{X}$ by the 4×4 transformation matrix $\mathbf{M}$ (the construction of which is detailed below) to obtain the transformed 4-vector $\mathbf{X}' = (X'_1, X'_2, X'_3, X'_4)^\mathrm{T}$.
3. Recover the transformed 3-vector $\mathbf{x}'$ by dividing each element of $\mathbf{X}'$ by X'_4, which resets the fourth element to unity. The first three components of $\mathbf{X}'$ then comprise $\mathbf{x}'$. That is, $\mathbf{x}' = (X'_1/X'_4, X'_2/X'_4, X'_3/X'_4)^\mathrm{T}$.

As will be seen below, X'_4 describes an overall scaling of the coordinates. If this is unity, $\mathbf{x}'$ may be recovered simply by taking the first three elements of $\mathbf{X}'$, avoiding three unnecessary divisions. The homogeneous transformation is thus represented as a single matrix multiplication [cf. Eq. (A.2)]

$$\mathbf{X}' = \mathbf{M} \cdot \mathbf{X} \tag{A.6}$$

An advantage of this formalism is that a composite transformation matrix, calculated by the matrix multiplication of its component transformation matrices, includes all the affine transformations, that is, including translations,

$$\mathbf{M}_{tot} = \mathbf{M}_3 \mathbf{M}_2 \mathbf{M}_1$$

A.2.2   Construction of the Homogeneous Transformation Matrix

The upper left 3×3 portion of the transformation matrix $\mathbf{M}$ has the same general form, and controls the same transformations, as the matrix $\mathbf{A}$ above (i.e., rotations, scalings, and shear transformations):

$$\mathbf{M} = \begin{pmatrix} A_{11} & A_{12} & A_{13} & t_{x1} \\ A_{21} & A_{22} & A_{23} & t_{x2} \\ A_{31} & A_{32} & A_{33} & t_{x3} \\ 0 & 0 & 0 & s \end{pmatrix}$$

The translation matrix $\mathbf{t}$ is now a part of the same 4×4 transformation matrix, and the lower-right element s denotes an overall scaling factor that affects the vector via the normalization division in step 3 above.

Example   Consider the same example as in Section A.1 that is, a rotation about the x_1 axis, followed by a translation parallel to the x_2 axis, then finally a rotation about the x_3 axis. In the homogeneous transformation case, the combined transformation simply involves concatenation of the individual transformation matrices, to produce an overall transformation matrix: $\mathbf{x}''' = \mathbf{M}_3 \mathbf{M}_2 \mathbf{M}_1 \mathbf{x} = \mathbf{M}_{tot} \mathbf{x}$. The reader may verify that in this case, the elements of the overall transformation matrix are

$$\mathbf{M}_{tot} = \begin{pmatrix} \cos\theta_{x3} & -\sin\theta_{x3}\cos\theta_{x1} & \sin^2\theta_{x1} & -t_{x2}\sin\theta_{x3} \\ \sin\theta_{x3} & \cos^2\theta_{x3} & -\sin\theta_{x1}\cos\theta_{x3} & t_{x2}\cos\theta_{x3} \\ 0 & \sin\theta_{x3} & \cos\theta_{x3} & 0 \\ 0 & 0 & 0 & 1 \end{pmatrix}$$

where θ_{x1} and θ_{x3} are the angles of rotation about the x_1 and x_3 axes, respectively, and t_{x2} is the distance of translation parallel to the x_2 axis [see also Vince (1995)].

GLOSSARY

Accommodation. A depth cue corresponding to the change in the geometry of the lens in the eye so as to focus on nearby objects. This is used by the brain as an indication of depth. Also known as "stereopsis".

Active. An active image space comprises a collection of discrete elements. These may take the form of a 3-D matrix or be mounted on a moving structure. Upon receipt of an appropriate stimulus each element moves from a passive to an active state. When active an element may, for example, act as a source of visible light.

Beam addressed. A volumetric display unit in which the voxel activation mechanism employs one or more directed energy beams (usually, lasers or electron beams).

Binocular parallax. A depth cue. The slightly different perspective seen by each eye gives rise to small differences in the retinal projection in each eye. These differences are interpreted by the brain as an indication of depth. Also known as "stereopsis".

Cathode ray tube (CRT). A display device in which one or more electron beams is rapidly scanned over a phosphor-coated screen. Usually, the screen takes the form of an essentially flat area occupying one end of an evacuated vessel which also includes the electron gun and (possibly) post deflection acceleration electrodes.

Dead image space. Region(s) of a volumetric image space within which it is impossible to activate voxels, whatever the arrangement of the voxel activation devices. This may be due to physical obstruction, or in the case of active matrix displays, to interstitial space between discontinuous subspaces.

Dead zone. A region of an image space within which image quality is compromised by a reduction in one or more image space characteristics. Specific subclasses discussed in this book include dead image space, distortional dead zone, voxel placement dead zone, shadowing dead zone, and visual dead zone.

Direct volume display device (DVDD). An alternative name for volumetric display devices, frequently encountered in the literature.

Directly addressed. A volumetric device in which the voxel generation mechanism employs a number of discrete elements (voxel generation centres). Each of these may be activated by electrical or optical signals.

Display system. The combination of the display unit and the graphics engine. It is therefore assumed to comprise not only the apparatus for the physical generation of the volumetric images but the software and hardware responsible for the manipulation (and possibly production) of the data.

Display unit. The apparatus responsible for the physical generation of the visible 3-D image. This comprises the image space creation, voxel generation, and voxel activation subsystems. The display unit excludes the graphics engine.

Distortional dead zone. A region of the image space in which the voxel shape or size is distorted beyond a specified tolerance.

Electron gun. A device for creating, gating, and deflecting beams of electrons in evacuated vessels (see Plate xiii).

Fill factor. The ratio of the voxel activation capacity to the voxel location capacity. Approximately equal to the fraction of the image space volume that can be filled with voxels in a single refresh period.

Flicker fusion frequency. The update frequency at which the visual system perceives a scene or depicted object as being continually present. In practice, this parameter depends on ambient lighting conditions, the brightness of the displayed image, and may vary among individuals. Moreover, prolonged viewing normally demands higher refresh frequencies to prevent eye strain or headaches. The flicker fusion frequency has yet to be fully researched in the context of volumetric displays.

Fluorescence efficiency. In stepwise excitation processes, a parameter denoting the fraction of all fluorescent centers pumped by the activation radiation into the fluorescent state that decay by the emission of visible photons. It takes a fractional value between zero and one.

Fluorescent center. In displays employing the stepwise excitation of fluorescence, a single ion or atom exhibiting a stepwise excitation process.

Graphics engine. The data manipulation and control systems (both software and hardware) that are responsible for transforming the initial data into the form required by the voxel activation subsystem. Other responsibilities may include tasks such as voxel ordering and calibration.

Homogeneous image space. An image space that demonstrates acceptable uniformity in the spatial positioning of voxels. This term is also used to refer to the optical characteristics of an image space.

Image refresh. Periodic reactivation of transient voxels within the image space in order that they appear continually present to the observer.

Image space. The volume within which volumetric images are physically depicted for viewing.

Image space creation subsystem. The mechanism by which the image space is created. For example, this may comprise the rapid motion of a suitable 2-D surface, or a static volume of material.

Inhomogeneous image space. An image space that demonstrates unacceptable nonuniformity in the spatial positioning of voxels. This term is also used to refer to the adverse optical characteristics of an image space.

Motion parallax. A depth cue. The continually changing perspective afforded by moving objects, or parts of an object, at different distances from the observer provides a cue to distance.

Multiplanar display system. An alternative name for certain volumetric display systems, frequently encountered in the literature.

Parallelism. The number of voxels that can be activated concurrently.

Passive. A passive image space is one in which voxels are generated by interaction between the material of the image space (or the surface of emission) and one or more directed energy beams (such as electron or laser beams).

Physiological depth cue. A depth cue that relies on feedback from the muscles of the eyes to provide information about the distance of an object.

Retinal image depth cue. A depth cue that employs information present in the projection of the scene being viewed on the retina.

Screen. In swept-volume volumetric display units, the combination of the surface of emission and its supporting structure.

Shadowing dead Zone. A region where voxel activation is precluded by a portion of the screen forming a physical obstruction and so preventing beam propagation.

Static-volume systems. A subclass of volumetric display in which image space creation does not involve mechanical motion. Rather, it comprises a static volume of a suitable material, or arrangement of materials, and may be solid, gaseous, or liquid.

Stepwise excitation of fluorescence. A physical, usually optical upconversion mechanism whereby two or more energy beams combine to produce visible light.

Stereopsis. An alternative name for binocular parallax.

Subspace. The subvolume of the image space in which voxels are created by a single voxel activation unit. The number of subspaces that comprise an image space corresponds to the number of voxel activation mechanisms.

Supporting structure. The structural portion of the screen in a swept-volume display unit, supporting the surface of emission.

Surface of emission. The surface on which voxels are generated while moving through the image space within a swept-volume display unit. A surface of emission may be passive (e.g., a light-scattering surface) or active (e.g., a matrix of voxel generation centers).

Swept-volume systems. A subclass of volumetric displays in which the image space is created by the rapid mechanical motion of a surface (screen).

System adjustment time. The time required for the voxel activation subsystem to adjust before activation of the next P voxels, where P is the parallelism of voxel activation.

Visual dead zone. A region of the image space, dependent on the position of the viewer, in which depicted data appear to be reduced in intensity.

Volumetric. A display device in which the images are cast within a physically 3-D volume. For the purposes of this book, systems employing a virtual image space are generally not considered.

Voxel. A single image element, or "particle," within the 3-D image space from which volumetric images may be constructed.

Voxel activation capacity. The maximum number of voxels that may be activated within the image space during a single image refresh period.

Voxel activation subsystem. The subsystem of the display unit responsible for causing the transition of each voxel from its inactive to its active state. This may take the form of optical or electronic signals to individual voxel generation centers, or a deflected laser or electron beam.

Voxel activation time. The time required for the activation stimulus to create voxels of sufficient brightness.

Voxel descriptor. A unit of data containing all the information required by the voxel activation subsystem pertaining to the position, intensity, color and perhaps opacity of the voxel.

Voxel generation centers. Individual voxel elements within an active image space or on an active surface of emission.

Voxel generation subsystem. The subsystem of the display unit comprising the physical process responsible for the conversion of the input from the voxel activation subsystem into visible light. For example, this may involve the generation of light in the presence of electrical signals (optoelectronic components) or beams of electrons (cathodoluminescence).

Voxel location capacity. The total number of voxel locations that are available within the image space during a single image refresh. Often, only a fraction of these may be activated during each refresh period.

Voxel placement dead zone. A region of the image space in which voxels cannot be generated to a sufficiently high spatial accuracy.

Voxel time. The total time required to create each voxel. This is composed of the system adjustment time and the voxel activation time.

Voxel visibility lifetime. The time scale over which a voxel remains visible after removal of the voxel activation stimulus. This is affected by the ambient lighting conditions.

REFERENCES

JOURNAL ARTICLES, BOOKS AND CONFERENCE PROCEEDINGS

Abrahamson, A., *The history of television, 1880 to 1941*, McFarland and Co., Jefferson, NC, 1987.

Akiyama, K., Tetsutani, N., Ishibashi, M., Ichinose, S., and Yasuda, H., "Consideration on three-dimensional visual communication systems," *IEEE Journal on Selected Areas in Communications*, **9**(4), 555–560 (1991).

Aviation Week [author unknown] "New display gives realistic 3-D effect," *Aviation Week*, October 31, 66–68 (1960).

Bahr, D., Langhans, K., Gerken, M., Vogt, C., Bezecny. D., and Homann, D., "Felix: a volumetric 3D laser display," *Projection Displays II: SPIE Proceedings*, **2650**, 265–273 (1996).

Bains, S., "Radial scanning produces 3-D image on a flat screen," *Laser Focus World*, January, 41–42 (1993).

Barnes, R. H., Moeller, C. E., Kircher, J. F., and Verber, C. M., "Two-step excitation of fluorescence in iodine monochloride vapour," *Applied Physics Letters*, **24**, 610–612 (1974).

Batchko, R. G., *Radial scanning in volumetry*, LAMDA Systems Corp., Chicago, 1992.

Batchko, R. G., "Three-hundred-sixty degree electro-holographic stereogram and volumetric display system," *Proceedings of the International Society for Optical Engineering*, **2176**, 30–41 (1994).

Benton, S. A., St. Hilaire, P., Lucente, M., Sutter, J. D., and Plesniak, W. J., "Real-time computer-generated 3D holograms," *SPIE Proceedings*, **1983**, 536–543 (1993).

Blondel, A., [Title unknown], *La lumière électrique*, **41**, 1891 [pages unknown].

Blundell, B. G., "On the development of higher performance volumetric display systems," *Spring Conference on Computer Graphics and its Applications*, Budmerice, Slovakia, April 1998.

Blundell, B. G., and King, W., "Outline of a low-cost prototype system to display three-dimensional images," *IEEE Transactions on Instrumentation and Measurement*, **40**(4), 792–793 (1991).

Blundell, B. G., and Schwarz, A. J., "A graphics hierarchy for the visualisation of 3D images by means of a volumetric display system," *Proceedings IEEE Tencon '94*, 1–5 (1994).

Blundell, B. G., Schwarz, A. J., and Horrell, D. K., "Volumetric three-dimensional display systems: their past, present and future," *IEE Engineering Science and Education*, **2**(5), 196–200 (1993a).

Blundell, B. G., Schwarz, A. J., and Horrell, D. K., "The Cathode Ray Sphere: a prototype volumetric display system," *Proceedings Eurodisplay '93 (Late News Papers)*, 593–596 (1993b).

Blundell, B. G., and Schwarz, A. J., "Volumetric three-dimensional displays," *McGraw-Hill Yearbook of Science and Technology*, McGraw-Hill, New York, 1995a, pp. 95–97.

Blundell, B. G., and Schwarz, A. J., "Visualization of complex system dynamics on a volumetric 3D display device," *Proceedings Visualisierung — dynamik und komplexität*, Bremen, Germany, 24–26 Sept., 1995b.

Blundell, B. G. and Schwarz, A. J. "An alternative approach to human–computer interaction," *Interfaces*, **41**, Sept/Oct 1999.

Blundell, B. G., Schwarz, A. J., and Horrell, D. K., "The Cathode Ray Sphere: a prototype system to display volumetric three-dimensional images," *Optical Engineering*, **33**(1), 180–186 (1994).

Bolas, M. T., Fisher, S. S., and Merritt, J. O. (eds.), *Stereoscopic displays and virtual reality systems V, SPIE Proceedings*, **3295** (1998).

Braun, F., "Uber ein Verfahren zur Demonstration und zum Studium des zeitlichen Verlaufes variabler Ströme," *Annalen der Physik (Leipzig) [3rd series]*, **60**, 552–559 (1897).

Brewster, D., *The stereoscope: its history, theory and construction (with its application to the fine and useful arts and to education)*, John Murray, London, 1856.

Brinkmann, U., "A laser-based three-dimensional display," *Lasers and Applications*, **2**(3), March, 55–56 (1983).

Brueche, E., and Scherzer, O., *Geometrische Elektronenoptik*, Springer-Verlag, Berlin, 1934.

Burckhardt, C. B., "Optimum parameters and resolution limitation of integral photography," *Journal of the Optical Society of America*, **58**(1), 71–76 (1968a).

Burckhardt, C. B., "Information reduction in holograms for visual display," *Journal of the Optical Society of America*, **58**(2), 241–246 (1968b).

Burton, H. E., "The optics of Euclid," *Journal of the Optical Society of America*, **35**(5), 357–372 (1945).

Busch, H., "Berechnung der Bahn von Kathodenstrahlen in axial symmetrischen Elektromagnetischen Felde," *Annalen der Physik (Leipzig)*, **8**, 974–993 (1926).

Busch, H., "Über die Wirkungweise der Konzentrierungs Spule bei der Braunschen Röhre," *Archiv für Elektrotechnik (Berlin)*, **18**, 583–594 (1927).

Caulfield, H. J. (ed.), *Handbook of optical holography*, Academic Press, San Diego, CA, 1979.

Clifton, T. E., and Wefer, F. L., "Direct volume display devices," *IEEE Computer Graphics and Applications*, **13**(4), 57–65 (1993a).

Clifton, T. E., and Wefer, F. L., "Functional architecture and component technologies for direct volume display devices," *SID'93 Digest*, 1031–1034 (1993b).

Coatrieux, J. L., Toumoulin, C., Hamon, C., and Luo, L., "Future trends in 3D medical imaging," *IEEE Engineering in Medicine and Biology*, **9**, December, 33–39 (1990).

Coddington, J. L., and Schipper, R. J., "Practical solid state three-dimensional (3-D) display," *IRE International Conference Record*, **10**(3), 177–184 (1962).

Crookes, W., "On the illumination of lines of molecular pressure and the trajectory of molecules," *Philosophical Transactions of the Royal Society of London*, **170**, 135–164 (1879).

Cross, M., "Lasers add a third dimension to graphics," *New Scientist*, November 11, 358, (1982).

Davisson, C. J., and Calbick, C. J., "Electron lenses," *Physical Review*, **42**, 580 (1932).

Dember, W. N., and Warm, J. S., *Psychology of depth perception*, 2nd ed., Holt, Rinehart and Winston, New York, 1979.

Downing, E. A., Hesselink, L., MacFarlane, R. M., and Barty, C. P. J., "Solid-state three-dimensional display," *CLEO'94 Proceedings*, 6–7 (1994).

Downing, E. A., Hesselink, L., Ralston, J., and Macfarlane, R., "A three-color, solid-state three-dimensional display," *Science*, **273**, 1185–1189 (1996).

Dudley, L. P., *Stereoptics*, MacDonald and Co., London, 1951.

Edgerton, S. Y., *The Renaissance rediscovery of linear perspective*, Basic Books, New York, 1975.

Edgerton, S. Y., *The heritage of Giotto's geometry: art and science on the eve of the scientific revolution*, Cornell University Press, Ithaca, NY, 1991.

Fisher, S. S., and Merritt, J. O. (eds.), *Stereoscopic displays and applications, SPIE Proceedings*, **1256** (1990).

Fisher, S. S., Merritt, J. O., and Bolas, M. T. (eds.), *Stereoscopic displays and virtual reality systems, SPIE Proceedings*, **2177** (1994).

Fisher, S. S., Merritt, J. O., and Bolas, M. T. (eds.), *Stereoscopic displays and virtual reality systems II, SPIE Proceedings*, **2409** (1995).

Fisher, S. S., Merritt, J. O., and Bolas, M. T. (eds.), *Stereoscopic displays and virtual reality systems III, SPIE Proceedings*, **2653** (1996).

Fisher, S. S., Merritt, J. O., and Bolas, M. T. (eds.), *Stereoscopic displays and virtual reality systems IV, SPIE Proceedings*, **3012** (1997).

Foley, J. D., van Dam, A., Feiner, S. K., and Hughes, J. F., *Computer graphics: principles and practice*, 2nd ed., Addison-Wesley, Reading, MA, 1990.

Foley, J. D., van Dam, A., Feiner, S. K., Hughes, J. F., and Phillips, R. L., *Introduction to computer graphics*, Addison-Wesley, Reading MA, 1994.

Fuchs, H., Pizer, S. M., Tsai, L. C., Bloomberg, S. H., and Heinz, E. R., "Adding a true 3-D display to a raster graphics system," *IEEE Computer Graphics and Applications*, **2**(5), 73–78 (1982).

Fuchs, H., Levoy, M., and Pizer, S. M., "Interactive visualization of 3D medical data," *IEEE Computer Graphics and Applications*, **9**(4), 46–51 (1989).

Fuchtbauer, C., "Excitation of mercury spectral lines," *Zeitschrift für Physik*, **21**, 635–638 (1920).

Gabor, D., "A new microscopic principle," *Nature*, **161**, 777–778 (May 15 1948).

Gibson, J. J., *The perception of the visual world*, Houghton Mifflin, Boston, 1950.

Golden, B. L., and Stewart, W. R., "Empirical analysis of heuristics," Chapter 7, 207–249, in Lawler et al. (1985), q.v.

Golden, B. L., Bodin, L., Doyle, T., and Stewart, W. R., "Approximate travelling salesman algorithms," *Operations Research*, **28**(3), 694–711 (1980).

Harris, L. D., Camp, J. J., Ritman, E. L., and Robb, R. A., "Three-dimensional display and analysis of tomographic volume images using a varifocal mirror," *IEEE Transactions on Medical Imaging*, **5**(2), 67–72 (1986).

Hart, S. J., and Dalton, M. N., "Display holography for medical tomography," *SPIE Proceedings*, **1212**, 116–135 (1990).

Hartwig, R., "A three dimensional computer display," invited presentation and demonstration, TELI — European Union of Science Journalists Associations, German Study Tour, Heidelberg, November 5, 1982.

Helmholtz, H., *Handbuch der Physiologischen Optik*, Vol. 3, Leopold Voss, Leipzig, 1909. [English translation by Southall (1962), q.v.]

Hesselink, L., Downing, E., and Akella, A., *Proposal to develop high efficiency material hosts and fully integrate a solid-state 3D display with RGB color capabilities*, research proposal, Stanford University, Stanford, CA, 1995.

Hill, F. S., *Computer graphics*, Macmillan, New York, 1990.

Hittorf, W., "Uber die Electricitätsleitung der Base," *Annalen der Physik (Leipzig) [2nd series]*, **136**, 1–31 (1869).

Honda, T., Doumuki, T., Akella, A., Galambos, L., and Hesselink, L., "One-color one-beam pumping of Er^{3+}-doped ZBLAN glasses for a three-dimensional two-step excitation display," *Optics Letters*, **23**(14), 1108–1110 (1998).

Hunter, S., Kiamilev, F., Essener, S., Parthenopoulos, D. A., and Rentzepis, P. M., "Potentials of two-photon based 3-D optical memories for high performance computing," *Applied Optics*, **29**(14), 2058–2066 (1990).

Isono, H., Yasuda, M., Takemori, D., Kanayama, H., Yamada, C., and Chiba, K., "Autostereoscopic 3-D TV display system with wide viewing angle," *Proceedings Eurodisplay '93*, 407–410 (1993).

Ittelson, W. H., *Visual space perception*, Springer-Verlag, New York, 1960.

Ives, H. E., "A camera for making parallax panoramagrams," *Journal of the Optical Society of America*, **17**, December, 435–439 (1928).

Ives, H. E., "Motion pictures in relief," *Journal of the Optical Society of America*, **18**, February, 118–122 (1929).

Ives, H. E., "Parallax panoramagrams made with a large diameter lens," *Journal of the Optical Society of America*, **20**, 332–342 (1930a).

Ives, H. E., "Parallax panoramagrams made with a large diameter concave mirror,"*Journal of the Optical Society of America*, **20**, 597–600 (1930b).

Ives, H. E., "Optical properties of a Lippmann lenticulated sheet," *Journal of the Optical Society of America*, **21**, 171–176 (March 1931).

Ivins, W. M., *On the rationalization of sight: with an examination of three Renaissance texts on perspective*, Da Capo Press, New York, 1973.

Jansson, D. G., Berlin, E. P., Straus, I., and Goodhue, X., "A three-dimensional computer display," Annual Conference on Computer Graphics and CAD/CAM Systems, MIT, Cambridge, MA (April 1979).

Julesz, B., "Binocular depth perception without familiarity cues," *Science*, **145**, 356–362 (1964).

Julesz, B., *Foundations of cyclopean perception*, University of Chicago Press, Chicago, 1971.

Kameyama, K., and Ohtomi, K., "A shape modelling system with a volume scanning display and multisensory input device," *Presence*, **2**(2), 104–111 (1993).

Kameyama, K., Ohtomi, K., and Fukui, Y., "Interactive volume scanning 3-D display with an optical relay system and multidimensional input devices," *Stereoscopic displays and applications IV: SPIE Proceedings*, **1915**, 12–20 (1993).

Kaufman, L., *Sight and mind: an introduction to visual perception*, Oxford University Press, New York, 1974.

Kennedy, D. N., and Nelson, A. C., "Three-dimensional display from cross-sectional tomographic images: an application to magnetic resonance imaging," *IEEE Transactions on Medical Imaging*, **6**(2), 134–140 (1993).

Ketchpel, R. D., "CRT provides three-dimensional display," *Electronics*, November 2, 54–57 (1962).

Ketchpel, R. D., "Direct-view three-dimensional display tube," *IEEE Transactions on Electron Devices*, **10**, 324–328 (1963).

Kim, I. I., Korevaar, E., and Hakakha, H., "Three-dimensional volumetric display in rubidium vapour," *Projection Displays II: SPIE Proceedings*, **2650**, 274–284 (1996).

Knoll, M., and Ruska, E., "Bietrag zur geometrischen Elektronenoptik," *Annalen der Physik (Leipzig)*, **12**(5), 607–640 (1932).

Lasher, M., Soltan, P., Dahlke, W., Acantilado, N., and MacDonald, M., "Laser projected 3-D volumetric displays," *Projection Displays II: SPIE Proceedings*, **2650**, 285–295 (1996).

Lawler, E. L., Lenstra, J. K., Rinnooy Kan, A. H. G., and Schmoys, D. B. (eds.), *The travelling salesman problem: a guided tour of combinatorial optimization*, Wiley, New York, 1985.

Leith, E. N., "White light holograms," *Scientific American*, **235**, 80–95 (October 1976).

Leith, E. N., and Upatnieks, J., "Wavefront reconstruction with diffused illumination and three-dimensional objects," *Journal of the Optical Society of America*, **54**, 1295–1301 (1964).

Leith, E. N., and Upatnieks, J., "Photography by laser," *Scientific American*, **212**, 24–35 (June 1965).

Leith, E. N., Upatnieks, J., Hildebrand, K., and Haines, K., "Requirements for a wavefront reconstruction television facsimile system," *Journal of the Society of Motion Picture and Television Engineers*, **74**, 893–896 (1965)

Lewis, J. D., Verber, C. M., and McGhee, R. B., "A true three-dimensional display," *IEEE Transactions on Electron Devices*, **18**, 724–732 (1971).

Lippmann, M. G., "Epreuves reversibles donnant la sensation du relief," *Journal of Physics [4th series]*, **7**, 821–825 (November 1908).

Lucente, M., "Interactive computation of holograms using a look-up table," *Journal of Electronic Imaging*, **2**(1), 28–34 (1993).

MacDonald, L. W., and Lowe, A. C., *Display systems: design and applications*, Wiley, New York, 1997.

MacFarlane, D. L., "A volumetric three dimensional display," *Applied Optics*, **33**(31), 7453–7457 (1994).

MacFarlane, D. L., Schultz, G. R., Higley, P. D., and Meyer, J., "A voxel based spatial display," *SPIE Proceedings*, **2177**, 196–202 (1994).

Mark, H., and Hull, F., "Three-dimensional viewing of tomographic data," *SPIE Proceedings*, **120**, 192–194 (1977).

Martin, A., "Cathode ray tubes for industrial and military applications," *Advances in Electronics and Electron Physics*, **67**, 183–328 (1986).

Mayo, H., *Outlines of human physiology*, 3rd ed., Burgess and Hill, London, 1833.

McAllister, D. F. (ed.), *Stereo computer graphics and other true 3D display technologies*, Princeton University Press, Princeton, NJ, 1993.

Merritt, J. O., "Common problems in the evaluation of 3D diplays," *SID '83 Digest*, 192–193 (1983).

Merritt, J. O., and Fisher, S. S. (eds.), *Stereoscopic displays and applications II, SPIE Proceedings*, **1457** (1991).

Merritt, J. O., and Fisher, S. S. (eds.), *Stereoscopic displays and applications III, SPIE Proceedings*, **1669** (1992).

Merritt, J. O., and Fisher, S. S. (eds.), *Stereoscopic displays and applications IV, SPIE Proceedings*, **1915** (1993).

Nithyanandum, N. "A three-dimensional digital image display system" *IEEE Transactions on Broadcasting*, **BC-21**(4), 53 (1975).

Nordin, G. P., Kulick, J. H., Jones, M., Nasiatka, P., Lindquist., R. G., and Kowel, S. T., "Demonstration of a novel three-dimensional autostereoscopic display," *Optics Letters*, **19**, 901–903 (June 1994).

Okoshi, T., *Three-dimensional imaging techniques*, Academic Press, San Diego, CA, 1976.

Okoshi, T., "Three dimensional displays," *Proceedings of the IEEE*, **68**(5), 548–564 (1980).

Onural, L., Bozdagi, G., and Atalar, A., "New high-resolution display device for holographic three-dimensional video: principles and simulations," *Optical Engineering*, **33**(3), 835–844 (1994).

Ozawa, L., *Cathodoluminescence*, Kodansha, Tokyo, 1990.

Parker, M. J., and Wallis, P. A., "Three-dimensional cathode-ray tube displays," *Journal of the IEE*, **95**, 371–390 (September 1948).

Parker, M. J., and Wallis, P. A., 'Discussion on "Three-dimensional cathode-ray tube displays," *Journal of the IEE*, **96**(III), no.42, 291–294 (1949).

Parthenopoulos, D. A., and Rentzepis, P. M., "Two-photon volume information storage in doped polymer systems," *Journal of Applied Physics*, **68**(11), 5814–5818 (1990).

Petty, M. C., Bryce, M. R., and Bloor, D., *An introduction to molecular electronics*, Oxford University Press, New York, 1995.

Plücker, J., "Uber die Einwirkung der Magneten auf die elektriscen entladungen inverdünten Gasen," *Annalen der Physik (Leipzig) [2nd series]*, **103**, 88–106 (1858).

Poole, H. H., *Fundamentals of display systems*, Macmillan, London, 1966.

Richards, W., "Stereopsis and stereoblindness," *Experimental Brain Research*, **10**, 380–388 (1970).

Sakamoto, K., Okamoto, M., Ueda, H., Takahashi, H., and Shimizu, E., "Real-time 3-D color display using a holographic optical element," *SPIE Proceedings*, **2652**, 124–131, 1996.

Schiffman, H. R., *Sensation and perception*, 3rd ed., Wiley, New York, 1990.

Schwarz, A. J., and Blundell, B. G., "Considerations regarding voxel brightness in volumetric displays utilizing two-step excitation processes," *Optical Engineering*, **32**(11), 2818–2823 (1993).

Schwarz, A. J., and Blundell, B. G., "Regions of extreme image distortion in rotating-screen volumetric display systems," *Computers and Graphics*, **18**(5), 643–652 (1994a).

Schwarz, A. J., and Blundell, B. G., "Considerations for accurate voxel positioning on a rotating-screen volumetric display system," *IEE Proceedings on Optoelectronics*, **141**(5), 336–344 (1994b).

Schwarz, A. J., and Blundell, B. G., "Optimisation of a dot-graphics plotting technique for volumetric graphics displays," *IEEE Computer Graphics and Applications*, **17**(3), 72–88 (1997).

Sheat, D., Gentry, P., McCartney, D., and Chamberlin, G., "3-D displays for videotelephone applications," *Proceedings Eurodisplay '93*, 411–413 (1993).

Sher, L. D., "The oscillating-mirror technique for realizing true 3-D," Chapter 11, 196–213, in McAllister (1993), q.v.

Sherr, S., *Electronic displays*, 2nd ed., Wiley, New York, 1992.

Shimada, S., "A new approach to the real-image 3D globe display," *SID'93 Digest*, 1001–1004 (1993).

Simon, W., "A method of producing a three-dimensional cathode ray tube display," *Behavior Research Methods Instrumentation*, **1**, 179 (1969).

Simon, W., and Walters, T., "A spinning mirror autostereoscopic display," *SPIE Proceedings*, **120**, 180–183 (1977).

Solomon, D., "Volumetric imaging launches graphics into a 3-D world," *Photonics Spectra*, 129–135 (June 1993).

Soltan, P., and Lasher, M., *Nonmoving 3D volumetric display using upconversion materials*, U.S. Navy Research and Development (NRaD) Report, April 11, 1996.

Soltan, P., Trias, J., Robinson, W., and Dahlke, W., "Laser-based 3D volumetric display system," *High-resolution displays and projection systems: SPIE Proceedings*, **1664** 177–192 (1992).

Soltan, P., Trias, J., Dahlke, W., Lasher, M., and MacDonald, M., "Laser-based 3D volumetric display system (2nd generation)," *SID'94 Proceedings* (1994).

Southall, J. P. C. (ed.), *Helmholtz's treatise on physiological optics*, Vol. 3, Dover, New York, 1962. [English translation of Helmholtz (1909), q.v., a reprint of the original 1925 Optical Society of America publication.]

Sproson, W. N., *Colour science in television and display systems*, Adam Hilger, Bristol, UK, 1983.

Szilard, J., "An improved three-dimensional display system," *Ultrasonics*, 273–276 (November 1974).

Toda, T., Takahashi, S., and Iwata, F., "Three-dimensional (3D) video system using grating image," *SPIE Proceedings*, **2652**, 54–61 (1996).

Traub, C., "Stereoscopic display using rapid varifocal mirror oscillations," *Applied Optics*, **6**(6), 1085–1087 (1967).

Travis, A. R. L., "The display of three-dimensional video images," *Proceedings of the IEEE*, **85**(11), 1817–1832 (1997).

Tricoles, G., "Computer-generated holograms: an historical review," *Applied Optics*, **26**(20), 4351–4360 (1987).

Valyus, N. A., *Stereoscopy*, Focal Press, London, 1966.

Verber, C. M., "Present and potential capabilities of three-dimensional displays using sequential excitation of fluorescence," *SPIE Proceedings*, **120**, 62–67 (1977).

Vince, J., *Virtual reality*, Addison-Wesley, Reading, Mass., 1995.

von Ardenne, M., "Negative lionenstrahlen bei der Formierung von Hochvakuum Elektronenstrahlröhren," *Archiv für Elektrotechnik (Berlin)*, **29**, 731–732 (1935).

von Ardenne, M., *Cathode ray tubes*, Pitman, London, 1939.

von Ardenne, M., *Elektron Ubermikroskopie*, Springer-Verlag, Berlin, 1940.

Wade, N. J. (ed.), *Brewster and Wheatstone on vision*, Academic Press, London, 1983.

Wade, N. J., "On the late invention of the stereoscope," *Perception*, **16**, 785–818 (1987).

Wade, N. J., and Swanston, M., *Visual perception: an introduction*, Routledge, London, 1991.

Wefer, F. L., *Computer graphics direct volume display devices*, Mitre Corp. Internal Report MTR 92W0000141, Mitre Corporation, McLean, VA, (1994).

Wehnelt, A., "Über der austritt negative ionen aus glühenden Metallverbindungen und damit zusammenhangende Erscheinenungen," *Annalen der Physik (Leipzig) [4th series]*, **14**, 425–468 (1904).

Wheatstone, C., "Contributions to the theory of vision: part the first, on some remarkable, and hitherto unobserved, phenomena of binocular vision," *Philosophical Transactions of the Royal Society*, **128**, 371–394 (1838).

Williams, R. D., and Donohoo, D., "Image quality metrics for volumetric laser displays," *Stereoscopic displays and applications II: SPIE Proceedings*, **1457**, 210–220 (1991).

Williams, R. D., and Garcia, F., "A real time autostereoscopic multiplanar 3D display system," *SID'88 Digest*, **19**, 91–94 (1988).

Withey, E. L., "Cathode-ray tube adds third dimension," *Electronics (Engineering Edition)*, May 23, 81–83 (1958).

Wood, R. W., "Controlled orbital transfers of electrons in optically excited mercury atoms," *Proceedings of the Royal Society, London*, **106**, 679–694 (1924).

Yamada, H., Masuda, C., Nozaki, T., Nishitani, K., and Miyaji, K., "A 3-D display using a laser and moving screens," *ICALEO* **48**, 71–77 (1984).

Yamanaka, R., Yamamoto, K., Handa, N., and Yoshikura, H., "A 3D display with a linearly moving mirror to reflect a series of 2D cross-sections and its application to noninvasive angiography," *IEEE Transactions on Medical Imaging*, **7**, 193–197 (1988).

Yeh, Y. Y. (1993) "Visual and perceptual issues in stereoscopic color displays," pp. 50–70, in McAllister (1993), q.v.

Zito, R., "Rate analysis of multiple-step excitation in mercury vapour," *Journal of Applied Physics*, **34**(5), 1535–1543 (1963).

Zito, R., and Schraeder, A. E., "Optical excitation of mercury vapour for the production of isolated fluorescence," *Applied Optics*, **2**(12), 1323–1328 (1963).

PATENTS

U.S. Patents

2,604,607; F.S. Howell, 1952, *Three-dimensional indicator tube and circuit therefor.*

2,749,480; M. Ruderfer, 1956, *Apparatus for producing three-dimensional visual patterns.*

2,762,031; R.A. Fryklund, 1956, *Three-dimensional position indicating system.*

2,967,905; M. Hirsch, filed January 13, 1958, granted January 10, 1961, *Three dimensional display apparatus.*

3,123,711; J. Fajans, 1964, *Luminous spot display device.*

3,140,415; R.D. Ketchpel, granted July 7, 1964, *Three-dimensional display cathode ray tube.*

3,204,238; A.M. Skellett, filed July 13, 1962, granted August 31, 1965, *Cathode ray tube for three-dimensional presentations.*

3,300,779; R. Sirkis, filed March 15, 1963, *Three dimensional pictorial displays.*

3,474,248; M.R. Brown and G.S. Waters, 1969, *Three-dimensional visual display systems.*

3,541,542; M.A. Dugay, J.A. Giordmaine, and P.M. Rentzepis, granted November 17, 1970, *Display system using two-photon fluorescent materials.*

3,609,706; Arthur W. Adamson, granted September 28, 1971, *Method and apparatus for generating three-dimensional patterns.*

3,609,707; J.D. Lewis and A.H. Adelman, 1971, *Method and apparatus for generating three-dimensional patterns.*

3,636,551; E.T. Maguire, granted January 18, 1972, *Computer-controlled three dimensional display.*

3,829,838; J.D. Lewis, C.M. Verber, and R.B. McGhee, 1974, *Computer-controlled three-dimensional pattern generator.*

4,063,233; W.G. Rowe, 1977, *Three-dimensional display devices.*

4,130,832; L.D. Sher, filed July 11, 1977, granted December 19, 1978, *Three-dimensional display.*

4,160,973; E.P. Berlin, Jr., granted July 10, 1979, *Three-dimensional display.*

4,870,485; E. Downing, granted September 26, 1989, *Three dimensional image generating apparatus having a phosphor chamber.*

4,922,336; R.R.A. Morton, granted May 1, 1990, *Three-dimensional display system.*

5,042,909; F. Garcia, Jr., P. Williamson, R.D. Williams, granted August 27, 1991, *Real time three dimensional display with angled rotating screen and method.*

5,148,310; R.G. Batchko, granted September 15, 1992, *Rotating flat screen fully addressable volume display.*

5,703,606; B.G. Blundell, granted December 30, 1997, *Three dimensional display system.*

5,745,197; M.S. Leung, N.A. Ives, and G. Eng, 1998, *Three-dimensional real image volumetric display system and method.*

5,813,742; R.S. Gold and J.E. Freeman, 1998, *Layered display system and method for volumetric presentation.*

5,854,613; P. Soltan, filed May 31, 1996, granted December 29, 1998, *Laser based 3D volumetric display system.*

European Patents

461.600 (French patent) [Brevet d'invention, demandé le 31 Octobre 1912, publié le 6 janvier 1914] Emile Luzy and Charles Dupuis, *Procédé pour obtenir des projections en relief.*

EP 0 310 928 A2; Felix Garcia, Jr. and Rodney D. Williams, filed September 28 1988, *Real time three dimensional display and method.*

EP 0 418 583 A2; Felix Garcia Jr. granted March 27 1991, *Real time three dimensional display.*

EP 0 491 284 A1; Rodney D. Williams, granted June 24 1992, *Volume display system and method for inside-out viewing.*

R. Hartwig, 1984. DE 26 22 802 C2 German patent.

Other Patents

PCT/NZ96/00028; B.G. Blundell, filed April 4, 1996, *Improvements in a three dimensional display system.*

NZ patent application 314700; B.G. Blundell, filed April 29, 1997, *A high definition volumetric display system.*

INDEX